Progress in Mathematics

Volume 93

Michèle Audin

Torus Actions on Symplectic Manifolds

Second revised edition

Springer Basel AG

Author:

Michèle Audin
Institut de Recherche Mathématique Avancée
Université Louis Pasteur
7 rue René-Descartes
67084 Strasbourg Cedex
France
e-mail: michele.audin@math.u-strasbg.fr

2000 Mathematics Subject Classification 14F43, 53Dxx, 70G45

A CIP catalogue record for this book is available from the Library of Congress, Washington D.C., USA

Bibliographic information published by Die Deutsche Bibliothek
Die Deutsche Bibliothek lists this publication in the Deutsche Nationalbibliografie; detailed bibliographic data is available in the Internet at <http://dnb.ddb.de>.

ISBN 978-3-0348-9637-5 ISBN 978-3-0348-7960-6 (eBook)
DOI 10.1007/978-3-0348-7960-6

Originally published by Birkhäuser Verlag in 2004
Softcover reprint of the hardcover 2nd edition 2004
Printed on acid-free paper produced of chlorine-free pulp. TCF ∞
ISBN 978-3-0348-9637-5

9 8 7 6 5 4 3 2 1 www.birkhauser-science.com

et la cheminée avec, entre les pots à pieds coniques, les deux gravures dans leurs cadres ovales, représentant l'une un morse sur son glaçon et l'autre un jeune garçon suspendu à un trapèze, et la table ronde avec le dessous de plat métallique en forme de losange, le poisson rouge dans son bocal sphérique, l'éventail en dentelle noire, le cendrier publicitaire hexagonal sur lequel est dessinée une chambre à air et la petite boîte à cigarettes en bois sculpté affectant la forme d'un tonneau G. P. [**116**]

CONTENTS

INTRODUCTORY PREFACE

How I have (re-)written this book

The book the reader has in hand was supposed to be a new edition of [**14**]. I have hesitated quite a long time before deciding to do the re-writing work—the first edition has been sold out for a few years.

There was absolutely no question of just correcting numerous misprints and a few mathematical errors. When I wrote the first edition, in 1989, the convexity and Duistermaat–Heckman theorems together with the irruption of toric varieties on the scene of symplectic geometry, due to Delzant, around which the book was organized, were still rather recent (less than ten years). I myself was rather happy with a small contribution I had made to the subject. I was giving a post-graduate course on all that and, well, these were lecture notes, just lecture notes.

By chance, the book turned out to be rather popular: during the years since then, I had the opportunity to meet quite a few people(1) who kindly pretended to have learnt the subject in this book.

However, the older book does not satisfy at all the idea I have now of what a good book should be. So that this "new edition" is, indeed, another book.

First of all, it is written more carefully. In particular, I have tried to follow the recommendations of [**16**]. The first edition, of course, contains only those results known(2) up through 1989. During the intervening fifteen years, I have enlarged both my knowledge of new results and my point of view, arising from work on more general integrable systems. I now find the relations among them to have gained in importance, and this has enlarged the book as well. It is not only a thicker book but also

(1) often former MIT students, so that I guess the advertisement was made by Guillemin

(2) To quote [**54**], "after a decade of dramatic progress, many chapters of the theory [had] reached the final stage".

- it contains more mathematics: I have split the previous chapter on the bases of symplectic geometry, adding more considerations on integrable systems, and I have (coherently) added a chapter on integrable systems and torus actions on certain moduli spaces (following Goldman and Jeffrey–Weitsman) because this is a beautiful example of use of the results of this book, but also because moduli spaces are the spaces of the future and cannot be ignored in a new book;
- besides, it contains more theorems, even on the previously treated subjects, for instance Atiyah's Kähler refinement of the convexity theorem and Delzant's uniqueness result;
- it contains much more relevant examples and many new exercises;
- it contains more proofs. Contrary to what happened in the old treatment, there is *no* proof "left as an exercise to the readers";
- it contains fewer misprints(3) (thanks to automatic spell-checking with *Excalibur*);
- it contains more figures and, this time, the figures are beautiful (but there will be no advertisement for expensive software in this book);
- the bibliography has been completed and up-dated;
- there is an index.

... Well, to make, eventually, a long story short, I have worked a lot on this new book, I have enjoyed it most of the time and I hope that this will be a genuinely good reference book. In any case, this is the last time I will write it.

Acknowledgements

I wish to thank for their help, conscious or not, all the people with whom I have had the opportunity to discuss the material presented here, before I wrote the first edition (Abdelkader Bouyakoub, Michel Brion, Thomas Delzant, Nicole Desolneux-Moulis, Patrick Iglesias, Paulette Libermann), during the time I was writing the first edition (Nicole Bopp, Julien Duval, Jacques Fougeront, Philippe Gaucher, Li Jie, Santiago Lopez de Medrano, Jean-Yves Mérindol, Jean-Marc Rinkel, Hubert Rubenthaler, Marcus Slupinski) and after I wrote it (Anton Alekseev, Ana Cannas da Silva, River Chiang, Vladimir Fock, Daniel Gatien, Sophie Gerardy, Viktor A. Ginzburg, Viktor G. Ginzburg, Leonor Godinho, Victor Guillemin, Lisa Jeffrey, Yael Karshon, Laurent Laeng, Dusa McDuff, Pierre Molino, Elisa Prato, Sébastien Racanière, Alexei Reyman, Bruno Sévennec, Holger Spielberg, Suzan Tolman, Pol Vanhaecke, Tilmann Wurzbacher).

(3) I have also suppressed some (old) bad jokes but could not resist adding new (bad) ones.

I also owe a special debt of gratitude to Leonor Godinho, Dusa McDuff and Sébastien Racanière for helping me to improve the penultimate version of this book, and especially to Edwin Beschler, copy-editor at Birkhäuser, who read it very carefully and made a lot of useful suggestions and to Claude Sabbah who helped me to hunt out widows and orphans.

What I have written in this book

At least since the Felix Klein Erlangen program, everybody knows that it is useful to make groups act on sets, the action providing information both on the group acting and on the set acted on.

In this book, we shall be interested in "differentiable" (*i.e.*, locally linear) objects: the set will be a smooth manifold and the group a Lie group, acting by diffeomorphisms.

The group action allows one to decompose the manifold into pieces (strata) corresponding to the various orbit types. When the manifold is not too complicated, it is possible to recover "everything" from this stratification: this is the case for example when the group is the circle $\mathbf{S}^1$ and the manifold a surface... but there are rather few examples of such a simple situation.

The two principal motivations to investigate group actions on *symplectic* manifolds might be:

(1) This is a natural (?) framework for the Hamiltonian mechanics: since Lagrange, we are able to consider the phase space of any mechanical system as a symplectic manifold, the group then represents the more or less hidden symmetries of the system[(4)].
(2) From the point of view of the group itself, what is called the "orbit method" (Kostant, Kirillov, *etc.*) and which is a tool to construct representations of the group, uses symplectic geometry in an essential way. The fundamental objects in the theory are the coadjoint orbits, which are naturally symplectic manifolds.

There is another good reason to investigate symplectic actions: this is *easier!* The so-called *Hamiltonian* actions are actually, by the very definition, those for which there is a *function* on the manifold, the critical points of which correspond to the fixed points of the group action. It is thus possible to use the well-understood methods (invented by Morse, Thom, and many others) of *Morse theory* to study the group action.

---

(4) A good reference for this point of view is the book [**123**].

For example, the central theorem in the first part of this book is the following statement (that any neophyte is allowed to find abstruse) due to Frankel [**51**] and Atiyah [**7**]:

***Theorem* (Corollary IV.3.2).** *Let X be an almost periodic vector field on a connected symplectic manifold (W,ω). Suppose H is a function $W \to \mathbf{R}$ such that $i_X\omega = dH$. Then all the levels $H^{-1}(t)$ are (empty or) connected.*

For those who felt the statement was indeed abstruse, here are a few applications which should be clearer.

A theorem of Kushnirenko. In the first application, we enumerate the solutions of a particular system of algebraic equations. Consider a finite subset $S \subset \mathbf{Z}^n$ of multi-exponents, and the system of n equations and n unknowns:

$$\sum_{\alpha \in S} c_\alpha^j z^\alpha = 0 \text{ for } 1 \leqslant j \leqslant n,$$

the parameters c_α^j being complex and the unknown z an element of the complex torus $(\mathbf{C}^\star)^n$.

For example,

- If $n = 1$ and $\inf S = k \geqslant 0$, the system consists of an equation
$$z^k \sum_{\ell=0}^{m-k} c_\ell z^\ell$$
and thus has, in general, $m-k$ (the length of the shortest interval containing S) nonzero solutions.
- If $S = \{e_1, \dots, e_n\}$ is the set of the vectors in the canonical basis of $\mathbf{Z}^n$, then our system is a homogeneous linear system and has in general no solution (in $(\mathbf{C}^\star)^n$).
- If we add 0 (then $S = \{e_1, \dots, e_n, 0\}$), the system is a linear system and has in general exactly one solution.

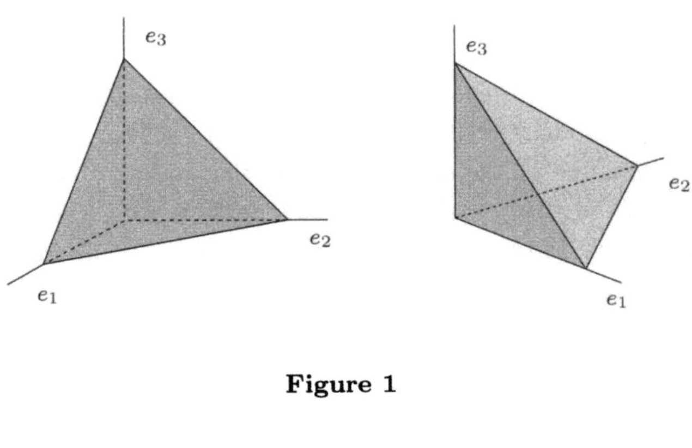

Figure 1

The following theorem, due to Kushnirenko [**90**], generalizes the observations made in these examples:

***Theorem* (Theorem IV.4.25).** *The number of solutions of the system*

$$\sum_{\alpha \in S} c_\alpha^j z^\alpha = 0 \text{ for } 1 \leqslant j \leqslant n$$

for general enough coefficients c_α^j is

$$N(S) = n! \operatorname{Vol}(\widehat{S})$$

where $\widehat{S}$ is the convex hull of S in $\mathbf{R}^n$.

The Toeplitz–Hausdorff theorem. This is a classical problem, the Toeplitz–Hausdorff problem, to determine the *numerical image* of an operator A on a Hilbert space $\mathcal{H}$. This is simply the set $W(A)$ of all the values taken by the inner product $\langle Av, v\rangle$, when v varies in the unit sphere of $\mathcal{H}$. Here, we will only consider matrices, that is, we will restrict ourselves to the finite dimensional case, $\mathcal{H} = \mathbf{C}^n$, endowed with the standard Hermitian form. In this case, A is an $n \times n$ complex matrix.

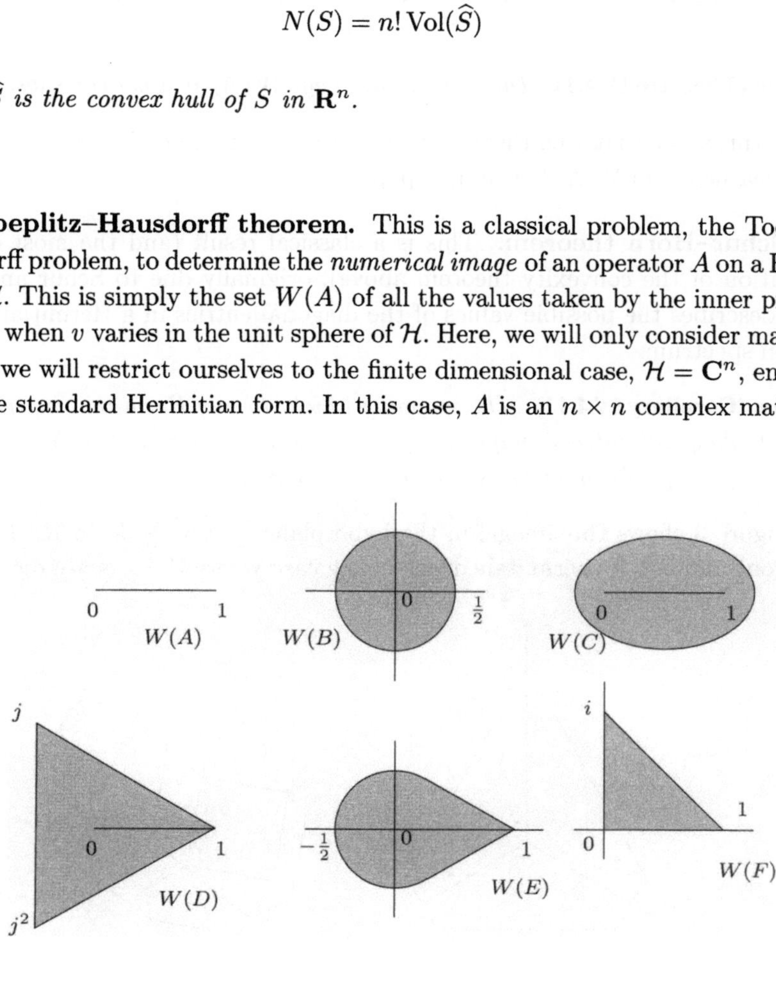

Figure 2

The numerical image can have rather different shapes, according to the specific chosen A. The examples given here are inspired by those in [**65**]; they correspond to the matrices

$$A = \begin{pmatrix} 1 & 0 \\ 0 & 0 \end{pmatrix}, \quad B = \begin{pmatrix} 0 & 0 \\ 1 & 0 \end{pmatrix}, \quad C = \begin{pmatrix} 0 & 0 \\ 1 & 1 \end{pmatrix},$$

$$D = \begin{pmatrix} 0 & 0 & 1 \\ 1 & 0 & 0 \\ 0 & 1 & 0 \end{pmatrix}, \quad E = \begin{pmatrix} 0 & 0 & 0 \\ 1 & 0 & 0 \\ 0 & 0 & 1 \end{pmatrix}, \quad F = \begin{pmatrix} 1 & 0 & 0 \\ 0 & i & 0 \\ 0 & 0 & 0 \end{pmatrix}$$

for which it is (more or less easily) checked that the numerical image is that shown in Figure 2. Notice that all these subsets are convex, this being a more general fact.

***Theorem* (Theorem IV.4.1).** *The numerical image $W(A)$ is a convex subset of* $\mathbf{C}$.

It turns out that this theorem is also a consequence of Corollary IV.3.2, as I have learned it in V. A. Ginzburg's paper [**53**].

The Schur–Horn theorem. This is a classical result (and the most classical application of the convexity theorem above), originally due to Schur and Horn, which describes the possible values of the diagonal entries in a Hermitian matrix of given spectrum.

***Theorem* (Corollary IV.4.11).** *The mapping which, to any Hermitian matrix, associates its diagonal entries, maps $\mathcal{H}_{(\lambda_1,\ldots,\lambda_n)}$ onto the convex hull in $\mathbf{R}^n$ of the set $(\lambda_{\sigma(1)},\ldots,\lambda_{\sigma(n)})$ (for σ in the symmetric group $\mathfrak{S}_n$).*

Figure 3 shows this image, in the hyperplane $\sum v_i = \sum \lambda_i$ in $\mathbf{R}^n$ for $n = 3$ (hexagon) and $n = 4$ (permutahedron) in the case where the λ_i's are distinct.

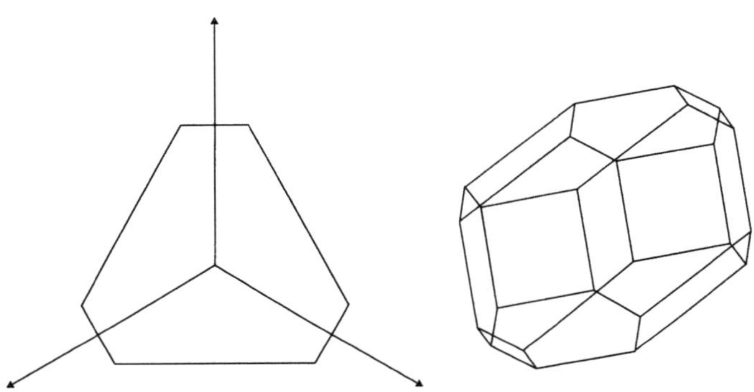

Figure 3

Although I did my best to present rather different applications of Corollary IV.3.2, I cannot hide that they have two important common points:

- in their assumptions: the manifolds $(\mathbf{C}^\star)^n$ and $\mathcal{H}_\lambda$ are both endowed with an action of the torus $\mathbf{T}^n$.
- in their conclusions: both contain the word "convex": actually, Corollary IV.3.2 is the essential step in the proof of the so-called *Atiyah, Guillemin and Sternberg convexity theorem.*

Our abstruse statement, in addition to giving unified proofs of previously different theorems, allows us to classify certain group actions on symplectic manifolds.

Contents of this book. As its title says, this book is mainly devoted to the study of torus actions in symplectic geometry. Hence, I start with some generalities on Lie group actions on manifolds. This is Chapter I, in which I spend some time on $\mathbf{S}^1$-actions on surfaces and on 3-manifolds, especially on Seifert manifolds. Then I introduce symplectic geometry (in Chapter II) and Hamiltonian group actions (in Chapter III), especially torus actions and action-angle variables.

Chapter IV is the core of the book. Devoted to some applications of Morse theory to Hamiltonian group actions, it contains all the convexity theorems mentioned in this introduction. I have tried to give a lot of examples, in the main text as well as in the exercises. The next chapter, Chapter V, describes a family of examples of symplectic manifolds (moduli spaces of flat connections) endowed with integrable systems (Goldman's functions), the action-angle variables (investigated by Jeffrey and Weitsman) of which give further examples of convex polyhedra appearing in symplectic geometry. Another reason to have included these examples here is the fact that this is an opportunity to introduce the basic definitions on connections, that I use to do the algebraic topology needed in Chapter VI. This one is devoted to the other beautiful achievement of the theory in the 1980s, namely, the Duistermaat–Heckman theorem. It happens that the language of equivariant cohomology fits very well with the study of Hamiltonian actions, since the existence of a momentum mapping for the G-action on the manifold W is equivalent to the existence of an extension of the symplectic form to the Borel construction W_{G} on W. This remark allowed Berline and Vergne [**22**] to give a lucid proof of the spectacular theorem of Duistermaat and Heckman [**44**]. I present here this perfect example of a theorem which becomes practically tautological once the right language to state it in is found.

In Chapter VII, I present a topological construction of the complex toric varieties. This comes from the work by Delzant [**39**], the seminal paper [**38**] by Danilov and the book [**85**] by Kirwan. In the last chapter (Chapter VIII), I discuss the Hamiltonian circle actions on 4-manifolds, around results due to Dusa

McDuff [**103**], myself [**13**] (and independently Ahara and Hattori [**1**]) and Yael Karshon [**83**]. The main reason to present these results *hic et nunc* is that the proofs given here are an illustration of all the methods introduced in the book, since I use Seifert manifolds, Morse theory, the Duistermaat–Heckman Theorem and toric manifolds.

Prerequisites and bibliography. The readers are assumed to know the basic definitions on manifolds (see [**124, 91**], as a good general reference for all these notes, [**84**, § 5] and, for the calculus used in this book, [**43**]). I also use, of course, the classification of surfaces [**101, 59**] and, to stay in the spirit of the book, [**58**], which might also be used as an introduction to Morse theoretical methods. As to algebraic topology: the reader is supposed to learn a few elementary notions as she or he goes along, being assumed to know nothing at the beginning and a little more at the end. There will always be a reference or a hint of proof for what I shall use, but not always a complete proof.

I have tried to make an extensive and rather complete bibliography. There are quite a few good books on symplectic geometry on the market. I like very much the concise and efficient survey [**6**] of Givental and Arnold. I also recommend, of course, among many others, [**96**] and [**106**].

CHAPTER I

SMOOTH LIE GROUP ACTIONS ON MANIFOLDS

In this chapter, we list the basic definitions and properties of Lie group actions. Then we investigate the actions of the circle $\mathbf{S}^1$ on surfaces and 3-manifolds.

I.1. Generalities

To make this book reasonably self-contained, we start with a few generalities. The material here is quite classical and can be found in numerous good books. This is why some proofs are only sketched. The books I have used are [**27, 32, 71, 91**].

I.1.a. Lie group actions. Let G denote a Lie group[(1)] the unit element of which is 1 and the Lie algebra of which is $\mathfrak{g}$.

G-*manifolds.* Let W be a smooth (*i.e.*, $\mathcal{C}^\infty$) manifold on which G acts by diffeomorphisms. The action is written

$$\begin{aligned} \mathrm{G} \times W &\longrightarrow W \\ (g, x) &\longmapsto g \cdot x. \end{aligned}$$

This is a left action, that is,

$$g \cdot (h \cdot x) = (gh) \cdot x \text{ for all } g, h \in \mathrm{G}, x \in W.$$

A manifold endowed with a G-action is called is a G-manifold.

Equivariant mappings. If the group G acts on the two manifolds V and W, and if $\varphi : V \to W$ is a smooth map, it is said that φ is an *equivariant* map if

$$\forall x \in V, \quad \forall g \in \mathrm{G}, \quad \varphi(g \cdot x) = g \cdot \varphi(x).$$

---

(1) The readers will find a summary of useful results on Lie groups, Lie algebras, the exponential mapping and all that in § I.4.

Clearly, if the equivariant mapping φ is a diffeomorphism, the inverse diffeomorphism φ^{-1} is also an equivariant mapping.

G-vector bundles. We shall call a G-*vector bundle* over a G-manifold W any vector bundle $E \to W$ endowed with a G-action which is linear in the fibers and such that the projection is an equivariant map.

For instance, the tangent bundle $\mathsf{T}W$ of a G-manifold has a natural structure of G-vector bundle, given by

$$g \cdot (x, v) = (g \cdot x, \mathsf{T}_x g(v)).$$

Orbits and stabilizers. If x is a point in W, its orbit will be denoted $\mathrm{G} \cdot x$,

$$\mathrm{G} \cdot x = \{y \in W \mid y = g \cdot x \text{ for some } g \in \mathrm{G}\},$$

and its stabilizer G_x,

$$\mathrm{G}_x = \{g \in \mathrm{G} \mid g \cdot x = x\}.$$

Stabilizers of points are obviously closed subgroups, and hence Lie subgroups (see Theorem I.4.6) of G. Notice that stabilizers of points in the same orbit are conjugate to each other, as $\mathrm{G}_{g \cdot x} = g^{-1}\mathrm{G}_x g$ (this equality also shows that all possible conjugates appear). With each orbit is associated a conjugation class of subgroups of G, called the *type* of the orbit. The conjugation class of H is denoted by (H). Thus the type of $\mathrm{G} \cdot x$ is (G_x).

Notice also that, if $\varphi : V \to W$ is an equivariant mapping, we have $\mathrm{G}_x \subset \mathrm{G}_{\varphi(x)}$.

I.1.b. Orbit mappings and fundamental vector fields. For each point $x \in W$, the mapping

$$\begin{aligned} f_x : \mathrm{G} &\longrightarrow W \\ g &\longmapsto g \cdot x \end{aligned}$$

is a smooth map, called the *orbit mapping*. Its image is the orbit of x. Its differential at 1 is the linear map:

$$\mathsf{T}_1 f_x : \mathsf{T}_1\mathrm{G} = \mathfrak{g} \longrightarrow \mathsf{T}_x W.$$

With each vector $X \in \mathfrak{g}$ is thus associated a vector

$$\mathsf{T}_1 f_x(X) = \underline{X}_x \in \mathsf{T}_x W$$

tangent to W at x. Allowing x to vary in W, we get a vector field, the *fundamental vector field* associated with X, denoted by ${}^W\underline{X}$, $\underline{X}$, or even X when there is no risk of confusion. By definition, the flow of $\underline{X}$ is $\exp(tX) \cdot x$. Notice that the Lie bracket is related to the bracket of vector fields by the simple formula

$$\underline{[X, Y]} = [\underline{X}, \underline{Y}].$$

The stabilizer of x is a Lie subgroup of G, the homogeneous space G/G_x is thus a smooth manifold (see Theorem I.4.8) and one shows

Theorem I.1.1. *The orbit map*

$$f_x : G/G_x \longrightarrow W$$

is an injective immersion.

Proof. We simply evaluate the kernel of the differential of the orbit map

$$\mathsf{T}_g f_x : \mathsf{T}_g G \longrightarrow \mathsf{T}_{g\cdot x} W.$$

By invariance, it is sufficient to study the case where $g = 1$. We are thus looking at the kernel of $\mathsf{T}_1 f_x$, that is $\{X \in \mathfrak{g} \mid \underline{X}_x = 0\}$. It is easily checked that this set is the Lie algebra $\mathfrak{g}_x$ of G_x. □

Thus, orbits are images of manifolds by injective immersions. It does not follow that they are submanifolds. The most classical (counter-)example is that of $\mathbf{R}^2$ acting on a torus by irrational translations (Exercise I.1). Recall however that a *proper* injective immersion is an embedding, in particular this problem does not occur when the group is compact.

Corollary I.1.2. *If* G *is a compact Lie group, its orbits are submanifolds of* W. □

Notice that, by the very definition, the fundamental vector fields generate the tangent space to the orbit.

I.1.c. Examples.

Example I.1.3. The linear group $GL(n, \mathbf{R})$ acts on $\mathbf{R}^n$, as does $GL(n; \mathbf{C})$ on $\mathbf{C}^n$.

We will use mainly compact subgroups of $GL(n; \mathbf{R})$ or $GL(n; \mathbf{C})$, as the orthogonal group $O(n)$ or the unitary group $U(n)$ and especially Abelian compact subgroups (namely, tori), but we will also need the complex torus $(\mathbf{C}^\star)^n$.

Example I.1.4. The circle $\mathbf{S}^1$ acts on $\mathbf{C}^n$ by complex multiplication

$$t \cdot (z_1, \ldots, z_n) = (tz_1, \ldots, tz_n).$$

The point 0 is fixed; all the other orbits are circles. The spheres

$$\mathbf{S}^{2n-1} = \left\{(z_1, \ldots, z_n) \mid \sum |z_i|^2 = 1\right\}$$

are stable under this action. This apparently trivial example will turn out to be the fundamental example for this book!

Example I.1.5. The circle $\mathbf{S}^1$ also acts on $\mathbf{S}^3 \subset \mathbf{C}^2$ (for instance) in a more sophisticated way

$$t \cdot (z_1, z_2) = (t^{m_1} z_1, t^{m_2} z_2)$$

(here m_1 and m_2 are any elements of $\mathbf{Z}$).

Example I.1.6. The unitary group $\mathrm{U}(n)$ acts on the real vector space of Hermitian $n \times n$-matrices by conjugation

$$A \cdot M = AMA^{-1}.$$

It follows from the diagonalizability of Hermitian matrices that the orbits are the manifolds $\mathcal{H}_\lambda$ mentioned in the introduction of this book (this is Exercise I.4).

I.1.d. More definitions. The G-action is said to be *effective* if each element $g \neq 1$ moves at least one x in W, that is to say, if:

$$\bigcap_{x \in W} \mathrm{G}_x = \{1\}.$$

Any action may be replaced by an effective one:

Proposition I.1.7. *The group $\bigcap \mathrm{G}_x$ is a closed normal subgroup in* G, *and the* G-*action in W induces an effective action of* $\mathrm{G}/\bigcap \mathrm{G}_x$ *in W.*

Proof. This an intersection of closed subgroups, hence a closed subgroup. Let now $g \in \bigcap \mathrm{G}_x$ and h be any element of G. For $x \in W$, we have

$$(h^{-1}gh) \cdot x = h^{-1}g(h \cdot x) = h^{-1}(h \cdot x)$$

since g fixes all elements, hence $h^{-1}gh \in \bigcap \mathrm{G}_x$. □

Example I.1.8. Let $\mathbf{S}^1$ act on $\mathbf{C}$ by $t \cdot w = t^2 w$. This is not an effective action, since the subgroup $\{\pm 1\}$ fixes all the points. However, the quotient $\mathbf{S}^1/\{\pm 1\}$ acts effectively.

Example I.1.9. The $\mathbf{S}^1$-action on $\mathbf{S}^3$ by

$$t \cdot (z_1, z_2) = (t^{m_1} z_1, t^{m_2} z_2)$$

is effective if and only if m_1 and m_2 are relatively prime (an exercise for the readers).

Example I.1.10. If the group G is *simple*, the action is either effective—if the normal subgroup $\bigcap \mathrm{G}_x$ is the trivial subgroup $\{1\}$—or trivial—if this is the whole group G. This is the case for instance for $\mathrm{G} = \mathrm{SO}(3)$.

The action is *transitive* if it has only one orbit, *free* if all the orbits have the trivial subgroup $\{1\}$ as stabilizer, *semi-free* if they have G or $\{1\}$ as stabilizer (in other words if it is free outside fixed points).

The last notion in this paragraph is that of *orbit space*. This is the quotient space W/G endowed with the quotient topology. Even for a smooth Lie group action on a smooth manifold, the quotient can be very bad. In general, it is not even Hausdorff (think of the example of the irrational flow on the torus $\mathbf{T}^2$, there is another example in Exercise I.2). Anyway:

Proposition I.1.11. *If* G *is compact,* W/G *is a Hausdorff space and the projection* $\pi : W \to W/\mathrm{G}$ *is a proper and closed mapping.*

Proof. Recall that the quotient map π for the quotient by a group action is an open mapping, since, for any open set $U \subset W$, we have

$$\pi^{-1}(\pi(U)) = \bigcup_{g\in \mathrm{G}} g \cdot U,$$

an open set as a union of open sets. It turns out that, when G is compact, the projection π is also closed,

$$\pi^{-1}(\pi(A)) = \Theta(\mathrm{G} \times A)$$

being closed for any closed subset A. The map Θ here is simply the action map

$$\Theta : \mathrm{G} \times W \longrightarrow W,$$

the point here being that Θ is a closed map when G is compact. To see this, let $C \subset \mathrm{G} \times W$ be any closed subset and let y be a point of the closure $\overline{\Theta(C)}$. There are sequences (g_n), (x_n) such that $y = \lim g_n \cdot x_n$. As G is compact, we may assume that (g_n) converges to some element g. Then $(x_n) = (g_n^{-1} \cdot (g_n \cdot x_n))$ tends to $g^{-1}(y)$, thus $(g_n, x_n) \to (g, g^{-1}(y)) \in C$, since C is a closed subset. Hence $y = g \cdot g^{-1}(y) \in \Theta(C)$, so that $\Theta(C)$ is closed.

We know that π is closed, its fibers are compact, hence π is a proper map. We still need to prove that W/G is Hausdorff. Let $\mathrm{G}\cdot x$ and $\mathrm{G}\cdot y$ be two distinct orbits. They are compact (as images of G). As W is Hausdorff, they can be separated by open subsets. There is an open neighborhood U of x such that $\overline{U} \cap \mathrm{G}\cdot y = \varnothing$. Now $\pi(U)$ and the complement of $\pi(\overline{U})$ are both open subsets, they do not intersect, the former contains $\pi(x)$ and the latter $\pi(y)$. □

I.2. Equivariant tubular neighborhoods and orbit types decomposition

I.2.a. The slice theorem (equivariant tubular neighborhood). Although this is not the most general possible assumption, we will restrict ourselves to the case where the group G is compact. In this case, according to Corollary I.1.2, its orbits are submanifolds of W. We now describe the manifold W in the neighborhood of an orbit.

Let $x \in W$. Let V_x denote the quotient vector space $\mathsf{T}_x W/\mathsf{T}_x(\mathrm{G}\cdot x)$. For $g \in \mathrm{G}_x$, the tangent map:

$$\mathsf{T}_x g : \mathsf{T}_x W \longrightarrow \mathsf{T}_{g\cdot x} W = \mathsf{T}_x W$$

is an isomorphism, sending the tangent space of the orbit into itself (by the identity map), and in particular it induces a linear isomorphism from V_x to itself. Thus, with each point x in W, is associated a *linear representation* of its stabilizer, *i.e.*, a group homomorphism:

$$\mathrm{G}_x \longrightarrow \mathrm{GL}(V_x).$$

Hence, the stabilizer G_x acts on $\mathrm{G} \times V_x$, by left multiplication on G and by the linear representation just described on V_x.

More generally, if H is a closed subgroup of G, and if V is a vector space endowed with an H-linear action, there is a *free* action of H on $\mathrm{G} \times V$, namely

$$h \cdot (g, v) = (gh^{-1}, h \cdot v).$$

The quotient is denoted $\mathrm{G} \times_{\mathrm{H}} V$, and $[g, v]$ represents the equivalence class of (g, v); this is a vector bundle on G/H with fiber V:

$$\begin{array}{ccc} \mathrm{G} \times V & \longrightarrow & \mathrm{G} \times_{\mathrm{H}} V \\ \downarrow & & \downarrow \\ \mathrm{G} & \longrightarrow & \mathrm{G}/\mathrm{H} \end{array}$$

and it is endowed with a G-action by

$$g' \cdot [g, v] = [g'g, v].$$

We may consider G/H as a submanifold in $\mathrm{G} \times_{\mathrm{H}} V$, the zero section of the vector bundle, namely

$$\{[g, 0] \mid g \in \mathrm{G}\} \subset \mathrm{G} \times_{\mathrm{H}} V.$$

The next theorem (the slice theorem) says that there exists an extension $\overline{f}_x$ of the orbit map:

$$\begin{array}{ccc} \mathrm{G}/\mathrm{G}_x & \longrightarrow & \mathrm{G} \times_{\mathrm{G}_x} V_x \\ {\scriptstyle f_x}\downarrow & & \downarrow{\scriptstyle \overline{f}_x} \\ \mathrm{G} \cdot x & \longrightarrow & W \end{array}$$

at least to a neighborhood of the zero section.

Theorem I.2.1 **(The "slice theorem" [89]).** *There exists an equivariant diffeomorphism from an equivariant open neighborhood of the zero section in* $\mathrm{G} \times_{\mathrm{G}_x} V_x$ *to an open neighborhood of* $\mathrm{G} \cdot x$ *in* W*, which sends the zero section* G/G_x *onto the orbit* $\mathrm{G} \cdot x$ *by the natural map* f_x*.*

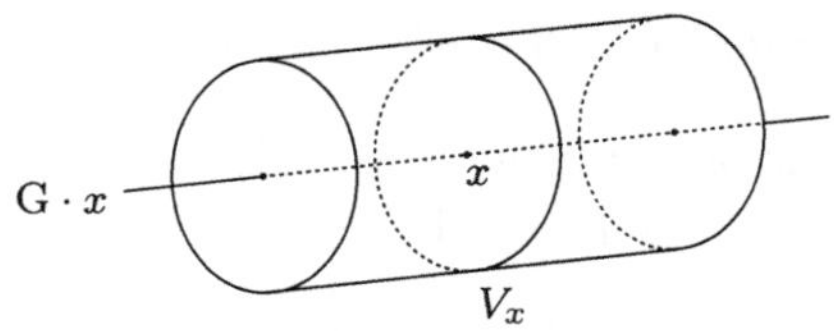

Figure 1

Sketch of a proof of theorem I.2.1. In the nonequivariant case, to prove the existence of a tubular neighborhood for a compact submanifold $Y \subset W$, one uses the *exponential map* of a *Riemannian metric*(2) on W which induces a diffeomorphism from a suitable neighborhood of the zero section in the normal bundle of Y onto a neighborhood of Y in W (see [**124**]).

To adapt this proof in the equivariant case, it suffices to choose the Riemannian metric in such a way that it is G-invariant (*i.e.*, such that G acts on W by isometries), thus making it possible to identify V_x with the orthogonal complement of $\mathsf{T}_x(\mathrm{G} \cdot x)$.

To be convinced of the existence of an invariant metric:

(1) The group G can be endowed with a Haar measure (this is to say, a measure invariant by left translations). As G is a Lie group, the proof of this fact is very easy: choose an n-linear skew-symmetric form (where n is the dimension of G) $\Omega \in \Lambda^n(\mathfrak{g})^\star$ on $\mathfrak{g} = \mathsf{T}_1\mathrm{G}$, define an invariant differential form on G by $\Omega_g = g \cdot \Omega$. This defines a measure. As G is compact, it is possible to normalize the measure in such a way that $\int_{\mathrm{G}} \Omega_g = 1$, multiplying Ω by an *ad hoc* scalar.

(2) Given a G-vector bundle $E \to W$, we have a G-action on the sections of E by:
$$(g \cdot s)(x) = gs(g^{-1}x).$$
Given a section s, associate to it an equivariant section $\overline{s}$ by:
$$\overline{s}(x) = \int_{\mathrm{G}} gs(g^{-1}x)\omega_g.$$

(2)The reader is invited to have a look at her or his favorite differential geometry textbook to find a proof of the existence of such a metric; it goes without saying that here all the manifolds are implicitly assumed to be paracompact (and Hausdorff)!

If E is the bundle of symmetric bilinear forms on the tangent bundle TW and if s is a Riemannian metric, it is easily checked that the invariant section $\overline{s}$ is still a Riemannian metric. □

I.2.b. Applications.

Proposition I.2.2. *The union of all the orbits of a given type is a submanifold of* W.

Proof. Fix a type, that is, a conjugacy class (H) of subgroups of G and let $W_{(\mathrm{H})}$ be the union of all the orbits of type (H):

$$W_{(\mathrm{H})} = \{x \in W \mid \mathrm{G}_x \in (\mathrm{H})\}.$$

Let $x \in W_{(\mathrm{H})}$. Let us prove that $W_{(\mathrm{H})}$ is a submanifold (this is a local property) in the neighborhood of the orbit of x. For this, we look at the orbits of type (H) in $\mathrm{G} \times_{\mathrm{H}} V$. If $g' \in \mathrm{G}_{[g,v]}$, we have: $g' \cdot [g, v] = [g'g, v]$ if and only if, for some $h \in \mathrm{H}$,

$$g'g = gh^{-1} \text{ and } v = h \cdot v.$$

Thus $\mathrm{G}_{[g,v]} = g\mathrm{H}_v g^{-1}$ is in the conjugation class of H_v and all conjugates do appear when g varies (and v is fixed).

The orbit of $[g, v]$ is of type (H), if and only if $\mathrm{H}_v = \mathrm{H}$, that is if and only if v is a fixed point of the H-action in V. Let

$$F = \{v \in V \mid h \cdot v = v, \forall\, h \in \mathrm{H}\}$$

be the set of these fixed points. This is a vector subspace of V and

$$(\mathrm{G} \times_{\mathrm{H}} V) \cap W_{(\mathrm{H})} = \left\{[g, v] \in \mathrm{G} \times_{\mathrm{H}} V \mid \mathrm{G}_{[g,v]} \in (\mathrm{H})\right\} = \mathrm{G} \times_{\mathrm{H}} F$$

is a subbundle of $\mathrm{G} \times_{\mathrm{H}} V$ and in particular a submanifold. □

The simplest example is that of the orbits of type (G):

Corollary I.2.3. *The set of fixed points of* G *is a submanifold of* W. □

Of course these submanifolds are not connected in general, and their components do not all have the same dimension. Anyway, the orbit types give a decomposition of W into submanifolds. This decomposition is *locally finite*:

Proposition I.2.4. *If* W *is compact, there are only finitely many orbit types.*

Proof. Let $n = \dim W$. If $n = 0$, the proposition is obviously true: W itself is a finite number of points. Suppose that the proposition is proved for all the G-manifolds of dimension $\leqslant n - 1$. Thanks to the compactness of W, it is sufficient to show that each tube $N = \mathrm{G} \times_{\mathrm{H}} V$ contains only a finite number of orbit types.

Choose an H-invariant metric on the H-vector space V, call SV the unit sphere for that metric and write $SN = \mathrm{G} \times_{\mathrm{H}} SV$ (this is the *sphere bundle* of N).

This is a G-manifold of dimension $n-1$, so that we can apply the induction hypothesis: in SN, there is only a finite number of orbit types.

We compare now the types in SN and in N. As the H-action on V is *linear*, for any nonzero λ, the orbit type of $[g, \lambda v] \in N$ is the same as that of $[g, v] \in SN$. Thus the types in N are: those in SN on the one hand, and that of the central orbit in G/H on the other hand. □

It is not only true that there are few orbit types. It happens, moreover, that one of the types takes up a lot of room:

Proposition I.2.5. *If W/G is connected, there is an orbit type* (H) *in W for which $W_{(\mathrm{H})}$ is a dense open subset of W. Moreover, $W_{(\mathrm{H})}/\mathrm{G}$ is a connected manifold.*

Such orbits are called *principal orbits.* If all the orbits are principal, the action is said to be *principal.*

Proof. Assume that G is connected. We prove the result by induction on $n = \dim W$. If $n = 0$ and W/G is connected, there is only one orbit and the proposition is obvious. As previously, assume the result proved for G-manifolds of dimension $\leqslant n-1$ and consider $N = \mathrm{G} \times_{\mathrm{H}} V$.

In order to apply the induction hypothesis to SN we would have to know that SN/G is connected. Nearly always, SN itself is connected (and thus *a fortiori* its quotient SN/G is): call π the projection of the vector bundle $N \to \mathrm{G/H}$; to connect two points a and b of SN by a path, choose a path from $\pi(a)$ to $\pi(b)$ in the connected orbit G/H and lift it to a path beginning at a in the boundary SN of the tube. The end b' of the lift is in the same fiber as b and to be able to connect it to b' (thus to a), it is sufficient that this fiber, which is a sphere in V, is connected.

The only possible problem occurs in the case where the unit sphere in V has dimension 0. In this case, $\dim V = 1$ and $SN \to \mathrm{G/H}$ is a twofold covering. If this covering is not trivial, it is connected and we are done; if it is trivial, it means that H acts *trivially* on the "slice" $V = \mathbf{R}$... in which case all the orbits contained in N are the same and we do not need the induction hypothesis to conclude that the proposition is true in N. In the other cases, the induction hypothesis applies to SN and there exists an orbit type (H) such that $(SN)_{(\mathrm{H})}$ is open and dense in SN. As in the proof of I.2.4, it is easily seen that the same happens in N.

Thus the proposition is proved in each equivariant tube. Choose a locally finite covering of W by such tubes and conclude, with the help of the connectedness of the orbit space, that the type (H) is the same in all the tubes. □

Example I.2.6. If G is *commutative*, let $(\mathrm{H}) = \{\mathrm{H}\}$ be the type of the principal orbits. The subgroup H fixes all the points of an open dense subset of W, and thus fixes all the points of W. If the action is effective, the principal orbits have type (1).

Remarks I.2.7. An orbit is principal if and only if the representation in V of its stabilizer is trivial. In this case, the quotient is a manifold. Notice that principal orbits have maximal dimension among all orbits.

An orbit is called *exceptional* if it is not principal but has the same dimension as the principal orbits, *singular* if its dimension is strictly less.

I.3. Examples: $\mathbf{S}^1$-actions on manifolds of dimension 2 and 3

In this section, we give examples of classification theorems. The most interesting one is that of fixed point free $\mathbf{S}^1$-actions on orientable 3-manifolds (Seifert manifolds), which will be used in the sequel to understand periodic Hamiltonians on 4-manifolds. Both as an introduction and a preliminary exercise, we consider first $\mathbf{S}^1$-actions on surfaces.

I.3.a. $\mathbf{S}^1$-actions on surfaces.

We begin with a list of examples.

The torus. The circle $\mathbf{S}^1$ acts on the torus $\mathbf{T}^2 = \mathbf{S}^1 \times \mathbf{S}^1$ by multiplication on one summand. This is the unique principal $\mathbf{S}^1$-action on a surface. The readers are invited to prove this, solving Exercise I.8) in the statement of which they will find some hints.

The sphere. The group $\mathbf{S}^1$ also acts on the unit sphere $\mathbf{S}^2$ of $\mathbf{R}^3$ as the group of rotations around a fixed axis. All the orbits are principal, except for the two fixed points where the axis meets the sphere. The orbit space is a closed interval (Figure 2).

The projective plane. The same circle also acts on the projective plane $\mathbf{P}^2(\mathbf{R})$. Here are three possible (but nevertheless equivalent) descriptions(3) of this action:

- If $\mathbf{P}^2(\mathbf{R})$ is the quotient of the sphere $\mathbf{S}^2$ by the antipodal map $(x \mapsto -x)$, the action previously defined on the sphere descends to the quotient.
- If $\mathbf{P}^2(\mathbf{R})$ is the "projective completion" of $\mathbf{R}^2$ (to which a line at infinity is added), then the linear $\mathbf{S}^1 = \mathrm{SO}(2)$-action on $\mathbf{R}^2$ may be extended to $\mathbf{P}^2(\mathbf{R})$.
- If $\mathbf{P}^2(\mathbf{R})$ is the set of lines in $\mathbf{R}^3$ and is described by homogeneous coordinates $[x, y, z]$, then $\mathbf{S}^1 = \mathrm{SO}(2)$ acts by $A \cdot [x, y, z] = [A \cdot (x, y), z]$.

(3)The reader wanting to know more about the projective plane might enjoy having a look at the nice book [**4**].

In each of the three descriptions given it is clear that all orbits are principal except for two of them. One is a fixed point and the other has a $\mathbf{Z}/2$ as stabilizer (it corresponds to the equatorial circle of $\mathbf{S}^2$ for example). The orbit space is a closed interval, each of its ends corresponding to a non-principal orbit (Figure 3).

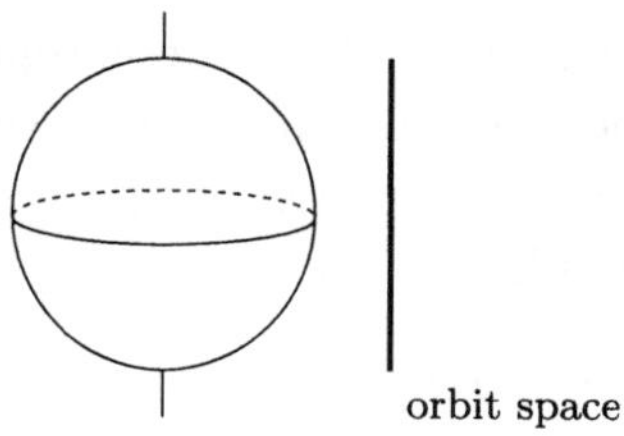

Figure 2

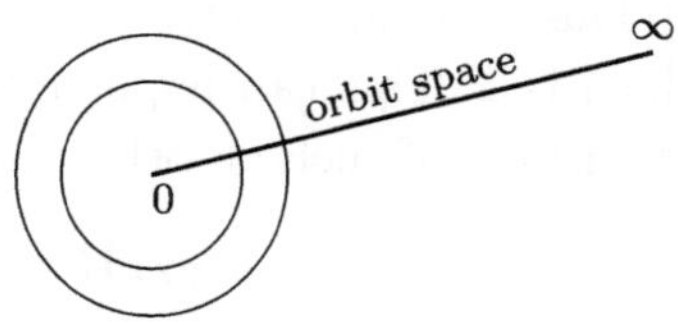

Figure 3

The Klein bottle. The circle also acts on the Klein bottle K. Recall (see Figure 4) that K is the quotient of $\mathbf{R}^2$ by the group of affine transformations generated by

$$t : (x, y) \longmapsto (x + 1, y) \text{ and } s : (x, y) \longmapsto (-x, y + 1).$$

We make $\mathbf{R}$ act on $\mathbf{R}^2$ by $(x, y) \mapsto (x, y + a)$. This induces an action of $\mathbf{R}/\mathbf{Z} = \mathbf{S}^1$

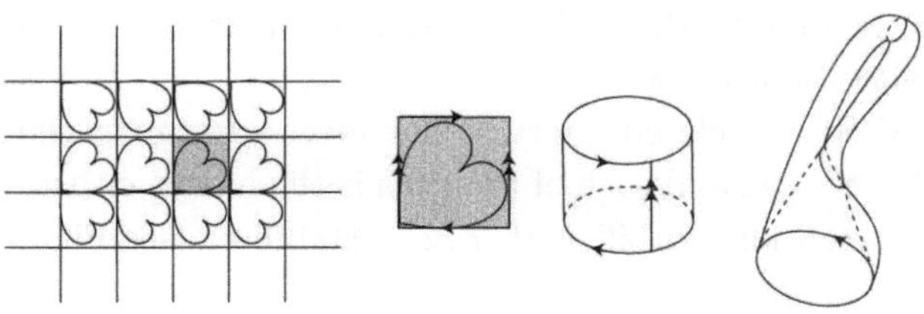

Figure 4

on K. The quotient map is identified with

$$\begin{aligned} K &\longrightarrow \mathbf{R}/(x \sim x + 1, x \sim -x) \\ [x, y] &\longmapsto [x] \end{aligned}$$

and this identifies the orbit space with the interval $[0, 1/2]$. All the orbits are principal, except those corresponding to 0 and $1/2$ which have a $\mathbf{Z}/2$ as stabilizer.

We prove now that this list exhausts all possible examples of closed surfaces endowed with effective $\mathbf{S}^1$-actions. Assume W is a closed (*i.e.*, compact and without boundary) connected surface.

The closed subgroups (and thus the possible orbit types) of $\mathbf{S}^1$ are: $\mathbf{S}^1$ itself and the finite cyclic groups $\{1\}$, $\mathbf{Z}/m$.

The principal orbits. As $\mathbf{S}^1$ is a commutative group and the action has been assumed to be effective, the principal orbits must have type $\{1\}$. In the neighborhood of a principal orbit, according to the slice theorem (Theorem I.2.1), the action is the product over $\mathbf{S}^1 \times \mathbf{R}$ of the multiplication and the trivial action. A neighborhood of such an orbit in the quotient space is thus homeomorphic to $\mathbf{R}$.

The exceptional orbits. The orbits of type $\mathbf{Z}/m$ $(m \geqslant 2)$ are exceptional. That group must have an effective action on the normal space to the orbit, the dimension of which is 1. This can happen only if $m = 2$, the action being $x \mapsto -x$. In the neighborhood of such an orbit, a tube looks like:

$$\mathbf{S}^1 \times_{\mathbf{Z}/2} \mathbf{R} = (\mathbf{S}^1 \times \mathbf{R})/(z,u) \sim (-z,-u).$$

This is a Möbius strip, thus the existence of an exceptional orbit forces W to be non-orientable. In the neighborhood of the orbit, the quotient map is $[z,t] \mapsto |t|$ and the quotient space is a half line.

The singular orbits. They have type $\mathbf{S}^1$: they are fixed points. Near such a point, the action linearizes as the standard $\mathbf{S}^1 = \mathrm{SO}(2)$-action on $\mathbf{R}^2$.

We have thus proved that the orbit space is a manifold of dimension 1 with boundary, it is of course compact and connected as W is. It suffices now to investigate all the possible cases.

(1) If $W/\mathbf{S}^1$ is a circle, the action is principal, and we have said that W is a torus, with the usual action.
(2) If not, $W/\mathbf{S}^1$ is a closed interval. The inverse image of the interior of this interval is $\mathbf{S}^1 \times \mathbf{R}$ and each of the ends is the image, either of a fixed point, or of an exceptional orbit, with $\mathbf{Z}/2$ as stabilizer (see Figure 5).

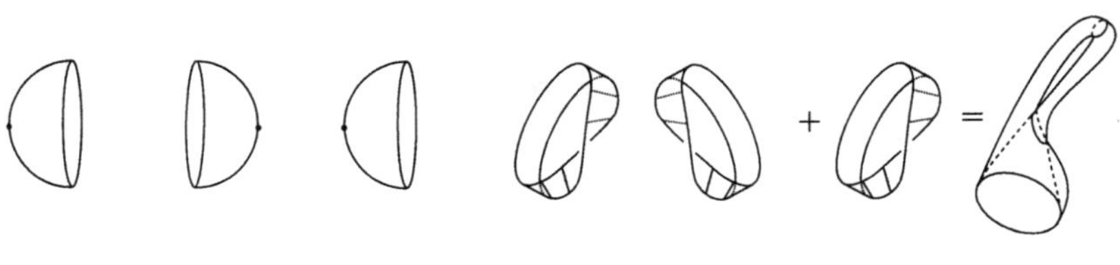

Figure 5

- If there are two fixed points, W is obtained by adding two points to $\mathbf{S}^1 \times \mathbf{R}$, and is thus a sphere.
- If there is only one fixed point, and one exceptional orbit, we must glue a disc to a Möbius strip along the boundary (which is a principal orbit). We thus get a projective plane.

– If both ends correspond to exceptional orbits, we must glue two Möbius strips along their common boundary, thus getting a Klein bottle.

I.3.b. Principal $\mathbf{S}^1$-actions on 3-manifolds. The study of 3-manifolds with fixed point free $\mathbf{S}^1$-action was initiated by Seifert during the 1930s [**122**]. The classification of these $\mathbf{S}^1$-actions is due to Raymond [**118**], who actually classified *topological* actions. Of course we shall restrict ourselves to the *differentiable* classification... However, both classifications turn out to be the same one.

We start with principal actions. The quotient space is then a closed surface B and the projection $\pi : W \to B$ is a "principal $\mathbf{S}^1$-bundle" over B (this is indeed a locally trivial bundle, according to the slice theorem).

In addition to the diffeomorphism type of B, we shall exhibit another invariant, a number called the *Euler class*. This is an integer when B is orientable (and an integer modulo 2 otherwise).

Figure 6

Assume the surface is orientable. Here is a description of the Euler class e. Choose a point in B and a disc D_0 around this point. The complement of the point has a wedge of $2g$ circles as deformation retract (see [**101**] and Figure 6). Note that g can be zero, the complement of D_0 is then a disc.

Recall (see, for instance, Exercise I.11) that a principal $\mathbf{S}^1$-bundle over a graph (or a topological space having the same homotopy type as a graph) is trivial. This allows us to choose a section $\sigma : \overline{B - D_0} \to \overline{W - \mathcal{V}_0}$. When restricted to D_0 the bundle is trivializable as well, thus $\mathcal{V}_0 = \pi^{-1}(D_0)$ is a solid torus

$$\mathcal{V}_0 = D_0 \times \mathbf{S}^1$$

and $W = (\overline{W - D_0}) \cup_\partial \mathcal{V}_0$. On the boundary $\partial\mathcal{V}_0$ we see two natural curves:

(1) The meridian a. By definition, this is a closed simple curve which bounds a disc in $\mathcal{V}_0$. Its homotopy class is (up to sign) well defined by this property: a is a generator of the kernel of

$$\pi_1(\partial\mathcal{V}_0) \cong \mathbf{Z}^2 \longrightarrow \pi_1(\mathcal{V}_0) \cong \mathbf{Z}$$

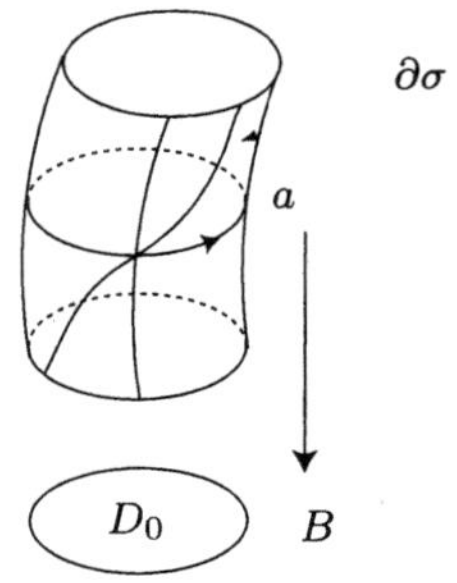

Figure 7

so that we can choose for a the boundary of a section of $\pi|_{D_0}$.

(2) Any orbit b of the $\mathbf{S}^1$-action. This is also a simple closed curve; with a chosen as above, the two curves meet at exactly one point.

As the surface B was supposed to be orientable and the action principal, it is easily checked that the manifold W is orientable as well. We choose an orientation of W, the induced orientation on $\mathcal{V}_0$, the boundary orientation[(4)] on $\partial\mathcal{V}_0$. We also orient b by the orientation of $\mathbf{S}^1$ and choose the orientation of a in such a way that $a \cdot b = +1$.

If $\partial\sigma$ is oriented *oppositely* to the boundary orientation, it meets b at one point and positively and we may write

$$\partial\sigma \sim a - eb$$

for some $e \in \mathbf{Z}$. This integer is the *Euler class* of the principal $\mathbf{S}^1$-action on W (or of the principal $\mathbf{S}^1$-bundle $W \to B$).

Proposition I.3.1. *The integer e is independent of all the choices made. Together with the diffeomorphism type of B, it determines the equivariant diffeomorphism type of W. For any orientable surface B and any integer e, there exists a manifold W endowed with a principal* $\mathbf{S}^1$*-action, the orbit space of which is B and the Euler class e.*

Proof. One constructs, given B and e, a manifold W with an $\mathbf{S}^1$-action the invariants of which are B and W:

We firstly choose a disc D_0 in B, then write $W_0 = \overline{(B - D_0)} \times \mathbf{S}^1$, endow it with the product action, thus $\sigma(x) = (x, 1)$ is a section. We consider also a solid torus $D^2 \times \mathbf{S}^1$ with the product action, and define an equivariant diffeomorphism

$$\varphi : \partial W_0 \longrightarrow \partial(D^2 \times \mathbf{S}^1)$$

(4) The convention we use to orient boundaries is the one in which the *outward* normal followed by the orientation of the boundary gives the orientation of the whole manifold.

reversing orientations. Make the choice of φ precise by prescribing which curve in ∂W_0 is sent to a meridian (*i.e.*, is homotopic to a constant in the solid torus). Take $\partial\sigma_0 + eb$, where b denotes the homotopy class of any orbit.

The manifold obtained this way has the right invariants. Conversely, if W is endowed with a principal $\mathbf{S}^1$-action with these invariants, it is easy to write down a diffeomorphism of W with the above model. □

Remark I.3.2. The Euler class e vanishes if and only if the fibration $W \to B$ has a section, in other words is trivializable.

Example I.3.3. The $\mathbf{S}^1$-action on $\mathbf{S}^3$ by $t\cdot(z_1,z_2) = (tz_1, tz_2)$ is principal and the quotient is a sphere $\mathbf{S}^2$ which we may consider here as the complex projective line $\mathbf{P}^1(\mathbf{C})$, the quotient mapping being the Hopf fibration:

$$\begin{aligned} \mathbf{S}^3 &\longrightarrow \mathbf{P}^1(\mathbf{C}) \\ (z_1,z_2) &\longmapsto [z_1,z_2] \end{aligned}$$

which associates with each unit vector the complex line it generates. The projective line $\mathbf{P}^1(\mathbf{C})$ is the union of the two discs

$$D_0 = \{[z_1,z_2] \mid |z_1| \leqslant |z_2|\} \cong \{t\in\mathbf{C} \mid |t|\leqslant 1\} \text{ by } t = \frac{z_1}{z_2}$$

and

$$D_\infty = \{[z_1,z_2] \mid |z_1| \geqslant |z_2|\} \cong \{u\in\mathbf{C} \mid |u|\leqslant 1\} \text{ by } u = \frac{z_2}{z_1},$$

neighborhoods respectively of $0 = [0,1]$ and of $\infty = [1,0]$. Hence the sphere decomposes into two solid tori on which the fibration is trivializable (see Figure 8).

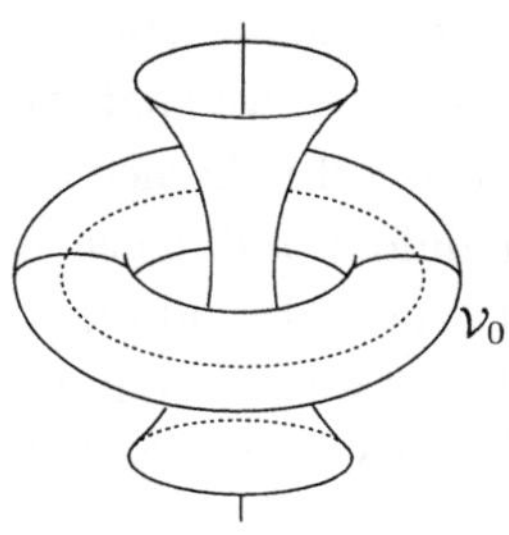

Figure 8

With the same notation as above, we can choose for a meridian a in $\partial\mathcal{V}_0$ the image of the section

$$t \longmapsto \frac{1}{\sqrt{2}}(t,1)$$

and for the curve b the orbit

$$\left\{\frac{1}{\sqrt{2}}(w,w) \mid w\in\mathbf{S}^1\right\} \text{ of the point } \frac{1}{\sqrt{2}}(1,1).$$

On the other side, $\partial\sigma$ is the image of the section

$$u \longmapsto \frac{1}{\sqrt{2}}(1, u)$$

(see Figure 9). Taking care of the orientations, we get

$$\partial\sigma \sim a + b$$

so that the Euler class of the Hopf bundle is -1.

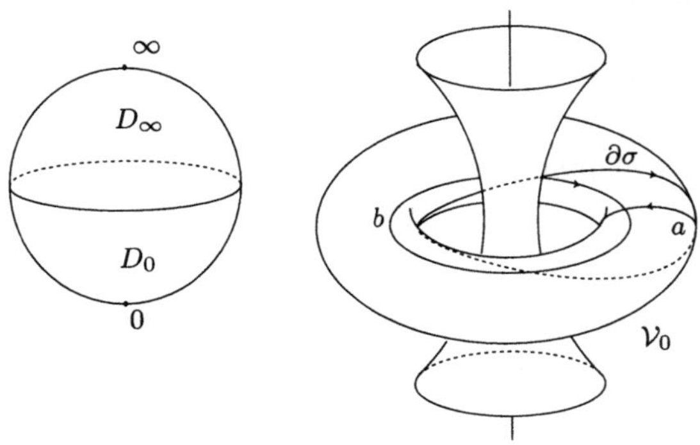

Figure 9

I.3.c. Oriented Seifert manifolds. We want now to generalize the previous study to the case of closed 3-manifolds endowed with an $\mathbf{S}^1$-action, still without fixed points but with possibly exceptional orbits. We shall restrict ourselves to the case of oriented manifolds. The proof we shall give of the classification theorem can be found in [**115**], another useful reference is [**24**].

Let W be an oriented manifold of dimension 3, endowed with an $\mathbf{S}^1$-action. There are principal orbits, of type $\{1\}$, and exceptional orbits, of type $\mathbf{Z}/m$ (for $m \geqslant 2$). As in the case of surfaces, we try to understand the local structure of the orbit space near an exceptional orbit with the help of the slice theorem I.2.1.

Orbits of type $\mathbf{Z}/m$. The slice has dimension 2 and $\mathbf{S}^1$ acts on $V = \mathbf{R}^2$ as a subgroup of O(2). If $m \geqslant 3$, $\mathbf{Z}/m$ sits in SO(2) as the subgroup of rotations of order m, but if $m = 2$, there are two different ways in which $\mathbf{Z}/2$ can be seen as a subgroup of O(2): it can be generated either by a reflection or by a rotation. Written with the help of a well-chosen complex coordinate in $\mathbf{R}^2$ and with $\xi = e^{2i\pi/m}$,

$$\xi \cdot v = \xi v$$

and possibly, if $m = 2$,

$$\xi \cdot v = \overline{v}.$$

In this case, the equivariant tube is

$$\mathbf{S}^1 \times \mathbf{C}/(z,v) \sim (-z,\overline{v}).$$

Writing $v = x + iy$, we see that this is the product of a Möbius strip and a line $\mathbf{R}$. This cannot happen since we have assumed that W was *orientable*.

Hence, all the orbits have tubular neighborhoods of the form $\mathbf{S}^1 \times_{\mathbf{Z}/m} \mathbf{C}$ where $\mathbf{Z}/m$ acts on $\mathbf{C}$ by rotations of order m.

Remark I.3.4. The exceptional orbits are isolated in W: in $\mathbf{S}^1 \times_{\mathbf{Z}/m} \mathbf{C}$ all the orbits are principal, except for the central ($v = 0$) one. In particular, as W is compact, there are only finitely many exceptional orbits.

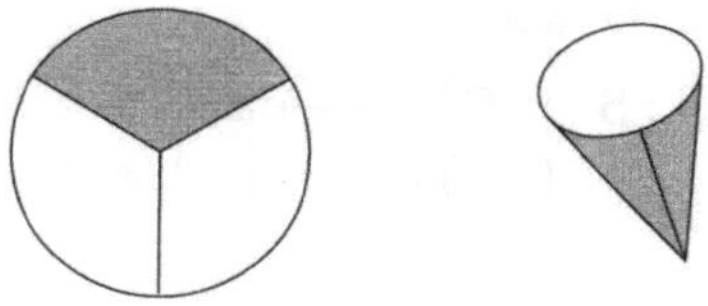

Figure 10

Near an orbit of type $\mathbf{Z}/m$, the orbit map is identified with

$$\begin{aligned} \mathbf{S}^1 \times_{\mathbf{Z}/m} \mathbf{C} &\longrightarrow \mathbf{C}/(\mathbf{Z}/m) \\ [z,v] &\longmapsto [v]. \end{aligned}$$

In particular, the quotient space is a closed topological surface B, together with the additional structure inherited from the differentiable structure of W: this is the structure of a differentiable manifold except at a finite number of points, where B has a conical "singularity". One says that B is an orbifold[5] (see § III.2.f). The additional structure is included in the $\mathbf{S}^1$-action on W. In this chapter we shall therefore concentrate on the *topological* surface B.

We shall only need to consider the case where B is an *orientable* surface: when these objects will appear again, in Chapter VIII, they will be symplectic (and hence oriented). In the sequel of this section, W is thus an oriented manifold and the quotient surface B is assumed to be orientable as well. The projection

$$W \longrightarrow B$$

is a *Seifert fibration* and the manifold W is a *Seifert manifold*.

---

[5] previously called a V-*manifold*.

The Seifert invariants of an exceptional orbit. Thanks to the slice theorem, a tubular neighborhood of an exceptional orbit looks like:

$$\mathcal{V} = \mathbf{S}^1 \times_{\mathbf{Z}/m} D^2$$

where $\mathbf{Z}/m$ acts as a subgroup on $\mathbf{S}^1$ and acts in a linear fashion on the disc $D^2 \subset \mathbf{R}^2 = \mathbf{C}$. This action is by rotations of order m, thus the quotient is given by

$$(z, v) \sim (\zeta^{-\beta} \cdot z, \zeta \cdot v) \quad \zeta = e^{2i\pi/m}$$

for some integer β, that is prime to m (and defined only modulo m).

The two relatively prime integers (m, β) with $0 < \beta < m$ are called the *Seifert invariants* of the exceptional orbit.

Let us show now that $\mathcal{V}$ is a solid torus. Let ν be an integer such that $\beta\nu = -1 \bmod m$. Consider the map

$$\begin{aligned} \widehat{\varphi} : \mathbf{S}^1 \times D^2 &\longrightarrow \mathbf{S}^1 \times D^2 \\ (z, v) &\longmapsto (z^m, z^{-\nu} v). \end{aligned}$$

We have

$$\widehat{\varphi}(\zeta^{-\beta} z, \zeta v) = (\zeta^{-m\beta} z^m, (\zeta^{-\beta} z)^{-\nu} \zeta v) = (z^m, z^{-\nu} v)$$

so that $\widehat{\varphi}$ factors through the $\mathbf{Z}/m$-action, defining a map

$$\varphi : \mathcal{V} \longrightarrow \mathbf{S}^1 \times D^2.$$

For $t \in \mathbf{S}^1$, we have

$$\varphi(t \cdot [z, v]) = \varphi([tz, v]) = (t^m z^m, t^{-\nu} z^{-\nu} v) = t \cdot \varphi(z, v)$$

for the $\mathbf{S}^1$-action on $\mathbf{S}^1 \times D^2$ defined by

$$t \cdot (Z, V) = (t^m Z, t^{-\nu} V).$$

In other words, φ is an equivariant mapping. Moreover, it is clearly bijective and an equivariant homeomorphism, hence the tube $\mathcal{V}$ is indeed a solid torus.

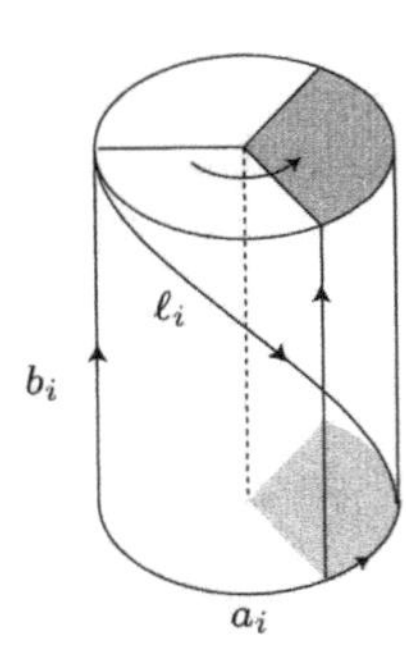

Figure 11

We have seen that any principal orbit turns m times around the exceptional orbit and thus meets any "meridian" at m points (see Figure 11 in which $m = 3$ and $\nu = 2$, where the principal orbit is b_i, composed of m segments in the figure, and where a_i is a meridian).

Examples I.3.5. Before giving the classification, let us investigate some examples.

(1) We have seen that, if m_1 and m_2 are relatively prime, the $\mathbf{S}^1$-action on $\mathbf{S}^3$ by
$$t \cdot (z_1, z_2) = (t^{m_1} z_1, t^{m_2} z_2)$$
is effective. The orbits $(z_1, 0)$ and $(0, z_2)$ are exceptional (if m_1 and m_2 are at least 2), with respective types $\mathbf{Z}/m_1$ and $\mathbf{Z}/m_2$.

A tubular neighborhood of the orbit $(z_1, 0)$ is a solid torus $\mathbf{S}^1 \times D^2$ on which $\mathbf{S}^1$ acts as above.

If α are β are chosen both by Bézout and by equation
$$\beta m_2 - \alpha m_1 = 1$$
with $0 < \beta < m_1$, $0 < \alpha < m_2$, then the orbit invariants are (m_1, β) and $(m_2, m_2 - \alpha)$.

(2) Let $\mathbf{Z}/p$ act on $\mathbf{S}^3$ by
$$\xi \cdot (z_1, z_2) = (\xi z_1, \xi^q z_2)$$
where $\xi = e^{2i\pi/p}$ and q is prime to p.

This is a *free* action of a finite group on a compact manifold. The quotient is a 3-manifold, denoted $L(p, q)$ and called a *lens space*(6) *of type* (p, q). Then the circle acts on $L(p, q)$ by projecting any action on the sphere. For instance, from the principal action we get an action with one exceptional orbit of type $\mathbf{Z}/q$ if $q \geqslant 2$.

Let us now come back to the classification problem: we have a finite number r of exceptional orbits in W, having $\mathbf{Z}/m_1, \dots, \mathbf{Z}/m_r$ as stabilizers and respective Seifert invariants $(m_1, \beta_1), \dots, (m_r, \beta_r)$. On the other hand, we have an orientable quotient surface B.

We remove r discs $D_1, \dots, D_r$ from B around the r points corresponding to the exceptional orbits, and we also remove a disc D_0 around a point corresponding to any principal orbit. This removes $r + 1$ solid tori $\mathcal{V}_0, \mathcal{V}_1, \dots, \mathcal{V}_r$ from W.

Thus, when restricted to $W - (\mathcal{V}_0 \cup \cdots \cup \mathcal{V}_r)$, the $\mathbf{S}^1$-action is principal and has a section (even if $r = 0$). We choose such a section
$$\sigma : \overline{B - (D_0 \cup \cdots \cup D_r)} \longrightarrow \overline{W - (\mathcal{V}_0 \cup \cdots \cup \mathcal{V}_r)}.$$

(6) This is an example of a manifold the fundamental group of which is $\mathbf{Z}/p$.

Its boundary $\partial\sigma$ has a component over each ∂D_i, that we call $\partial_i\sigma$. Consider the curve $\partial_i\sigma$. We know that it meets b_i at exactly one point, thus $(\partial_i\sigma, b_i)$ is a basis of $\pi_1(\partial\mathcal{V}_i)$. Choose a meridian a_i as above and a "parallel" ℓ_i in such a way that (a_i, ℓ_i) is a basis of $\pi_1(\partial\mathcal{V}_i)$ (oriented as $(\partial_i\sigma, b_i)$) (see Figure 11).

We have $a_i \cdot b_i = m_i$ and $\ell_i \cdot b_i = \nu_i$, and thus

$$\begin{cases} \partial_i\sigma = x a_i + y\ell_i \\ b_i = -\nu_i a_i + m_i \ell_i \end{cases}$$

that is

$$(\partial_i\sigma, b_i) = (a_i, l_i)\begin{pmatrix} x & -\nu_i \\ y & m_i \end{pmatrix}$$

where $xm_i + \nu_i y = 1$. Invert the matrix to get

$$(a_i, l_i) = (\partial_i\sigma, b_i)\begin{pmatrix} m_i & \nu_i \\ -y & x \end{pmatrix}.$$

Hence

$$a_i = m_i\partial_i\sigma - y b_i$$

with $y\nu_i = 1 - xm_i$ that is $-y \equiv \beta_i \bmod m_i$. The meridian a_i is thus written

$$a_i = m_i\partial_i\sigma + n_i b_i$$

where m_i is actually the order of the central orbit stabilizer in $\mathcal{V}_i$ and n_i is congruent to $\beta_i \bmod m_i$.

We can now modify the section σ in such a way that n_i is exactly β_i: any map $\varphi : (B - \cup D_i) \to \mathbf{S}^1$ can be used to modify σ just by multiplication

$$\sigma_\varphi(x) = \varphi(x) \cdot \sigma(x).$$

If the degree of $\varphi|_{\partial D_i}$ is d_i, then $\partial_i\sigma_\varphi \sim \partial_i\sigma + d_i b_i$ and

$$a_i = m_i\partial_i\sigma_\varphi + (n_i - d_i m_i) b_i.$$

To replace n_i by β_i in the decomposition of the meridian above it is thus sufficient to find a map $\varphi : B - \cup D_i \to \mathbf{S}^1$, such that $\deg\varphi|_{\partial D_i} = d_i$ and $0 < n_i - d_i m_i < m_i$. Now use the lemma:

Lemma I.3.6. *Let V be a connected surface the boundary of which has $r+1$ (circle) components $\partial V = \bigcup_{i=0}^r C_i$. Let $\psi : \partial V \to \mathbf{S}^1$ be a continuous map and let $d_i =$ $\deg(\psi|_{C_i})$. There exists an extension of ψ to V if and only if $\sum d_i = 0$.*

Proof. Stokes formula shows that the condition is necessary: we can assume that ψ is differentiable, if there exists an extension, there exists a smooth extension $\widetilde{\psi} : V \to \mathbf{S}^1$. If $d\theta$ denotes the volume form on $\mathbf{S}^1$, we compute

$$\sum_{i=0}^r d_i = \int_{\partial V} \psi^\star d\theta = \int_V d(\widetilde{\psi}^\star d\theta) = 0.$$

We prove the reciprocal by induction on r. If $r = 0$ (the case where the boundary is connected) and $d_0 = 0$, then ψ is homotopic to the constant map. We can use the homotopy to construct the extension $\widetilde{\psi}$: take it to be constant outside a collar (see Figure 12). To proceed from r to $r+1$, we cut the surface along the boundary

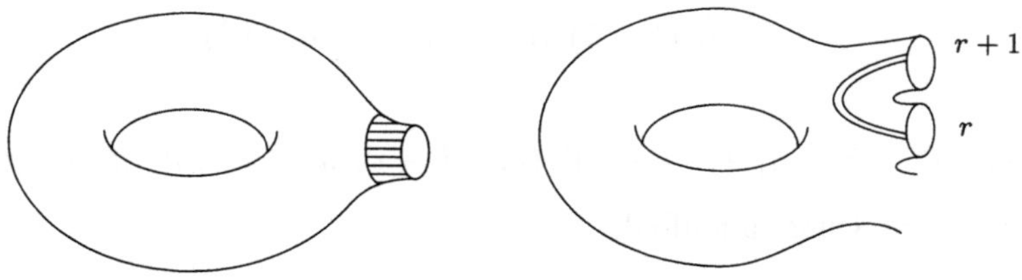

Figure 12

of a small strip connecting the components of the boundary labelled r and $r+1$ (as shown on the right part of the same figure). The boundary of the new surface has r connected components. Notice that the closed surface obtained from this new surface is the same as before. Moreover, the degree along the new boundary component is the sum $d_r + d_{r+1}$, so that the total sum is zero and the induction hypothesis gives the desired extension $\widetilde{\psi}$. □

Using the lemma, one can find a map φ which can be used to modify the n_i's over all the ∂D_i's for $1 \leqslant i \leqslant r$. This forces

$$\deg \varphi|_{\partial D_0} = -(d_1 + \cdots + d_r)$$

and at the same time forces an invariant β such that $\partial_0\sigma + \beta b_0$ is the meridian in $\mathcal{V}_0$ to appear and explains why we needed the "extra" tube $\mathcal{V}_0$. This way, we have determined a list of (normalized) invariants:

$$(g \mid \beta, (m_1, \beta_1), \dots, (m_r, \beta_r))$$

and we have, just as in Proposition I.3.1:

Theorem I.3.7. *Let W be a fixed point free oriented $\mathbf{S}^1$-manifold the orbit space of which is an oriented surface of genus g. The diffeomorphism type of W is determined by the list*

$$(g \mid \beta, (m_1, \beta_1), \dots, (m_r, \beta_r))$$

in which r is the number of exceptional orbits, β_i is a number which is prime to m_i and is chosen such that $0 < \beta_i < m_i$ and (m_i, β_i) are the Seifert invariants of the orbit numbered i. Reciprocally, given such a list one can find an $\mathbf{S}^1$-manifold with these invariants. □

Up to the order, the list is unique once we require the Seifert invariants to be normalized. Previous remarks (how to get normalized invariants) show moreover that the gluing data

$$(g \mid \beta, (m_1, \beta_1), \ldots, (m_r, \beta_r))$$

or

$$(g \mid 0, (1, -\beta), (m_1, \beta_1), \ldots, (m_r, \beta_r))$$

or even

$$(g \mid 0, (1, -\beta - (d_1 + \cdots + d_r)), (m_1, \beta_1 + d_1 m_1), \ldots, (m_r, \beta_r + d_r m_r))$$

define the same Seifert manifold.

I.3.d. Associated principal actions and Euler class. If W is a Seifert manifold, then for any integer n, one can make $\mathbf{Z}/n$ act on W as a subgroup of $\mathbf{S}^1$.

Proposition I.3.8. *Let W be an oriented Seifert manifold. Let n be a common multiple of the orders of the exceptional orbits. The manifold W', quotient of W by the $\mathbf{Z}/n$-action, is endowed with an action of $\mathbf{S}^1/(\mathbf{Z}/n) \cong \mathbf{S}^1$ which is both* effective *and* principal.

Proof. We look at the neighborhood of an orbit (principal or exceptional). We will prove that the action on W' is effective and principal, and, at the same time, that W' is a manifold.

Consider a tube $\mathcal{V}$, neighborhood of an orbit with stabilizer $\mathbf{Z}/m$ (m might be equal to 1), and write $n = mk$. On $\mathcal{V} \cong \mathbf{S}^1 \times D^2$, the circle $\mathbf{S}^1$ acts by

$$t \cdot (z, u) = (t^m z, t^{-\nu} u).$$

Take the quotient of the tube by $\mathbf{Z}/n \subset \mathbf{S}^1$:

$$\begin{aligned} \mathbf{S}^1 \times D^2 &\longrightarrow \mathbf{S}^1 \times D^2 \\ (z, u) &\longmapsto (z^k, u^{mk}). \end{aligned}$$

The quotient $\mathcal{V}'$ is a solid torus as well and the quotient map an n-fold covering (branched along the central orbit when this orbit is exceptional, that is, when $m \geqslant 2$). Looking at each tube, we see that the quotient W' of W by $\mathbf{Z}/n$ is a manifold. Let us look now at the $\mathbf{S}^1$-action. The diagram

$$\begin{array}{ccc} \mathcal{V} & \longrightarrow & \mathcal{V}' \\ \downarrow & & \downarrow \\ \mathcal{V} & \longrightarrow & \mathcal{V}' \end{array} \qquad \begin{array}{ccc} (z, u) & \longmapsto & (z^k, u^{mk}) = (Z, U) \\ \downarrow & & \downarrow \\ (t^m, t^{-\nu} u) & \longmapsto & (t^{mk} z^k, t^{-\nu mk} u^{mk} = (t^n Z, t^{-\nu n} U) \end{array}$$

shows that the $\mathbf{S}^1$-action on $\mathcal{V}$ (*resp.* W) induces an action of $\mathbf{S}^1/(\mathbf{Z}/n)$ on $\mathcal{V}'$ (*resp.* W') which is indeed effective and principal: after identification of $\mathbf{S}^1$ and $\mathbf{S}^1/(\mathbf{Z}/n)$ by $t \mapsto t^n$, this can be written $(tZ, t^{-\nu} U)$ on tubes. □

Notice that the quotient of W' by this principal action is the same surface B that was the quotient of W: each orbit of the action on W gives one and only one orbit in W'.

To define an *Euler class* for the $\mathbf{S}^1$-action on W, one can try to divide by n the Euler class of the principal action just defined on W'. We need then to verify that the rational number obtained in this way does not depend on the choice of n. It is of course sufficient to understand what happens when a multiple n' of n is chosen.

Proposition I.3.9. *Let W be a manifold of dimension 3 endowed with an action of the circle $\mathbf{S}^1$. Let W' be the quotient of W by the subgroup $\mathbf{Z}/k$ of $\mathbf{S}^1$. Make $\mathbf{S}^1 \cong \mathbf{S}^1/(\mathbf{Z}/k)$ act on W'. The Euler class e' of the new action is $e' = ke$.*

Proof. Use the same notation as in I.3.b. Let p be the projection of W on W'. If σ is a section of

$$\overline{W - \mathcal{V}_0} \longrightarrow \overline{B - D_0},$$

then $p \circ \sigma$ is a section of

$$\overline{W' - \mathcal{V}'_0} \longrightarrow \overline{B - D_0}.$$

On Figure 13, a and b represent a meridian and an orbit in $\partial\mathcal{V}_0$, a' and b' represent

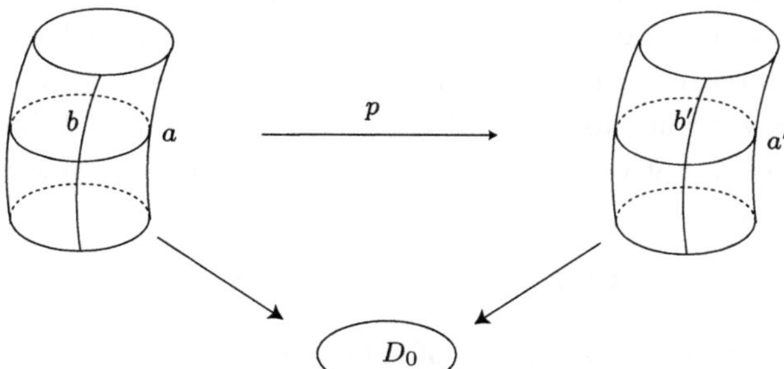

Figure 13

analogous objects in $\partial\mathcal{V}'_0$. Now,

$$p_\star : \pi_1\partial\mathcal{V}_0 \longrightarrow \pi_1\partial\mathcal{V}'_0$$

sends the class of a to that of a' and the class of b to that of kb'. It sends of course the class of $\partial\sigma$ to that of $\partial(p \circ \sigma)$. Thus:

$$a' - e'b' = [\partial\sigma'] = [\partial(p \circ \sigma)] = p_\star(a - eb) = a' - keb'. \qquad \square$$

Thus we get a well-defined Euler class $e \in \mathbf{Q}$ for all (oriented, with oriented base space) Seifert fibrations.

There exist other definitions of the Euler class for a Seifert fibration. We shall have the opportunity to meet one more later on. The interested readers may consult [**24**] for a rather complete list of definitions and references.

I.4. Appendix: Lie groups, Lie algebras, homogeneous spaces

Recall that a group G is said to be a *Lie group* if this is both a group and a manifold, the two structures being compatible in the sense that the group operations are differentiable mappings. This structure is particularly interesting, because it allows one to use the infinitesimal calculus.

The main examples are the ones used in this book, the (compact) torus $\mathbf{T}^n = \mathbf{R}^n/\mathbf{Z}^n$, the (complex) torus $(\mathbf{C}^\star)^n$, the linear groups $\mathrm{GL}(n;\mathbf{R})$, $\mathrm{GL}(n;\mathbf{C})$ and their subgroups, $\mathrm{O}(n)$, $\mathrm{U}(n)$, and so on.

The Lie algebra. Let G be a Lie group with unit 1. Denote by $\mathfrak{g}$ its tangent vector space $\mathsf{T}_1\mathrm{G}$ at the point 1. Recall that the map which, to a vector field tangent to G, associates its value at the point 1, defines an isomorphism from the vector space of left invariant vector fields on G to the tangent space $\mathfrak{g}$ (this is obviously a linear map, and it is easy to check that it is one-to-one).

We can thus think of the elements of $\mathfrak{g}$ as being vector fields on the group G. Moreover, the bracket of two left invariant vector fields is again a left invariant vector field (easy verification) so that the bracket of vector fields induces a bracket on $\mathfrak{g}$. As the bracket of vector fields, this bracket is skew-symmetric and satisfies the Jacobi identity

$$[[X,Y],Z]+[[Y,Z],X]+[[Z,X],Y]=0,$$

so that $\mathfrak{g}$ has, indeed, the structure of a *Lie algebra*, this is the[(7)] Lie algebra of the group G.

Here is an alternative description of the Lie bracket, through the useful notions of "Adjoint" and "adjoint", representations. Firstly, the group acts on itself by conjugation

$$\forall g \in \mathrm{G},\ \mathrm{G} \longrightarrow \mathrm{G}$$
$$h \longmapsto ghg^{-1}.$$

[(7)] Any Lie algebra will define (more or less uniquely, with some additional assumptions) a Lie group. When speaking of a Lie algebra $\mathfrak{g}$, I will always mean "the Lie algebra of the Lie group G" (whether the group G is explicitly mentioned or not).

We may differentiate this map at 1, this is the *Adjoint action*

$$\begin{aligned} \forall g \in \mathrm{G}, \mathrm{Ad}_g : \mathfrak{g} &\longrightarrow \mathfrak{g} \\ X &\longmapsto \mathrm{Ad}_g(X), \end{aligned}$$

which we may perfectly well consider as a map

$$\begin{aligned} \mathrm{G} &\longrightarrow \mathrm{End}(\mathfrak{g}) \\ g &\longmapsto \mathrm{Ad}_g \end{aligned}$$

and differentiate at 1

$$\begin{aligned} \mathfrak{g} &\longrightarrow \mathrm{End}(\mathfrak{g}) \\ X &\longmapsto \mathrm{ad}_X . \end{aligned}$$

The Lie bracket can also be defined by

$$\begin{aligned} \mathfrak{g} \times \mathfrak{g} &\longrightarrow \mathfrak{g} \\ (X, Y) &\longmapsto \mathrm{ad}_X(Y), \end{aligned}$$

namely $[X, Y] = \mathrm{ad}_X(Y)$, as is easily checked from the definitions.

Another viewpoint on the same object: ad_X is a vector field on $\mathfrak{g}$ (which value at Y is $[X, Y]$), this is the fundamental vector field of the adjoint action associated with X.

Examples I.4.1

(1) If the Lie group G is the torus $\mathbf{T}^n$, the Lie algebra is $\mathbf{R}^n$, as a vector space, with the trivial Lie bracket $[X, Y] \equiv 0$, this reflecting the fact that the group is Abelian and thus the Adjoint action trivial.

(2) Consider the group $\mathrm{G} = \mathrm{SO}(n)$ of the positive isometries in the Euclidean space $\mathbf{R}^n$. Differentiating the relation ${}^tAA = \mathrm{Id}$ at Id gives that the tangent space is the vector space $\mathfrak{g} = \mathfrak{so}(n)$ of $n \times n$ skew-symmetric matrices. Using the definitions above, one sees that the group $\mathrm{SO}(n)$ acts on $\mathfrak{so}(n)$ by conjugation (this is the Adjoint action here) and that the Lie bracket on $\mathfrak{g}$ is simply the commutator $[X, Y] = XY - YX$ (this will always be the case with a group of matrices, notice that in this case, the Jacobi identity for the Lie bracket becomes obvious).

There are other examples in the exercises at the end of this chapter.

The exponential mapping. Let $X \in \mathfrak{g}$. We can consider X as a vector field on G and, as such, integrate it: there exists a map

$$\begin{aligned}]-\varepsilon, \varepsilon[&\longrightarrow \mathrm{G} \\ t &\longmapsto \varphi_t(X) \end{aligned}$$

defined for ε small enough, and such that

$$\frac{d}{dt}\varphi_t(X) = X(\varphi_t(X)) \text{ and } \varphi_0(X) = 1.$$

Using the fact that G is a group, it is checked that φ_t is indeed defined on $\mathbf{R}$ (that is, for all values of t) and is a group homomorphism, namely

$$\varphi_{s+t}(X) = \varphi_s(X)\varphi_t(X).$$

This allows us to define a mapping

$$\begin{aligned} \mathbf{R} \times \mathfrak{g} &\longrightarrow \mathrm{G} \\ (t, X) &\longmapsto \varphi_t(X) \end{aligned}$$

and the *exponential mapping:*

$$\begin{aligned} \mathfrak{g} &\longrightarrow \mathrm{G} \\ X &\longmapsto \exp(X) = \varphi_1(X). \end{aligned}$$

Example I.4.2. Let G be the torus of dimension 1, $\mathrm{G} = \mathbf{R}/\mathbf{Z}$, so that $\mathfrak{g} = \mathbf{R}$ $\exp : \mathbf{R} \to \mathbf{S}^1$ is the... exponential mapping $t \mapsto e^{2i\pi t}$. Notice that we can also consider this torus as being the unit circle

$$\mathbf{S}^1 = \{u \in \mathbf{C} \mid |u| = 1\}$$

and the exponential mapping $\theta \mapsto e^{i\theta}$. They are not exactly the same, since the former has period 1 while the latter has period 2π. We shall try to be precise, using the notation t for the 1-periodic coordinate and θ for the 2π-periodic one.

The main properties of the exponential mapping are listed in the next proposition.

Proposition I.4.3. *The exponential mapping is a differentiable mapping that satisfies*

(1) $\exp(t+u)X = \exp tX \cdot \exp uX$,
(2) $\exp(-tX) = (\exp tX)^{-1}$.

Moreover, this is a diffeomorphism from a neighborhood of $0 \in \mathfrak{g}$ *onto a neighborhood of* $1 \in \mathrm{G}$.

Proof. The proof of the two equalities is straightforward, starting from the definition. The differentiability properties also come from the definition, which implies both the fact that exp is differentiable and that $(d\exp)_0 = \mathrm{Id}$. □

Remark I.4.4. The exponential mapping is not in general a *global* diffeomorphism. Think for instance of the example of the circle above.

Notice that it follows from the proposition that there is a whole neighborhood of 1 in the image of the exponential mapping. From this the next proposition is deduced.

Proposition I.4.5. *In a connected Lie group, any element can be written as a product of exponentials.*

Proof. A connected Lie group is path-connected. Let g be any element of the group G and let $\gamma : [0,1] \to G$ be a path from 1 to g. Let U be an open neighborhood of 1 that is included in the image of the exponential mapping. Let $g_1 = \gamma(t_1)$ be an element of $U \cap \gamma([0,1])$, for some $t_1 > 0$. We select then $g_2 = \gamma(t_2)$ (for some $t_2 > t_1$) in $g_1 \cdot U \cap \gamma([0,1])$, and so on. We thus get a covering of the compact set $\gamma([0,1])$ by the open subsets $g_k \cdot U$, which we can assume to be finite. Hence, g belongs to, say, $g_{n-1} \cdot U$, so that $g = g_{n-1} \exp(X_n)$, but g_{n-2} belongs to $g_{n-2} \cdot U$, and so on, so that eventually

$$g = \exp(X_1) \cdots \exp(X_n). \qquad \square$$

Closed subgroups of Lie groups. A classical and very useful theorem of Cartan asserts:

Theorem I.4.6. *A closed subgroup of a Lie group is a Lie subgroup.*

The main points are, $H \subset G$ being a closed subgroup, to show that

$$\mathfrak{h} = \{X \in \mathfrak{g} \mid \exp(tX) \in H \quad \forall t \in \mathbf{R}\}$$

is a vector space (this will be the Lie algebra of H), and to prove the following lemma.

Lemma I.4.7. *Let $\mathfrak{h}'$ be a vector subspace supplementary to $\mathfrak{h}$ in $\mathfrak{g}$. The mapping*

$$\begin{aligned} \varphi : \mathfrak{h} \oplus \mathfrak{h}' &\longrightarrow G \\ (X, X') &\longmapsto \exp X' \exp X \end{aligned}$$

is a diffeomorphism from a neighborhood of 0 *onto its image.*

The lemma itself is a consequence of the inverse function theorem and of the fact that $(d\varphi)_0(X + X') = X + X'$.

Another very important property is that homogeneous spaces of Lie groups are manifolds, this is what allows us to work with the orbit mappings as in § I.1.b.

Theorem I.4.8. *Let* H *be a closed subgroup of the Lie group* G *and let*

$$\pi : G \longrightarrow G/H$$

be the quotient map. There is a unique manifold structure on G/H *such that*

- *the quotient map π is a differentiable mapping,*
- *for any point $[x]$ of the quotient, there exists a neighborhood U of $[x]$ and a differentiable local section $\sigma : U \to G$ of π.*

Notice that π is then a submersion.

Proof. The quotient space G/H is endowed with the quotient topology. The Lie group G is Hausdorff and the subgroup H is closed, hence G/H is Hausdorff. Let us now look at the differentiability properties.

The proof of the uniqueness property is fairly standard. Assume $(\mathrm{G/H})'$ is the same set but with a (maybe) different differentiable structure. Let $[x] \in \mathrm{G/H}$, we have a neighborhood U and a differentiable section

$$\sigma : U \longrightarrow \mathrm{G}$$

so that $\mathrm{Id}_U = \pi \circ \sigma$ is a differentiable mapping, thus proving that

$$\mathrm{Id} : \mathrm{G/H} \longrightarrow (\mathrm{G/H})'$$

is differentiable at $[x]$. The same reasoning for the "other" identity map

$$\mathrm{Id} : (\mathrm{G/H})' \longrightarrow \mathrm{G/H}$$

shows that Id is a diffeomorphism, which is the expected uniqueness property.

Let us now prove the existence. To begin with, we prove the result in a neighborhood of $[1] \in \mathrm{G/H}$. Let us write, as in Lemma I.4.7, $\mathfrak{g} = \mathfrak{h} \oplus \mathfrak{h}'$, so that the mapping

$$\begin{aligned} \varphi : \mathfrak{h} \oplus \mathfrak{h}' &\longrightarrow \mathrm{G} \\ (X, X') &\longmapsto \exp(X')\exp(X) \end{aligned}$$

is a diffeomorphism from a neighborhood of 0, say V onto its image. We can assume that $V = V_1 \times V_1'$, for some $V_1 \subset \mathfrak{h}$, $V_1' \subset \mathfrak{h}'$. Let $U = \pi(\varphi(V))$. The set

$$\pi^{-1}(U) = \bigcup_{[x]\in U} \pi^{-1}([x]) = \bigcup_{y\in\varphi(V)} y \cdot \mathrm{H} = \bigcup_{h\in\mathrm{H}} \varphi(V) \cdot h$$

is an open subset of G, as $\varphi(V) \cdot h$ is open. Hence U is open, by definition of the quotient topology. Now the mapping

$$\begin{aligned} \tau : U &\longrightarrow \mathfrak{h}' \\ [x] &\longmapsto X' \text{ where } x = \exp(X')\exp(X) \end{aligned}$$

is both

- well-defined: let $x \in \varphi(V)$, $x = \exp(X')\exp(X)$ and $y \in \mathrm{H} \cap \exp(V)$, $y = \exp(Y)$, then

  $$xy = \exp(X')\exp(X)\exp(Y) = \exp(X')\exp(Z) \text{ for some } Z \in \mathfrak{h} \cap V$$

  (because φ is a diffeomorphism) and thus $\tau([xy]) = \tau[x]$,
- and injective: if $x = \exp(X')\exp(X)$ and $y = \exp(X')\exp(Y)$, we have

  $$y^{-1}x = (\exp(Y))^{-1}(\exp(X'))^{-1}\exp(X')\exp(X) = (\exp(Y))^{-1}\exp(X) \in \mathrm{H},$$

  thus $[x] = [y]$.

We can then use τ to put a differentiable structure on U, defined so that τ is a diffeomorphism (a local chard). Let us now prove that this structure satisfies the expected properties.

The mapping

$$\begin{aligned} \mathfrak{g} = \mathfrak{h} \oplus \mathfrak{h}' &\xrightarrow{\ p\ } \mathfrak{h}' \\ (X, X') &\longmapsto X' \end{aligned}$$

is differentiable, and so is the mapping

$$\varphi^{-1} : \varphi(V) \longrightarrow V.$$

It follows that the composition

$$\begin{aligned} \rho = p \circ \varphi^{-1} : \varphi(V) &\longrightarrow V_1' \\ \exp(X') \exp(X) &\longmapsto X' \end{aligned}$$

is differentiable and that $\tau^{-1} \circ \rho : \varphi(V) \to U$ is a differentiable mapping as well. But

$$\tau^{-1} \circ \rho(x) = \tau^{-1} \circ \rho(\exp(X') \exp(X)) = \tau^{-1}(X') = X$$

so that $\tau^{-1} \circ \rho = \pi$, hence π is differentiable.

Now, the expected section $\sigma : U \to \mathrm{G}$ is defined as $\sigma = \exp \circ \tau$, an obviously differentiable mapping, that satisfies

$$\pi \circ \sigma([x]) = \pi \circ \exp \circ \tau([x]) = \pi(\exp(X')) = [\exp(X')] = [\exp(X') \exp(X)] = [x]$$

so that $\pi \circ \sigma = \mathrm{Id}_U$.

We have thus proved the theorem in a neighborhood of $[1] \in \mathrm{G}/\mathrm{H}$. To prove it at the other points, we just need to use the left translations

$$L_x[y] = [xy]. \qquad \square$$

Exercises

Exercise I.1. Fix a real number α and let $\mathbf{R}$ act on the torus $\mathbf{T}^2 = \mathbf{R}^2/\mathbf{Z}^2$ by

$$t \cdot (x, y) = (x + t, y + \alpha t).$$

Find the orbits and show that these are submanifolds of $\mathbf{T}^2$ if (and only if) α is rational (hint: if α is irrational, all the orbits are dense).

Exercise I.2. Let $\mathrm{G} = \mathbf{C}^\star$ act on $\mathbf{C}^{n+1}$ by

$$\lambda \cdot v = \lambda v.$$

Let X be the orbit space. Prove that, for the quotient topology, the only open subset containing the image $[0]$ of 0 is X itself. Hence, X is not Hausdorff. Prove that $X - \{[0]\}$ is Hausdorff (and even compact). This is of course the complex projective space $\mathbf{P}^n(\mathbf{C})$.

Exercise I.3. Consider the circle acting on the 3-sphere by

$$t \cdot (z_1, z_2) = (t^{m_1} z_1, t^{m_2} z_2).$$

Find the orbits (they depend on m_1 and m_2). Imagine "analogous"(?) actions of the torus $\mathbf{T}^p$ in $\mathbf{C}^n$.

Exercise I.4. Prove that the orbits of the $\mathrm{U}(n)$-action on the real vector space of Hermitian matrices are the manifolds $\mathcal{H}_\lambda$ mentioned in the introduction of this book. Given a vector $\lambda \in \mathbf{R}^n$ what is the stabilizer of $\mathcal{H}_\lambda$? (hint: if $\lambda_1, \ldots, \lambda_n$ are distinct real numbers, the stabilizer of the diagonal matrix is the torus $\mathbf{T}^n$ of all *diagonal* unitary matrices.) See also Proposition II.1.15 below.

Exercise I.5. Prove that the $\mathbf{S}^1$-action on $\mathbf{C}^2$ is semi-free. When is the $\mathbf{S}^1$-action on $\mathbf{S}^3$ by $(t^{m_1} z_1, t^{m_2} z_2)$ free?

Exercise I.6. Let H and K be closed subgroups of G. Prove that there exists an equivariant map

$$\varphi : \mathrm{G/H} \longrightarrow \mathrm{G/K}$$

if and only if H is conjugate to a subgroup of K.

Exercise I.7. A Lie group G acts on a manifold V. Assume there exists a nonempty open subset on which the action is not effective. Prove that the action on V is not effective.

***Exercise I.8* (Surfaces endowed with principal $\mathbf{S}^1$-actions).** Assume that W is a closed surface endowed with a *principal* $\mathbf{S}^1$-action. Prove that the orbit space is a circle and that the equivariant tubular neighborhoods have the form $\mathbf{S}^1 \times \mathbf{R}$.

Prove that it is possible to glue all these tubes, writing $\mathbf{S}^1 = [0,1]/(0 \sim 1)$, and that W is obtained from $\mathbf{S}^1 \times [0,1]$ by the identification of $\mathbf{S}^1 \times 0$ and $\mathbf{S}^1 \times 1$ through an equivariant diffeomorphism φ.

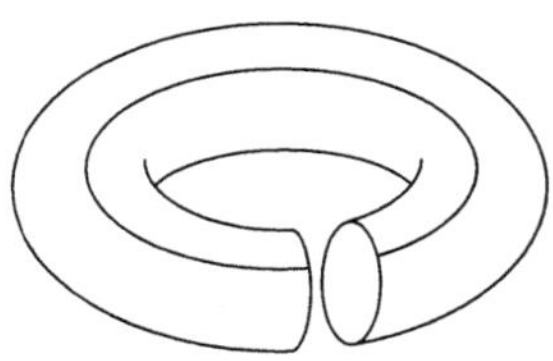

Figure 14

Show that φ is a rotation $z \mapsto az$ and that W is diffeomorphic, by an equivariant diffeomorphism, to $\mathbf{S}^1 \times \mathbf{S}^1$.

Exercise I.9. Check that the three $\mathbf{S}^1$-actions on $\mathbf{P}^2(\mathbf{R})$ described in § I.3 are the same.

Exercise I.10. Recall (or prove!) that a closed subgroup of SO(3) either is conjugate to SO(2) or O(2) or is finite. Use this fact to determine the surfaces that can be endowed with (a nontrivial!) SO(3)-action.

Exercise I.11. Recall (or prove) that any locally trivial ($\mathbf{S}^1$-)bundle over the interval $[0,1]$ is globally trivializable. Prove that a locally trivial $\mathbf{S}^1$-bundle over $\mathbf{S}^1$ is isomorphic to

$$\begin{aligned} [0,1]\times \mathbf{S}^1/(0,z)\sim(1,\varphi(z)) &\longrightarrow \mathbf{S}^1 \\ (t,z) &\longmapsto e^{2i\pi t} \end{aligned}$$

for some diffeomorphism $\varphi : \mathbf{S}^1 \to \mathbf{S}^1$. Assume now that this is a principal $\mathbf{S}^1$-bundle. Check that $z^{-1}\varphi(z)$ is a constant $a = \exp(2i\pi t_0) \in \mathbf{S}^1$ and that the map

$$\begin{aligned} [0,1]\times \mathbf{S}^1 &\longrightarrow [0,1]\times \mathbf{S}^1 \\ (t,z) &\longmapsto (t,\exp(-2i\pi t t_0)z) \end{aligned}$$

defines a global trivialization of our bundle. Deduce that, if the orbit space of a principal $\mathbf{S}^1$-action is either a wedge of circles, or more generally a graph, or the complement of a point in a surface or a disc, then the fibration is trivializable.

Exercise I.12. Assume p and q are coprime integers, with $2 \leqslant q < p$. Find the quotient and Seifert invariants for the exceptional orbit of the $\mathbf{S}^1$-action on the lens space $L(p,q)$. Give a description of $L(p,q)$ by gluing of two solid tori.

Exercise I.13. Prove that the Euler class of the Seifert manifold with invariants $(g \mid \beta, (m_1,\beta_1),\ldots,(m_r,\beta_r))$ is

$$e = \beta - \sum_{i=1}^{r} \frac{\beta_i}{m_i}.$$

Hint: both sides are equal in the case of a principal action.

Exercise I.14. Show that the action of $\mathbf{S}^1$ on $\mathbf{S}^3$ by

$$t\cdot(z_1,z_2) = (t^{m_1}z_1, t^{m_2}z_2) \quad (m_1 \text{ and } m_2 \text{ coprime}),$$

has invariants: $(0 \mid 1, (m_1,\beta)(m_2, m_2-\alpha))$ with $\beta m_2 - \alpha m_1 = 1$. Hint: the orbit invariants having been calculated in § I.3.5 and it is easy to compute the Euler class, then everything follows from the previous exercise.

Exercise I.15. The unitary group $\mathrm{U}(n)$ is the group of all complex $n\times n$-matrices satisfying ${}^t\overline{A}A = 1$. Check that its Lie algebra is

$$\mathfrak{u}(n) = \left\{X \in M_n(\mathbf{C}) \mid {}^t\overline{X} + X = 0\right\},$$

that is, the real vector space of skew-Hermitian matrices, matrices the real part of which is skew-symmetric and the imaginary part symmetric.

Exercise I.16. Prove that the map

$$\varphi : (x, y, z) \longmapsto \begin{pmatrix} 0 & -z & y \\ z & 0 & -x \\ -y & x & 0 \end{pmatrix}$$

is a Lie algebra isomorphism from $\mathbf{R}^3$ (endowed with the vector product) to $\mathfrak{so}(3)$ (endowed with the Lie bracket) and that conjugation of skew-symmetric matrices by rotations amounts to the canonical action of the rotations on $\mathbf{R}^3$, namely that, for $A \in \mathrm{SO}(3)$, the vector associated with the skew-symmetric matrix

$$A \begin{pmatrix} 0 & -z & x \\ z & 0 & -y \\ -y & x & 0 \end{pmatrix} A^{-1} \text{ is the vector } A \begin{pmatrix} x \\ y \\ z \end{pmatrix}.$$

Determine the adjoint orbits. Check that the map

$$\begin{aligned} \mathbf{R}^3 &\longrightarrow [0, +\infty[\\ v &\longmapsto \|v\|^2 \end{aligned}$$

is the quotient map $\mathfrak{so}(3) \to \mathfrak{so}(3)/\mathrm{SO}(3)$.

Exercise I.17. Let $\mathrm{G} = \mathrm{SO}(2,1)$ be the group of isometries of the quadratic form $x^2 + y^2 - z^2$ on $\mathbf{R}^3$. Let J be the matrix of this form. Check that $A \in \mathrm{SO}(2,1)$ if and only if ${}^tAJA = J$. Deduce that the Lie algebra of $\mathrm{SO}(2,1)$ is that of all matrices X such that ${}^tXJ + JX = 0$. Prove that that $\mathfrak{so}(2,1)$ is isomorphic to $\mathbf{R}^3$ (with the Lie algebra structure given by the vector product) by the map

$$\varphi : (x, y, z) \longmapsto \begin{pmatrix} 0 & -z & y \\ z & 0 & x \\ y & x & 0 \end{pmatrix}$$

and that the adjoint action is the standard action of $\mathrm{SO}(2,1)$ on $\mathbf{R}^3$. Determine the adjoint orbits.

Exercise I.18. The special unitary group $\mathrm{SU}(n)$ is the group of unitary matrices of determinant 1. Prove that the Lie algebra $\mathfrak{su}(n)$ is

$$\mathfrak{su}(n) = \left\{ X \in M_n(\mathbf{C}) \mid {}^t\overline{X} + X = 0 \text{ and } \det X = 1 \right\}.$$

Check that the Lie algebra $\mathfrak{su}(2)$ is

$$\mathfrak{su}(2) = \left\{ \begin{pmatrix} ia & b + ic \\ -b + ic & -ia \end{pmatrix} \mid a, b, c \in \mathbf{R} \right\}$$

and that the map

$$\varphi : \mathfrak{su}(2) \longrightarrow \mathbf{R}^3$$
$$\begin{pmatrix} ia & b+ic \\ -b+ic & -ia \end{pmatrix} \longmapsto \begin{pmatrix} a \\ b \\ c \end{pmatrix}$$

satisfies $\varphi([A, A']) = 2\varphi(A) \times \varphi(A')$. Deduce that $\mathfrak{su}(2)$ also[8] is isomorphic to $\mathbf{R}^3$ with the vector product.

Exercise I.19. Check that any vector in $\mathbf{R}^3$ is the vector product of two vectors of $\mathbf{R}^3$. Deduce that the Lie algebras $\mathfrak{so}(3)$, $\mathfrak{so}(2,1)$, $\mathfrak{su}(2)$ satisfy

$$\mathfrak{g} = [\mathfrak{g}, \mathfrak{g}].$$

The groups SO(3), SO(2, 1), SU(n), are said to be *semi-simple.*

***Exercise I.20* (Adjoint orbits of SO(4)).** In this exercise, we consider $\mathbf{R}^4$ with its canonical basis (e_1, e_2, e_3, e_4) and its canonical Euclidean structure (for which the canonical basis is orthonormal). The special orthogonal group SO(4) acts on $\mathbf{R}^4$ by isometries and on its Lie algebra $\mathfrak{so}(4)$ (the Lie algebra of 4×4-skew-symmetric matrices) by the adjoint action.

(1) A skew-symmetric matrix A defines a skew-symmetric bilinear form on $\mathbf{R}^4$ by

$$A \longmapsto ((x,y) \longmapsto {}^t x A y).$$

Check that this defines an isomorphism of vector spaces

$$\mathfrak{so}(4) \longrightarrow \bigwedge^2 (\mathbf{R}^4)^\vee.$$

Prove that the adjoint action defines an action of SO(4) on $\bigwedge^2(\mathbf{R}^4)^\vee$ by

$$(g \cdot \alpha)(x,y) = \alpha(g^{-1}x, g^{-1}y).$$

(2) The Hodge star operator acts on $\bigwedge^2(\mathbf{R}^4)^\vee$ by

$$(\star\alpha) \wedge \beta = (\alpha, \beta) \det \quad \forall \beta \in \bigwedge^2(\mathbf{R}^4)^\vee$$

where (,) denotes the Euclidean inner product on $\bigwedge^2(\mathbf{R}^4)^\vee$ for which the basis $(e_i^\star \wedge e_j^\star)_{i<j}$ is orthonormal and $\det = e_1^\star \wedge e_2^\star \wedge e_3^\star \wedge e_4^\star$. Prove that $\star$ is an involution and that

$$\star(e_1^\star \wedge e_2^\star) = e_3^\star \wedge e_4^\star, \quad \star(e_1^\star \wedge e_3^\star) = -e_2^\star \wedge e_4^\star, \quad \star(e_1^\star \wedge e_4^\star) = e_2^\star \wedge e_3^\star.$$

Determine the eigenspaces $\bigwedge_+$ and $\bigwedge_-$ of $\star$ for the eigenvalues $+1$ and -1 (respectively!), proving in particular that both have dimension 3.

[8] Using the quaternions, it is classical and easy to prove that the group SU(2) is a two-fold covering of the group SO(3), *the* good reason why SU(2) and SO(3) have isomorphic Lie algebras.

(3) Hence we have a decomposition

$$\bigwedge^2(\mathbf{R}^4)^\vee = \bigwedge_+ \oplus \bigwedge_- .$$

Prove that SO(4) acts on $\bigwedge_+$ and $\bigwedge_-$ by isometries. Deduce that the generic adjoint orbits of SO(4) are diffeomorphic with $\mathbf{S}^2 \times \mathbf{S}^2$.

(4) A skew-symmetric matrix can be written, uniquely, in the decomposition above

$$A = A_+ + A_-.$$

The adjoint orbit of A is determined by the two numbers $\|A_+\|^2$ and $\|A_+\|^2$ or, equivalently, by the two numbers $\|A_+\|^2 + \|A_+\|^2$ and $\|A_+\|^2 - \|A_+\|^2$. Check that $\|A_+\|^2 + \|A_+\|^2 = \|A\|^2$ and that $\|A_+\|^2 - \|A_+\|^2$ is (half) the *Pfaffian* of A, namely, if $A = (a_{i,j})$, that

$$\|A_+\|^2 - \|A_+\|^2 = \frac{1}{2}(a_{1,2}a_{3,4} - a_{1,3}a_{2,4} + a_{1,4}a_{2,3}) = \frac{1}{2}\operatorname{Pf}(A).$$

***Exercise I.21* (Small orbits of** SO(n)**).** Let the Lie group SO(n) act on its Lie algebra, the vector space $\mathfrak{so}(n)$ of skew-symmetric matrices, by conjugation. Let

$$A = \left(\begin{array}{cc|c} 0 & -a & \\ a & 0 & 0 \\ \hline & & \\ \multicolumn{2}{c|}{0} & 0 \\ & & \end{array}\right) \quad \text{for some real } a \neq 0.$$

Prove that the orbit of A consists of all the skew-symmetric matrices of rank 2 with square norm equal to $\|A\|^2 = a^2$. Determine the stabilizer of A and deduce that

$$\operatorname{SO}(n) \cdot A \cong \operatorname{SO}(n)/\operatorname{SO}(2) \times \operatorname{SO}(n-2).$$

This homogeneous space can also be interpreted as being the Grassmannian $\widetilde{G}_2(\mathbf{R}^n)$ of all oriented 2-planes in $\mathbf{R}^n$. What is the dimension of this orbit?

Assume that $n = 4$. Prove that the Grassmannian $\widetilde{G}_2(\mathbf{R}^4)$ is diffeomorphic to $\mathbf{S}^2 \times \mathbf{S}^2$ (see Exercise I.20).

Let now the group SO(n) act as usual on $\mathbf{R}^n$, and (diagonally) on $\mathbf{C}^n$. Check that this defines an action of SO(n) on $\mathbf{P}^{n-1}(\mathbf{C})$ and that the orbit of the point $[1, i, 0, \ldots, 0]$ is the quadric hypersurface of equation

$$\sum_{i=1}^{n} z_i^2 = 0$$

and is equivariantly diffeomorphic to our Grassmannian.

CHAPTER II

SYMPLECTIC MANIFOLDS

In this chapter, we define symplectic manifolds, give their basic properties, and give a few examples of such objects, among which the $\mathcal{H}_\lambda$'s we have already met: $\mathcal{H}_\lambda$ is the set of all $n \times n$ Hermitian matrices with given spectrum $\lambda = (\lambda_1, \ldots, \lambda_n) \in \mathbf{R}^n$. We already know that $\mathcal{H}_\lambda$ is indeed a manifold, as this is an orbit of the compact group $\mathrm{U}(n)$ (see Example I.1.6) acting by conjugation on the vector space $\mathcal{H}$ of all Hermitian matrices. We will prove that they are *symplectic* manifolds.

II.1. What is a symplectic manifold?

II.1.a. Symplectic vector spaces. A *symplectic* vector space is a real vector space endowed with a *nondegenerate* skew-symmetric bilinear form.

Example II.1.1. Consider a vector space F, its dual $F^\star$, and the direct sum $F \oplus F^\star$. The bilinear form

$$\begin{aligned}(F \oplus F^\star) \times (F \oplus F^\star) &\longrightarrow \mathbf{R}\\ ((x, \varphi), (x', \varphi')) &\longmapsto \varphi(x') - \varphi'(x)\end{aligned}$$

is obviously skew-symmetric and nondegenerate, and thus gives $F \oplus F^\star$ the structure of a symplectic vector space.

Example II.1.2. Consider the complex vector space $\mathbf{C}^n$, endowed with its standard Hermitian form, denoted

$$(X, Y) \longmapsto \langle X, Y \rangle.$$

We look at $\mathbf{C}^n$ as a *real* $2n$-dimensional vector space and consider the real bilinear form

$$\omega(X, Y) = \operatorname{Im}\langle X, Y \rangle.$$

This is skew-symmetric

$$\operatorname{Im}\langle X, Y \rangle = \operatorname{Im} \overline{\langle X, Y \rangle} = -\operatorname{Im}\langle X, Y \rangle,$$

and nondegenerate: if $\mathrm{Im}\langle X, Y\rangle = 0$ for all Y, then $\mathrm{Im}\langle X, iY\rangle = 0$ for all Y, and $\langle X, Y\rangle = 0$ for all Y, hence $X = 0$ since the Hermitian form is nondegenerate.

These two examples are not very different. They are actually isomorphic, since there is a unique (up to isomorphism) symplectic vector space in any (even) dimension. More precisely, the dimension of a symplectic vector space is even and is the only invariant of its isomorphism type. This is a consequence of the existence of symplectic bases, asserted by the following proposition.

Proposition II.1.3. *Let ω be a nondegenerate skew-symmetric bilinear form on a finite dimensional real vector space E. There exists a symplectic basis $(e_1, \dots, e_n, f_1, \dots, f_n)$, namely a basis of E such that $\omega(e_i, f_j) = \delta_{i,j}$, and $\omega(e_i, e_j) = \omega(f_i, f_j) = 0$.*

Proof. As ω is nondegenerate, it is not identically zero. Hence, there exist two vectors e_1 and f_1 such that $\omega(e_1, f_1) = 1$. One then checks that the restriction of ω to the orthogonal complement (with respect to ω) of the plane $\langle e_1, f_1\rangle$ is nondegenerate. One eventually concludes by induction on the dimension—once noticing that an alternated bilinear form on a vector space of dimension 1 is zero. □

Example II.1.4. Let $(x_1, \dots, x_n)$ be any basis of the vector space F, $(x_1^\star, \dots, x_n^\star)$ be the dual basis of $F^\star$. Then the basis

$$((x_1, 0), \dots, (x_n, 0), (0, -x_1^\star), \dots, (0, -x_n^\star))$$

is a symplectic basis for the symplectic vector space $F \oplus F^\star$ of Example II.1.1.

Example II.1.5. Let $(e_1, \dots, e_n)$ be a unitary basis of the complex vector space $\mathbf{C}^n$. Write $f_j = -ie_j$, then

$$\mathrm{Im}\langle e_j, f_k\rangle = -\mathrm{Im}\langle e_j, ie_k\rangle = -\mathrm{Im}(-i\langle e_j, e_k\rangle) = \langle e_j, e_k\rangle = \delta_{j,k}$$

and, analogously, $\omega(e_j, e_k) = \omega(f_j, f_k) = 0$, thus the basis

$$(e_1, \dots, e_n, -ie_1, \dots, -ie_n)$$

is a symplectic basis of the symplectic vector space $\mathbf{C}^n$ of Example II.1.2.

The symplectic group. This is the group $\mathrm{Sp}(2n; \mathbf{R}) \subset \mathrm{GL}(2n; \mathbf{R})$ of all "isometries" of the symplectic form. This is of course a Lie subgroup of $\mathrm{GL}(2n; \mathbf{R})$. If J denotes the skew-symmetric matrix

$$J = \begin{pmatrix} 0 & \mathrm{Id} \\ -\mathrm{Id} & 0 \end{pmatrix}$$

(so that $\omega(x, y) = {}^t x J y$),

$$\mathrm{Sp}(2n; \mathbf{R}) = \left\{ g \in \mathrm{GL}(2n; \mathbf{R}) \mid {}^t g J g = J \right\}.$$

Its Lie algebra is the space of matrices X such that

$$^{t}XJ + JX = 0,$$

the linear equation obtained as the differential, at the point Id, of the equation $^{t}gJg - J = 0$.

The symplectic form as a differential form. Let E be a symplectic vector space of dimension $2n$. We choose a symplectic basis, according to Proposition II.1.3 and write the coordinates of a vector

$$x = \sum_{i=1}^{n}(x_i e_i + y_i f_i)$$

in this basis, so that

$$\omega(x, x') = \sum_{i=1}^{n}(x_i y_i' - x_i' y_i).$$

One can thus write ω as a differential form

$$\omega = \sum_{i=1}^{n} dx_i \wedge dy_i.$$

This is an *exact* differential form (the differential of a degree 1-form):

$$\omega = d(\sum_{i=1}^{n} x_i dy_i) = -d\alpha,$$

where the form α is $\sum_{i=1}^{n} y_i dx_i$. The 1-form α is called the *Liouville form.*

II.1.b. The definition of a symplectic manifold. Let us try now to understand what could be the definition of a *symplectic* manifold W. One would first ask that each tangent space is endowed with the structure of a symplectic vector space as defined above. For any $x \in W$, we should be given a skew-symmetric bilinear form ω_x, say, on $\mathsf{T}_x W$. Of course, ω_x should depend smoothly on x, in other words, the collection of all the ω_x's should be a differential 2-form on W. We should ask, moreover, that the skew-symmetric form ω_x is nondegenerate, which can be expressed by writing, if the dimension (obviously even) of the manifold W is denoted $2n$,

$$\bigwedge^{n} \omega_x \neq 0 \in \bigwedge^{2n} \mathsf{T}_x^\star W$$

for all x, that is to say that the $2n$-form $\omega^{\wedge n}$ is a volume form on W. Note in particular that, $\omega^{\wedge n}$ being a volume form on W, this manifold must be orientable.

In fact this is not sufficient: the calculus that ω allows on W should be locally the same as that on our standard linear examples. Recall from above that the bilinear form on a symplectic vector space is an exact 2-form (since $\omega = -d\alpha$). We could try to require that the symplectic form ω on our supposed symplectic manifold is an exact form. Unfortunately, this would prevent the manifold W from being compact (which is a little embarrassing if one wants the $\mathcal{H}_\lambda$'s to be symplectic manifolds) because of the following proposition.

Proposition II.1.6. *Let W be a compact manifold. There exists no 2-form on W which is both nondegenerate and exact.*

Proof. We have said that the 2-form ω is nondegenerate if and only if $\omega^{\wedge n}$ is a volume form. Now, if $\omega = d\beta$ is exact, then

$$\omega^{\wedge n} = d(\beta \wedge \omega^{\wedge(n-1)})$$

is also an exact form. This is forbidden by the Stokes formula, the integral on W of the volume form is positive, hence this would give:

$$0 \neq \int_W \omega^{\wedge n} = \int_W d(\beta \wedge \omega^{\wedge(n-1)}) = \int_{\partial W} \beta \wedge \omega^{\wedge(n-1)} = 0$$

since W is a closed manifold. □

In fact, this was not a very good idea, in order to be able to calculate *locally* as in a symplectic vector space, to use a... *global* property such as of the exactness of a differential form. The local notion corresponding to exactness is the notion of closeness... and eventually we can define a *symplectic manifold* as a pair (W, ω) where W is a manifold and ω a closed nondegenerate 2-form.

We shall very often call the manifold W itself a symplectic manifold (where no ambiguity is possible).

Examples II.1.7

(1) Any symplectic vector space, with its "constant" symplectic form $\sum dx_j \wedge dy_j$ is a symplectic manifold.

(2) If V is a manifold, consider the total space of its cotangent bundle, with the projection:

$$\pi : \mathsf{T}^\star V \longrightarrow V.$$

On $\mathsf{T}^\star V$, there is a canonical[(1)] differential 1-form, the *Liouville form*, α, defined by

$$\alpha_{(x,\varphi)}(X) = \varphi(\mathsf{T}_x\pi(X)),$$

where $x \in V$, $\varphi \in \mathsf{T}^\star_x V$ is a linear form on $\mathsf{T}_x V$, and $X \in \mathsf{T}_{(x,\varphi)}(\mathsf{T}^\star V)$. The 2-form $\omega = -d\alpha$ is exact, hence closed. If $(x_1, \ldots, x_n)$ are local coordinates on V, and if $(y_1, \ldots, y_n)$ are the "dual" coordinates, then $(x_1, \ldots, x_n, y_1, \ldots, y_n)$ is a system of local coordinates on $\mathsf{T}^\star V$ in which $\alpha = \sum y_i dx_i$ and $\omega = \sum dx_i \wedge dy_i$. This shows that ω is nondegenerate.

---

(1) See Exercise II.12 for a justification of the word *canonical*.

We shall often refer to this form ω as *the* symplectic form on the cotangent bundle[(2)]. The example of $F \oplus F^\star$ above may be considered as the special case where $V = F = \mathbf{R}^n$.

(3) If W is a surface, then the notion of symplectic form coincides with that of volume form: all the orientable surfaces may thus be potentially considered as symplectic manifolds. Here are two examples, that are also shown in Figure 1.

(4) Consider, in particular, the unit sphere $\mathbf{S}^2$ in $\mathbf{R}^3$. The tangent space of $\mathbf{S}^2$ at a point v is the plane orthogonal to the unit vector v. Put

$$\omega_v(X, Y) = v \cdot (X \times Y) = \det(v, X, Y).$$

This is a nondegenerate 2-form and thus a symplectic form.

(5) The 2-torus is the quotient of $\mathbf{R}^2$ by its subgroup $\mathbf{Z}^2$. The "constant" symplectic form $dx \wedge dy$ is invariant by the integral translations, hence it descends to a symplectic form on the torus.

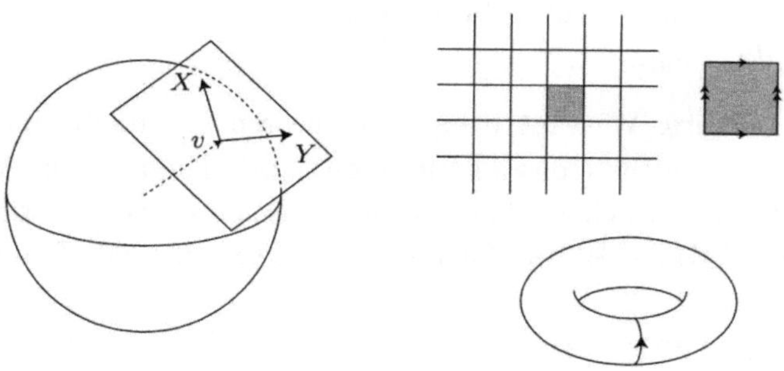

Figure 1

***Remark II.1.8* (The symplectic volume).** As we have noticed, if ω is a symplectic form, $\omega^{\wedge n}$ is a volume form. The $2n$-form $\omega^{\wedge n}/n!$ is called the *symplectic volume form*. Notice that

$$\left(\sum dx_i \wedge dy_i\right)^{\wedge n} = n! dx_1 \wedge dy_1 \wedge \cdots \wedge dx_n \wedge dy_n,$$

hence the normalization, which gives volume 1 to the torus $\mathbf{R}^{2n}/\mathbf{Z}^{2n}$.

(2) One more example of an exact symplectic form, on a noncompact manifold.

II.1.c. The Darboux theorem. The fact that all the symplectic vector spaces of the same dimension are isomorphic is the linear version of a more general result, the Darboux theorem, which asserts that, locally, all the symplectic manifolds are isomorphic. We will use a more general form, due to Weinstein, which describes symplectic manifolds in the neighborhood of certain submanifolds.

***Theorem II.1.9* ([128]).** *Let ω_0 and ω_1 be two symplectic forms on a manifold W. Assume they coincide on a closed submanifold V. Then there exists a neighborhood U_0 of V in W and a map*

$$\psi : U_0 \longrightarrow W$$

such that

$$\psi|_V = \mathrm{Id}_V \text{ and } \psi^\star\omega_1 = \omega_0.$$

Remark II.1.10. The map ψ given by this statement is automatically a local diffeomorphism because $\omega_0^{\wedge n} = \psi^\star\omega_1^{\wedge n}$ and because these two $2n$-forms are volume forms.

***Corollary II.1.11* (Darboux).** *Let x be a point in a symplectic manifold (W, ω). There exists a system of local coordinates $(x_1, \ldots, x_n, y_1, \ldots, y_n)$ centered at x in which $\omega = \sum dx_i \wedge dy_i$.*

Proof of the corollary. We want to apply the theorem to the submanifold $V = \{x\}$, by comparing ω with the (constant) form on T_xW. The theorem does not allow this, the two forms not being defined on the same space. We must first use a local diffeomorphism $\mathsf{T}_xW \to W$ and for instance the exponential of any Riemannian metric will do.

Thus let $\varphi = \exp_x : V_0 \to U_0$, where $V_0 \subset \mathsf{T}_xW$ and $V_0 \subset W$ are open subsets, so small that φ is a diffeomorphism. Define $\omega_0 = (\varphi^{-1})^\star\omega_x$. Now ω_0 and ω are two symplectic forms on U_0, which coincide at x by definition. We can thus apply the theorem: if necessary by restricting U_0, there exists a diffeomorphism ψ such that $\psi^\star\omega = \omega_0$. If $(x_1, \ldots, x_n, y_1, \ldots, y_n)$ are coordinates with respect to a symplectic basis of the vector space T_xW, the local chart $\psi \circ \varphi$ transforms them into a system of local coordinates in which $\omega = \sum dx_i \wedge dy_i$. □

Proof of the theorem. We apply the path method due to Moser [**111**]. This means that we consider the form $\omega_t = \omega_0 + t(\omega_1 - \omega_0)$. This is a closed form, and, as ω_0 and ω_1 coincide on V, it is nondegenerate on V and thus on a tubular neighborhood of V (nondegeneracy is clearly an open property). The segment $[0, 1]$ being compact, there exists a tubular neighborhood U of V on which all the forms ω_t are symplectic.

As ω_0 and ω_1 are closed, so is $\omega_0 - \omega_1$ and it is thus possible to find, on a neighborhood of V, a 1-form β such that $d\beta = \omega_0 - \omega_1$. Indeed, U is diffeomorphic with an open neighborhood of the zero section of a vector bundle on V, hence retracts on V. The Poincaré lemma applies to $\omega_0 - \omega_1$. It gives us the form β, that can be chosen so that $\beta_x = 0$ for all $x \in V$.

The 2-form ω_t being symplectic, it defines a pairing between tangent and cotangent spaces, we thus get, for all t, a vector field X_t dual to β through ω_t, that is, such that $i_{X_t}\omega_t = \beta$. We consider now X_t as a vector field depending on time, vanishing on V for all t. Its flow φ_t keeps V fixed, thus one can find a neighborhood U_0 of x where φ_t is defined and such that $\varphi_t(U_0) \subset U_0$.

We thus get:

$$\frac{d}{dt}\left[\varphi_t^\star \omega_t\right] = \varphi_t^\star \left[\frac{d\omega_t}{dt} + \mathcal{L}_{X_t}\omega_t\right] = \varphi_t^\star \left[\omega_1 - \omega_0 + \omega_0 - \omega_1\right] = 0$$

because $\mathcal{L}_{X_t}\omega_t = d i_{X_t}\omega_t + i_{X_t} d\omega_t = d\beta$ using the Cartan formula and the definition of φ. The form $\varphi_t^\star\omega_t$ does not depend on t, it equals ω_0 for $t = 0$ and we deduce the result with $\psi = \varphi_1$. □

The theorem has corollaries which describe a *normal form* for a symplectic manifold in the neighborhood of a submanifold on which the symplectic form has constant rank. We will use the case where the submanifold is *symplectic* (this is Corollary II.1.12), but notice that it describes, for instance, tubular neighborhoods of *isotropic* submanifolds (those on which ω vanishes(3)) of W. It asserts in particular that if L is an isotropic submanifold of maximal dimension (then $\dim L = \frac{1}{2}\dim W$ and L is said to be *Lagrangian*), a tubular neighborhood of L in W is (in a symplectic fashion) isomorphic to a neighborhood of the zero section in $\mathsf{T}^\star L$ (with the canonical symplectic form).

These corollaries are deduced from the theorem as the Darboux Theorem is, comparing a neighborhood of the submanifold under consideration with some model space (the tangent space $\mathsf{T}_x W$ in the case where the submanifold V was a point). We will need the notion of the *symplectic normal bundle* to a submanifold $V \subset W$ on which the symplectic form has constant rank. This is simply the vector bundle

$$(\mathsf{T}V)^\circ / (\mathsf{T}V) \cap (\mathsf{T}V)^\circ$$

denoting by $^\circ$ the orthogonality, in $\mathsf{T}W$, with respect to ω. This is indeed a vector bundle (this is the constant rank assumption) and it is symplectic in the sense that the fibers are symplectic vector spaces. Notice that this is indeed a *normal* bundle, namely a supplement of the tangent bundle $\mathsf{T}V \subset j^\star \mathsf{T}W$ if and only if V is a symplectic submanifold.

(3) This is a good opportunity to define a *coisotropic submanifold*, namely a submanifold V such that, for any point x, the orthogonal of $\mathsf{T}_x V$ is contained in $\mathsf{T}_x V$.

Corollary II.1.12. *Let (W,ω) and (W',ω') be symplectic manifolds. Suppose that V is a manifold that embeds both in W and W', that the pull backs of ω and ω' have constant rank and such that the symplectic normal bundles of V in W and W' are isomorphic as symplectic vector bundles. Then the map identifying the images of V in W and W' extends to a diffeomorphism φ between neighborhoods of these images such that $\varphi^\star\omega' = \omega$.*

Proof. The proof is the same as that of Corollary II.1.11, using tubular neighborhoods of V. See, for instance, [**96**], for detail. □

Remark II.1.13. In case we need them later, notice all the proofs above become easily "equivariant proofs" if there is a Lie group G acting on W, preserving the symplectic forms and the submanifolds[(4)]. This is the reason why we used $\varphi = \exp_x$ in the above proof instead of any local chart: it suffices now to choose an invariant metric to get equivariant local coordinates.

Remark II.1.14. A statement analogous to II.1.9 for Riemannian metrics would be definitely false: in the Riemannian case, there is a *local* invariant which distinguishes between the neighborhood of a point and the tangent space at that point, namely the curvature. We have just proved that there exist no local invariants in symplectic geometry.

II.1.d. Examples: U(n)-orbits in Hermitian matrices. Consider the group $\mathrm{U}(n)$ acting by conjugation on the vector space $\mathcal{H}$ of Hermitian matrices

$$\mathcal{H} = \left\{X \in M_n(\mathbf{C}) \mid {}^t\overline{X} = X\right\}.$$

Recall that Hermitian matrices have real eigenvalues and are diagonalizable in a unitary basis, so that the orbits of this action are the submanifolds

$$\mathcal{H}_\lambda = \{A \in \mathcal{H} \mid \text{spectrum of } A = (\lambda_1, \ldots, \lambda_n)\} \text{ for } \lambda = (\lambda_1, \ldots, \lambda_n) \in \mathbf{R}^n.$$

Let us describe first these orbits as manifolds.

Proposition II.1.15. *Assume that λ consists of k distinct values with multiplicities $n_1, \ldots, n_k$. The orbit $\mathcal{H}_\lambda$ is diffeomorphic with the homogeneous space $\mathrm{U}(n)/\mathrm{U}(n_1) \times \cdots \times \mathrm{U}(n_k)$.*

Proof. We put the eigenvalues in increasing order $\mu_1 < \cdots < \mu_k$. Then any element in $\mathcal{H}_\lambda$ is determined by the sequence $Q_1, Q_2, \ldots, Q_k$ of its eigenspaces, Q_i having dimension the multiplicity n_i of μ_i (recall once again that Hermitian matrices are diagonalizable) and the Q_i's being mutually orthogonal (Hermitian matrices are diagonalizable in a *unitary* basis). The stabilizer of the matrix corresponding to $Q_i = \{0\} \times \mathbf{C}^{n_i} \times \{0\}$ is $\mathrm{U}(n_1) \times \cdots \times \mathrm{U}(n_k)$. This gives the proposition. And this

---

(4) The equivariant form of Corollary II.1.12 is due to Marle [**100**].

shows that the orbit under consideration is diffeomorphic with the manifolds of flags

$$0 \subset P_1 \subset \cdots \subset P_k = \mathbf{C}^n$$

(with $P_i = Q_1 \oplus \cdots \oplus Q_i$). □

Examples II.1.16

(1) Complete flags. If all the eigenvalues $\lambda_1, \ldots, \lambda_n$ are distinct, any element in $\mathcal{H}_\lambda$ defines (and is defined by) the family of its eigenspaces, here a family of pairwise orthogonal lines in $\mathbf{C}^n$. One thus defines a bijection from $\mathcal{H}_\lambda$ onto the manifold of complete flags in $\mathbf{C}^n$,

$$\mathcal{D}(\mathbf{C}^n) = \{0 = P_0 \subset P_1 \subset \cdots \subset P_n = \mathbf{C}^n\},$$

associating to a matrix h the subspaces $P_i = \ell_1 \oplus \cdots \oplus \ell_i$ where ℓ_i is the eigenspace of h corresponding to λ_i. The dimension of these orbits is $n^2 - n$.

(2) Trivial case. On the other hand, if all the eigenvalues are equal, there is only one point in the orbit. There is not much to say here.

(3) Complex projective space. Suppose now that $\lambda_1 \neq \lambda_2 = \cdots = \lambda_n$. Then the eigenspace associated with λ_1 is a line and the one associated with λ_2 is the orthogonal hyperplane. The map: $\Psi_\lambda : \mathcal{H}_\lambda \to \mathbf{P}^{n-1}(\mathbf{C})$ which associates the λ_1-eigenspace with h is a diffeomorphism.

Notice that the diffeomorphism type of the orbit depends only on the multiplicities of the eigenvalues, not on their actual values.

We check now what was announced at the very beginning of this chapter: the orbits $\mathcal{H}_\lambda$ are symplectic manifolds. To begin with, some (bi)linear algebra in the space of all Hermitian matrices.

The mapping $h \mapsto ih$ is an isomorphism from the real vector space $\mathcal{H}$ onto the real vector space $\mathfrak{u}(n)$ of skew-Hermitian matrices, which is indeed the Lie algebra of the group $\mathrm{U}(n)$ (Exercise I.15). Now we have a symmetric bilinear form, defined on $\mathcal{H}$:

$$(X, Y) \longmapsto \operatorname{tr}(XY) = \operatorname{tr}({}^t\overline{X}Y).$$

This is an inner product, since

$$\operatorname{tr}({}^t\overline{X}Y) = \sum_{i,j} \overline{x}_{i,j} y_{i,j} = 2 \sum_{i<j} \operatorname{Re}(\overline{x}_{i,j} y_{i,j}) + \sum_k x_{k,k} y_{k,k}.$$

The same formula defines a (negative definite) bilinear form on $\mathfrak{u}(n)$. Notice that these forms are invariant under conjugation. They allow us to identify $\mathcal{H} = i\mathfrak{u}(n)$ with the dual vector space $\mathfrak{u}(n)^\star$.

Define now, for any matrix $h \in \mathcal{H}$, a skew-symmetric bilinear form ω_h on $\mathfrak{u}(n) = \mathsf{T}_1 \mathrm{U}(n) = i\mathcal{H}$ by

$$\omega_h(X, Y) = \operatorname{tr}([X, Y] ih).$$

Notice that $\omega_h(X,Y) = \operatorname{tr}(X(Yih - ihY))$ and $Yh - hY \in \mathcal{H}$. The kernel of ω_h is thus the subspace

$$K_h = \{Y \in \mathfrak{u}(n) \mid [Y, ih] = 0\}.$$

If we differentiate the relation $ghg^{-1} = h$ with respect to g, we see that K_h is nothing other than the Lie algebra of the stabilizer of h for the action of $\mathrm{U}(n)$ on $\mathcal{H}$ by conjugation.

Let us consider now the orbit map $f_h : \mathrm{U}(n) \to \mathcal{H}$ and its derivative

$$\mathsf{T}_1 f_h : \mathfrak{u}(n) \longrightarrow \mathsf{T}_h\mathcal{H}_\lambda \subset \mathcal{H}.$$

We know that its image is indeed $\mathsf{T}_h\mathcal{H}_\lambda$, that is, it is generated by the fundamental vector fields $[X,h]$, and its kernel is the Lie algebra K_h of the stabilizer. Hence, ω_h defines a nondegenerate skew-symmetric bilinear form $\widetilde{\omega}_h$ on $\mathsf{T}_h\mathcal{H}_\lambda$ by the formula

$$\widetilde{\omega}_h([X,h],[Y,h]) = \operatorname{tr}([X,Y]ih) \text{ for } X, Y \in \mathfrak{u}(n).$$

Thus we have defined, on any orbit $\mathcal{H}_\lambda$, a nondegenerate 2-form $\widetilde{\omega}$. The only thing which is left to prove is the fact that $\widetilde{\omega}$ is indeed a *closed* form. It is enough to prove that

$$d\widetilde{\omega}_h([X,h],[Y,h],[Z,h]) = 0 \text{ for all } X, Y, Z \in \mathfrak{u}(n).$$

Now, if $\underline{X}$ denotes the fundamental vector field associated with X, namely $\underline{X}_h = [X,h]$, one has

$$\begin{aligned} 3d\widetilde{\omega}(\underline{X},\underline{Y},\underline{Z}) &= \underline{X}\cdot\widetilde{\omega}(\underline{Y},\underline{Z}) - \underline{Y}\cdot\widetilde{\omega}(\underline{X},\underline{Z}) + \underline{Z}\cdot\widetilde{\omega}(\underline{X},\underline{Y}) \\ &\quad + \widetilde{\omega}([\underline{X},\underline{Y}],\underline{Z}) + \widetilde{\omega}([\underline{Y},\underline{Z}],\underline{X}) + \widetilde{\omega}([\underline{Z},\underline{X}],\underline{Y}). \end{aligned}$$

Now the form $\widetilde{\omega}$ is invariant, by definition, and the vector fields are fundamental ones, so that the first three terms vanish. And the sum which is left vanishes because of the Jacobi identity in the Lie algebra $\mathfrak{u}(n)$ (there is indeed a close relation between the closedness of symplectic forms and the Jacobi identity, see Remark II.3.4 and Exercise II.27).

Thus all the orbits $\mathcal{H}_\lambda$ under consideration are compact symplectic manifolds. Notice that, if the diffeomorphism type of the manifold $\mathcal{H}_\lambda$ depends only on the multiplicities of the eigenvalues, its symplectic form depends on the actual values of the λ_i's.

For instance, we have thus exhibited a lot of symplectic forms on $\mathbf{P}^{n-1}(\mathbf{C})$, one for each pair of distinct real numbers. See Exercise II.13.

II.2. Calibrated almost complex structures

Recall that $\mathbf{C}^n$ is a model for the symplectic vector spaces of dimension $2n$. This is not accidental: on any symplectic manifold, there are almost complex structures.

A (linear) complex structure on a real vector space E is an endomorphism J of E such that $J^2 = -1$. Analogously, an *almost complex structure* on a manifold W is a section J of the bundle $\operatorname{End} \mathsf{T}W$ such that $J_x^2 = -\operatorname{Id}_{\mathsf{T}_x W}$ for all x in W. For instance, if W is a complex manifold (namely a manifold with complex holomorphic change of local coordinates), its tangent space $\mathsf{T}_x W$ at any point has a natural structure of complex vector space and the multiplication by i is an almost complex structure. This is the situation we are trying to mimic: J plays the role of multiplication by i. Note that J gives $\mathsf{T}W$ the structure of a complex vector bundle, each $\mathsf{T}_x W$ becoming a complex vector space by

$$(a+ib)\cdot\xi = a\xi + bJ\xi.$$

Let, for instance, W be an oriented surface, endowed with a Riemannian metric, in such a way that we have a notion of rotation by $+\pi/2$ in every tangent plane. The family of all such rotations defines an almost complex structure on W, thus any oriented surface admits almost complex structures. Of course, not any manifold, even of even dimension, even oriented, may be endowed with an almost complex structure[5].

In $\mathbf{C}^n$, the standard Hermitian product decomposes into its real and imaginary parts:

$$\langle u,v\rangle = (u,v) + i\omega(u,v)$$

the Euclidean inner product of $\mathbf{R}^{2n}$ and its symplectic form, respectively. One easily checks that $\omega(iu,v)$ is the Euclidean inner product and moreover that $\omega(iu,iv) = \omega(u,v)$: multiplication by i is an isometry of ω. We now try to mimic this situation in the following definition.

An almost complex structure J on a symplectic manifold (W,ω) will be called *calibrated*[6] if J is an "isometry" of ω and if the symmetric bilinear form $\omega(JX,Y)$ is positive-definite at each point.

II.2.a. The linear case. The linear group $\mathrm{GL}(2n,\mathbf{R})$ acts on the space of complex structures on $\mathbf{R}^{2n}$ by conjugation: $g\cdot J = gJg^{-1}$. Using a J-complex basis, one sees that this action is transitive and that the stabilizer of a given almost complex structure J_0 is the group of all *complex* (for J_0) automorphisms. In other words, the set of all linear almost complex structures on $\mathbf{R}^{2n}$ can be identified with the homogeneous space

$$\mathcal{J}_n = \mathrm{GL}(2n,\mathbf{R})/\mathrm{GL}(n,\mathbf{C}).$$

---

(5) For instance, it can be shown that this is not the case for the 4-sphere.

(6) implicitly: "by the symplectic form".

Let us now show that there exist calibrated complex structures on any symplectic vector space (E, ω). First choose a Euclidean inner product (X, Y). As $(\,,)$ and ω are nondegenerate, the relation $(X, AY) = \omega(X, Y)$ defines an isomorphism $A : E \to E$. Write the polar decomposition of A, namely $A = BJ$, were B is symmetric positive definite, A and B commute and J is isometric.

Lemma II.2.1. *The endomorphism A is skew-symmetric, J is an isometry for ω, it satisfies $J^2 = -1$ and $JB = BJ$.*

Proof. We have

$$(X, AY) = \omega(X, Y) = -\omega(Y, X) = -(AX, Y)$$

thus A is skew-symmetric. Writing $J = B^{-1}A$, we get

$${}^tJ = {}^tA{}^tB^{-1} = -AB^{-1} = -B^{-1}A = -J.$$

But J is an isometry, so that ${}^tJJ = 1$, and J is indeed a complex structure. Moreover

$${}^t(BJ) = {}^tJ{}^tB = {}^tJB = -BJ$$

hence B and J commute, thus A and J commute as well, and we are done:

$$\omega(JX, JY) = (JX, AJY) = (JX, JAY) = (X, AY) = \omega(X, Y). \qquad \square$$

Now the inner product defined by B is

$$((X, Y)) = (BX, Y) = \omega(JX, Y).$$

We have shown the existence, on every symplectic vector space, of calibrated complex structures. Moreover, the form $((\,,)) + i\omega(\,,)$ is Hermitian.

Remark II.2.2. Doing the construction of J above fiberwise gives the fact that any symplectic vector bundle has the structure of a complex vector bundle.

Now that we know that there exist calibrated complex structures, we show that the set of all such structures is contractible. The proof given here is due to Sévennec (see [**12**]), another proof is sketched in Exercise II.16. Denote by $\mathcal{J}_c(\omega)$ the set of almost complex structures calibrated by ω.

Proposition II.2.3. *Let J_0 be a complex structure on $\mathbf{R}^{2n}$ calibrated by the form ω. The map*

$$J \longmapsto (J + J_0)^{-1} \circ (J - J_0)$$

is a diffeomorphism from $\mathcal{J}_c(\omega)$ onto the open unit ball in the vector space of symmetric matrices S such that $J_0S + SJ_0 = 0$.

Corollary II.2.4. *The space $\mathcal{J}_c(\omega)$ is contractible.* $\square$

Proof. The endomorphism $J + J_0$ is invertible, since, for $x \neq 0$,

$$\omega((J + J_0)x, x) = \omega(Jx, x) + \omega(J_0 x, x) > 0.$$

Thus the map in the statement is well-defined. Now, we put $A = J_0^{-1} \circ J$, so that

$$\begin{aligned} S &= (J + J_0)^{-1} \circ (J - J_0) \\ &= (J_0 \circ (J_0^{-1} + J))^{-1} \circ (J_0 \circ (J - J_0)) \\ &= (A + \mathrm{Id})^{-1} \circ (A - \mathrm{Id}). \end{aligned}$$

We want to prove that $\|S\| < 1$, that is, that $\|Ax - x\|^2 < \|Ax + x\|^2$ for all $x \neq 0$. But,

$$\begin{aligned} \|Ax + x\|^2 - \|Ax - x\|^2 &= 4(Ax) \cdot x \\ &= 4\omega(J_0 \circ Ax, x) \\ &= 4\omega(Jx, x) > 0. \end{aligned}$$

Now, J is calibrated if and only if for any x, $y \in \mathbf{R}^{2n}$,

$$\omega\left((\mathrm{Id} - S)\, x, (\mathrm{Id} - S)\, y\right) = \omega\left((\mathrm{Id} + S)\, x, (\mathrm{Id} + S)\, y\right),$$

which is equivalent to

$$\omega\left(Sx, y\right) + \omega\left(x, Sy\right) = 0.$$

Hence J is calibrated if and only if S is symmetric. Moreover, J is an almost complex structure ($J^{-1} = -J$) if and only if $J_0 S + S J_0 = 0$.

Conversely, let S be a symmetric matrix such that $\|S\| < 1$, which implies that $\mathrm{Id} - S$ is invertible and that the endomorphism

$$J = J_0 \circ (\mathrm{Id} + S) \circ (\mathrm{Id} - S)^{-1}$$

is well-defined. The two maps $J \mapsto S$ and $S \mapsto J$ are obviously inverse of each other. □

II.2.b. Calibrated almost complex structures on a symplectic manifold. Let W be a $2n$-dimensional manifold endowed with a symplectic form ω. With its tangent bundle are associated the bundles

$$\begin{aligned} \mathcal{J}(\mathsf{T}W) &\longrightarrow W \quad \text{with fibre } \mathcal{J}_n \\ \mathcal{J}_c(\mathsf{T}W, \omega) &\longrightarrow W \quad \text{with fibre } \mathcal{J}_c(\omega) \end{aligned}$$

of almost complex structures, *resp.* calibrated almost complex structures on W. A calibrated complex structure is simply a section of the bundle $\mathcal{J}_c(\mathsf{T}W, \omega) \to W$. As this bundle has a nonempty contractible fiber, we get:

Proposition II.2.5. *The space of almost complex structures calibrated by ω is nonempty and contractible.* □

This proof gives in particular the existence of a calibrated almost complex structure on any symplectic manifold. This can be proved directly, in the very same way that we have proven the existence of a calibrated (linear) complex structure on $\mathbf{R}^{2n}$. We just have to replace the inner product by a Riemannian metric and to notice that the map $A \mapsto B = \sqrt{{}^tAA}$ in the polar decomposition is smooth.

Remark II.2.6. The symplectic form ω induces a *symplectic* form on any submanifold of W which is preserved by a calibrated almost complex structure. Indeed, if $j : Z \subset W$ is the inclusion of such a submanifold, $j^\star\omega$ is nondegenerate:

$$\forall\, X \in \mathsf{T}_z Z - \{0\}\,, \exists\, Y \in \mathsf{T}_z Z \text{ such that } \omega(X, Y) \neq 0$$

because $Y = JX$ works.

The equivariant case. Assume now that our symplectic manifold W is acted on by a compact Lie group G. Assume that the G-action preserves the symplectic form in the sense that

$$\forall g \in \mathrm{G} \qquad g^\star\omega = \omega.$$

Due to the compactness of G, we know that there exists an invariant Riemannian metric $\langle\,,\rangle$ on W (see § I.2.a). The calibrated almost complex structure J obtained by the method above (polar decomposition) is invariant. To prove this, it suffices to prove that the field of endomorphisms A defined by

$$\langle X, A_x Y\rangle_x = \omega_x(X, Y)$$

is invariant. Let $g \in \mathrm{G}$, $x \in W$, $X \in \mathsf{T}_{g\cdot x}W$, $Y \in \mathsf{T}_x W$. Then

$$\begin{aligned}
\langle X, \mathsf{T}_x g \circ A_x(Y)\rangle_{g\cdot x} &= \langle (\mathsf{T}_x g^{-1})^{-1}(X), A_x(Y)\rangle_x \text{ by invariance of } \langle\,,\rangle \\
&= \omega_x(\mathsf{T}_x g^{-1})(X), Y) \text{ by definition of } A_x \\
&= \omega_{g\cdot x}(X, \mathsf{T}_x g(Y)) \text{ by invariance of } \omega \\
&= \langle X, A_{g\cdot x} \circ \mathsf{T}_x g(Y)\rangle_{g\cdot x}.
\end{aligned}$$

This shows that $A_{g\cdot x} \circ \mathsf{T}_x g = \mathsf{T}_x g \circ A_x$, that is, that A is invariant.

Examples. Once the orbits $\mathcal{H}_\lambda$ of $\mathrm{U}(n)$ have been described as flag manifolds of $\mathbf{C}^n$, they clearly have a natural additional structure: they are *complex* manifolds. As above, the complex diffeomorphism type depends only on the multiplicities. Let us describe the underlying almost complex structure. One can write $\mathfrak{u}(n) = \mathfrak{u}(n_1) \oplus \cdots \oplus \mathfrak{u}(n_k) \oplus \mathfrak{m}$ where $\mathfrak{m}$ is a *complex* vector space. One simply decomposes in blocks a matrix A in an orbit with multiplicities $n_1, \dots, n_k$:

$$A = \begin{pmatrix} A_1 & X_{1,2} & X_{1,3} & \cdots & \\ -{}^t\overline{X}_{1,2} & A_2 & X_{2,3} & \cdots & \\ \cdots & & & & \\ & & & & X_{k-1,k} \\ & & & & A_k \end{pmatrix}$$

where $A_i \in \mathfrak{u}(n_i)$ and $X_{i,j}$ is a *complex* matrix. The complex structure on $\mathfrak{m}$ is given by the complex structure of the spaces where the blocks $X_{i,j}$ live, not by the complex structure of the space of big matrices: if

$$X = \begin{pmatrix} 0 & X_{1,2} & X_{1,3} & \cdots \\ -{}^t\overline{X}_{1,2} & \cdots & \cdots & \cdots \\ -{}^t\overline{X}_{2,3} & \cdots & \cdots & \cdots \\ \vdots & \vdots & \vdots & \vdots \end{pmatrix} \in \mathfrak{m}$$

call $j(X)$ the same matrix with each $X_{k,l}$ replaced by $iX_{k,l}$: this is the complex structure of $\mathfrak{m}$. Of course, $\mathfrak{m}$ is not a Lie subalgebra of $\mathfrak{u}(n)$, but $\mathfrak{g} = \mathfrak{u}(n_1) \oplus \cdots \oplus \mathfrak{u}(n_k)$ is.

Now $\mathfrak{m}$ is a good representative of the tangent space to the homogeneous space $\mathrm{U}(n)/\mathrm{U}(n_1) \times \cdots \times \mathrm{U}(n_k)$ at the point image of $\mathrm{Id} \in \mathrm{U}(n)$, and it has a natural complex structure. Everything is invariant enough to define something on $\mathcal{H}_\lambda$.

Fix a diagonal matrix $D = (\mu_1, \dots, \mu_1, \dots, \mu_k, \dots, \mu_k)$ in the orbit $\mathcal{H}_\lambda$ with $\mu_1 < \cdots < \mu_k$ (so that the stabilizer of D is indeed $\mathrm{U}(n_1) \times \cdots \times \mathrm{U}(n_k)$). Any element in $\mathsf{T}_d\mathcal{H}_\lambda$ can be written in a unique way as $[X, D]$ for some $X \in \mathfrak{m}$ (the tangent space to the orbit is generated by fundamental vector fields, the uniqueness comes from the specification $X \in \mathfrak{m}$, this subspace being a supplement of the Lie algebra of the stabilizer).

Define now $J_D[X, D] = [j(X), D]$ in such a way that J_D is a complex structure on $\mathsf{T}_D\mathcal{H}_\lambda$. Let now $h = gDg^{-1}$ be any element in the orbit. Any element in $\mathsf{T}_h\mathcal{H}_\lambda$ can be written uniquely (once g is chosen) as $[gXg^{-1}, h]$ with $X \in \mathfrak{m}$. Put

$$J_h\left[gXg^{-1}, h\right] = \left[gj(X)g^{-1}, h\right]$$

thus defining an almost complex structure J on the orbit $\mathcal{H}_\lambda$. This complex structure is calibrated by the symplectic form $\widetilde{\omega}$, as is shown in Exercise II.19.

Kähler manifolds. Assume now that W is a *complex* manifold. That is to say, the almost complex structure J is the multiplication by i. Suppose that $\langle\ ,\ \rangle$ is a Hermitian metric on W. Its imaginary part is a (type $(1,1)$) nondegenerate 2-form ω. If ω is symplectic, that is to say, if it is closed, one says that the metric is *Kähler*. The manifold W endowed with this metric and with the form ω is also called *Kähler*. By its very definition, the complex structure is calibrated by the Kähler form. Moreover, in a Kähler manifold, the symplectic volume form $\omega^{\wedge n}/n!$ associated with the Kähler form is also the Riemannian volume.

Kähler manifolds are very common: any complex submanifold of a Kähler manifold is Kähler (for the induced $(1,1)$-form), thus all projective complex algebraic manifolds are Kähler—it suffices to prove that the projective space is, and it is (see Exercise II.21).

II.3. Hamiltonian vector fields and Poisson brackets

II.3.a. Hamiltonian vector fields. Being nondegenerate, the symplectic form ω defines a pairing between the tangent and cotangent spaces of W (already used in the proof of II.1.9). In particular, for any function $H : W \to \mathbf{R}$, there is a symplectic analogue of the gradient, namely a vector field X_H defined by

$$\omega(Y, X_H(x)) = (dH)_x(Y) \quad \forall Y \in \mathsf{T}_x W,$$

that is, by $i_{X_H}\omega = -dH$. We shall call X_H the symplectic gradient of H or the *Hamiltonian vector field* associated with H. If X is the Hamiltonian vector field associated with a function H, we shall also say that H is a *Hamiltonian* for X.

Remark II.3.1. As ω is nondegenerate, the zeros of the vector field X_H are precisely the critical points of the function H.

Example II.3.2. Consider, on the symplectic manifold $\mathbf{C}^n$ with the standard (linear) symplectic structure, the function

$$H = \frac{1}{2}\sum |z_i|^2 .$$

From

$$dH = \sum_{j=1}^{n} x_j dx_j + y_j dy_j \text{ and } \omega = \sum_{j=1}^{n} dx_j \wedge dy_j,$$

one gets

$$X_H = \sum_{j=1}^{n} \left(-y_j \frac{\partial}{\partial x_j} + x_j \frac{\partial}{\partial y_j}\right).$$

Remark II.3.3. If J is an almost complex structure calibrated by ω, and if grad denotes the gradient for the Riemannian metric $\omega(JX, Y)$, then $X_f = J\operatorname{grad} f$.

A vector field X on W will be said to be *Hamiltonian* if the 1-form $i_X\omega$ is exact, *locally Hamiltonian* if this 1-form is closed. One writes $\mathcal{H}(W)$ and $\mathcal{H}_{\mathrm{loc}}(W)$ respectively for the spaces of Hamiltonian and locally Hamiltonian vector fields on W.

II.3.b. The Poisson bracket. Assume now that H and K are two functions on W. Their *Poisson bracket* $\{H,K\}$ is the function defined by the formula

$$\{H,K\} = X_H \cdot K = dK(X_H).$$

It follows from this definition that

$$\begin{aligned}\{H,K\} &= dK(X_H) = \omega(X_H,X_K)\\ &= -\omega(X_K,X_H) = -dH(X_K)\\ &= -X_K\cdot H = -\{K,H\}.\end{aligned}$$

The Poisson bracket is thus skew-symmetric in H and K. This is also a derivation, this is to say that it satisfies the *Leibniz identity*:

$$\{H,K_1K_2\} = \{H,K_1\}K_2 + K_1\{H,K_2\}.$$

Using the general formula

$$\mathcal{L}_X i_Y - i_Y\mathcal{L}_X = i_{[X,Y]}$$

(see Exercise II.24) and the Cartan formula

$$\mathcal{L}_X = di_X + i_X d,$$

one gets

$$\begin{aligned} i_{[X_H,X_K]}\omega &= \mathcal{L}_{X_H} i_{X_K}\omega - i_{X_K}\mathcal{L}_{X_H}\omega\\ &= di_{X_H}i_{X_K}\omega + i_{X_H}di_{X_K}\omega - i_{X_K}di_{X_H}\omega - i_{X_K}i_{X_H}d\omega\\ &= di_{X_H}i_{X_K}\omega = d(\omega(X_K,X_H)) = -d\{H,K\},\end{aligned}$$

that is, the fact that

$$[X_H,X_K] = X_{\{H,K\}}.$$

Hence we also have, for any three functions f, g and h,

$$[X_f,X_g]\cdot h = \{\{f,g\},h\}.$$

From this, we shall deduce that the Poisson bracket also satisfies the Jacobi identity

$$\{f,\{g,h\}\} + \{g,\{h,f\}\} + \{h,\{f,g\}\} = 0$$

and thus that it defines the structure of a Lie algebra on $\mathcal{C}^\infty(W)$. Moreover, the mapping

$$\begin{aligned}\mathcal{C}^\infty(W) &\longrightarrow \mathcal{X}(W)\\ H &\longmapsto X_H\end{aligned}$$

is a morphism of Lie algebras from $\mathcal{C}^\infty(W)$ (endowed with the Poisson bracket) into the Lie algebra of the vector fields (endowed with the Lie bracket).

Proof of the Jacobi identity. By definition of the Lie bracket of vector fields, one has

$$[X_f, X_g] \cdot h = X_f \cdot (X_g \cdot h) - X_g \cdot (X_f \cdot h).$$

Using the definition of the Poisson bracket and the previous inequality, one gets

$$\begin{aligned}\{\{f,g\},h\} &= [X_f, X_g] \cdot h \\ &= X_f \cdot (X_g \cdot h) - X_g \cdot (X_f \cdot h) \\ &= X_f \cdot \{g,h\} - X_g \cdot \{f,h\} \\ &= \{f,\{g,h\}\} - \{g,\{f,h\}\},\end{aligned}$$

which is equivalent to the expected identity. □

Remark II.3.4. The fact that the Poisson bracket satisfies the Jacobi identity is equivalent to the fact that the symplectic form is *closed.* This is one more good reason to insist that only *closed* 2-forms are *symplectic.* See Exercise II.27.

II.3.c. Poisson manifolds. More generally, a *Poisson bracket* on a manifold V is a Lie algebra structure, denoted $\{\ ,\ \}$, on the vector space $\mathcal{C}^\infty(V)$ of functions, satisfying the Leibniz identity. This notion was first investigated by Lichnerowicz [**97**] and by Weinstein [**130**].

It is strictly larger than the notion of symplectic manifold. Here is a family of examples of Poisson manifolds which are not symplectic.

The dual vector space to a Lie algebra. Let $\mathfrak{g}$ be a Lie algebra and let $\mathfrak{g}^\star$ be the dual vector space. Notice that, up to now, this is only a vector space. In particular, the Lie algebra structure of $\mathfrak{g}$ defines no (canonical) Lie algebra structure on $\mathfrak{g}^\star$. What it does define is a canonical *Poisson structure*, created by Kirillov, Kostant and Souriau [**84, 87, 123**] and which I describe now.

For $f, g \in \mathcal{C}^\infty(\mathfrak{g}^\star)$ and $\xi \in \mathfrak{g}^\star$ one writes:

$$\{f,g\}(\xi) = \langle \xi, [df_\xi, dg_\xi] \rangle$$

where $df_\xi : \mathsf{T}_\xi \mathfrak{g}^\star = \mathfrak{g}^\star \to \mathbf{R}$ is identified to an element of $\mathfrak{g}$ (by biduality). The Jacobi identity for $\{\,,\}$ is easily deduced from that for $[\,,]$ in $\mathfrak{g}$.

There is no reason why $\mathfrak{g}^\star$ should be a symplectic manifold. For instance, nothing prevents its dimension from being odd.

The perspicacious reader will already have realized that there is a relationship between this general construction and that of the symplectic form on $\mathcal{H}_\lambda$ in § II.1.d. There we have considered the Lie group $\mathrm{G} = \mathrm{U}(n)$, its Lie algebra $\mathfrak{g} = \mathfrak{u}(n)$, the dual being identified to $\mathcal{H} = i\mathfrak{u}(n)$ by the invariant inner product $\operatorname{tr}(XY)$. Modulo these identifications, the formal analogy with the definition of $\{\,,\}$ is only a special case of a forthcoming result (see § II.3.c): the coadjoint orbits are *symplectic submanifolds* of $\mathfrak{g}^\star$, in the sense that, when restricted to any orbit, the Poisson bracket is the same as the one defined by the symplectic form of the orbit.

Adjoint and coadjoint actions. As we have seen in the Appendix to Chapter I, any Lie group G acts on its Lie algebra $\mathfrak{g}$ through the adjoint action, the fundamental vector field ${}^{\mathfrak{g}}\underline{X}$ associated with $X \in \mathfrak{g}$ being

$$^{\mathfrak{g}}\underline{X}_Y = [X, Y].$$

To define the *coadjoint* action we merely have to transpose the adjoint G-action on $\mathfrak{g}$:

$$\langle \mathrm{Ad}_g^\star \xi, X\rangle = \langle \xi, \mathrm{Ad}_{g^{-1}} X\rangle$$

where Greek letters denote elements of $\mathfrak{g}^\star$, Latin ones elements of $\mathfrak{g}$, and brackets duality. Thus $\mathrm{Ad}^\star$ is a (left) G-action on $\mathfrak{g}^\star$. Of course, the fundamental vector field on $\mathfrak{g}^\star$ associated with $X \in \mathfrak{g}$ by the coadjoint action is defined by

$$\langle {}^{\mathfrak{g}^\star}\underline{X}_\xi, Y\rangle = \langle \xi, [Y, X]\rangle.$$

Remark II.3.5. The principal orbits of the coadjoint representation of a compact connected Lie group have a maximal torus as stabilizer.

Indeed, as the group G is compact, it is easy to define an invariant inner product on its Lie algebra $\mathfrak{g}$, and this allows us to identify $\mathfrak{g}$ with its dual. The assertion is then equivalent to the fact that the stabilizer of any $X \in \mathfrak{g}$ for the adjoint representation contains a maximal torus. The infinitesimal analogue of this assertion—that is, the centralizer of any element in $\mathfrak{g}$ contains the Lie algebra of a maximal torus—is an easy consequence of the fact that all adjoint orbits meet the Lie algebra of any maximal torus (see **[25]**): if $X \in \mathfrak{t}$, its centralizer contains $\mathfrak{t}$.

The symplectic form on the orbits. Just rewriting what we did for $\mathcal{H}_\lambda$ in § II.1.d, one defines for any $\xi \in \mathfrak{g}^\star$ a skew-symmetric bilinear form on $\mathfrak{g}$ by

$$\omega_\xi(X, Y) = \langle \xi, [X, Y]\rangle.$$

Lemma II.3.6. *The kernel of ω_ξ is the Lie algebra $\mathfrak{g}_\xi$ of the stabilizer of $\xi \in \mathfrak{g}^\star$ for the coadjoint representation.*

Proof. Let $X, Y \in \mathfrak{g}$. By the very definition of $\mathrm{Ad}^\star$, we have:

$$\langle \mathrm{Ad}^\star_{\exp(-tX)} \xi, Y\rangle = \langle \xi, \mathrm{ad}_{\exp(tX)} Y\rangle.$$

Differentiating at 0 with respect to t:

$$\begin{aligned}\langle \frac{d}{dt} \mathrm{Ad}^\star_{\exp(-tX)} \xi|_{t=0}, Y\rangle &= \langle \xi, \frac{d}{dt} \mathrm{ad}_{\exp(tX)} Y\rangle \\ &= \langle \xi, [X, Y]\rangle = \omega_\xi(X, Y).\end{aligned}$$

Thus the kernel of ω_ξ is $\mathfrak{g}_\xi$. □

In particular, ω_ξ defines a nondegenerate skew-symmetric bilinear form on $\mathfrak{g}/\mathfrak{g}_\xi$, a vector space of which we know (from § I.1.b) that it is identified with $\mathsf{T}_\xi(\mathrm{G}\cdot\xi) \subset \mathfrak{g}^\star$. The proof that there exists a canonical symplectic structure on each coadjoint orbit, induced by the canonical Poisson structure of $\mathfrak{g}^\star$, proceeds exactly as the proof we have written for the orbits $\mathcal{H}_\lambda$ in § II.1.d (Exercise II.29).

The symplectic foliation. In general the Poisson bracket of a Poisson manifold W defines nothing on a submanifold $V \subset W$. If it defines a Poisson bracket, that is, if $\{f,g\}\big|_V$ depends only on $f|_V$ and $g|_V$, then V is said to be a *Poisson submanifold*. If moreover, the restricted Poisson bracket on V can be defined by a symplectic form, then V is a symplectic submanifold.

Now, any Poisson manifold has a *symplectic foliation*, a (singular) foliation the leaves of which are the maximal symplectic submanifolds. I will not discuss this in great generality, but refer the reader to [**96**] or [**125**] for instance. However, in the case of the dual $\mathfrak{g}^\star$ of the Lie algebra of a group G, this is quite simple: the symplectic leaves are just the coadjoint orbits.

Exercises

Exercise II.1. Let ω be a skew-symmetric bilinear form on a real vector space E. Prove that there exists a basis $e_1,\dots,e_r,f_1,\dots,f_r,g_1,\dots,g_k$ of E such that

$$\omega(e_i,f_j)=\delta_{i,j},\quad \omega(e_i,e_j)=\omega(e_i,g_\ell)=\omega(f_i,g_\ell)=\omega(g_\ell,g_m)=0$$

for all indices i, j, ℓ and m.

Assume that $\dim E = 2n$. Prove that $\omega^{\wedge n} \neq 0$ if and only if ω is nondegenerate.

Exercise II.2. Check that the formula

$$\omega = \frac{1}{2i}\sum_{j=1}^{n} dz_j \wedge d\overline{z}_j$$

defines a symplectic form on $\mathbf{C}^n$.

***Exercise II.3* (The symplectic group).** Prove that $\mathrm{Sp}(2;\mathbf{R}) = \mathrm{SL}(2;\mathbf{R})$. What is the dimension of the kernel of the linear map $A \mapsto {}^tAJ + JA$? Deduce that $\mathrm{Sp}(2n;\mathbf{R})$ has dimension $2n^2+n$. In a symplectic basis of $\mathbf{R}^{2n}$, write a matrix by $n\times n$ square blocks:

$$M = \begin{pmatrix} A & B \\ C & D \end{pmatrix}.$$

To which conditions on A, B, C and D does this matrix belong to the Lie algebra $\mathfrak{sp}(2n;\mathbf{R})$? to the Lie group $\mathrm{Sp}(2n;\mathbf{R})$?

Exercise II.4. Let E be an even dimensional vector space and let ω, ω' be two symplectic forms on E. Prove that the symplectic groups $\mathrm{Sp}(E,\omega)$ and $\mathrm{Sp}(E,\omega')$ are conjugated subgroups of $\mathrm{GL}(E)$.

Let $\Omega(E)$ be the space of all symplectic forms on the vector space E. Prove that the linear group of E acts on this space by

$$(g\cdot\omega)(X,Y)=\omega(gX,gY).$$

Deduce that $\Omega(E)$ is in one-to-one correspondence[7] with the homogeneous space $\mathrm{GL}(E)/\mathrm{Sp}(E)$, where $\mathrm{Sp}(E)$ is the symplectic group $\mathrm{Sp}(E,\omega_0)$ for a given form ω_0 on E.

Exercise II.5. Let $\mathbf{C}^n=\mathbf{R}^{2n}$ be endowed with the standard symplectic structure. Consider the subgroups $\mathrm{O}(2n)$, $\mathrm{Sp}(2n;\mathbf{R})$, $\mathrm{GL}(n,\mathbf{C})$, $\mathrm{U}(n)$ and compute the intersections of any two of them.

Exercise II.6. Let $A\in\mathrm{Sp}(2n)$. Check that the matrices ${}^tA$ and A^{-1} are similar[8]. Show that λ is an eigenvalue of A if and only if λ^{-1} is also an eigenvalue, and that both occur with the same multiplicity.

Exercise II.7. Let $A\in\mathrm{Sp}(2n)$. Prove that it can be written in a unique way as a product

$$A=S\cdot\Omega$$

where S is the positive definite symmetric and symplectic matrix $S=\sqrt{A^tA}$ and Ω is the orthogonal matrix $\Omega=S^{-1}A$. Prove that, in addition

$$\Omega=S^{-1}A\in\mathrm{Sp}(2n)\cap\mathrm{O}(2n)=\mathrm{U}(n).$$

Deduce that, as a manifold, the symplectic group $\mathrm{Sp}(2n;\mathbf{R})$ is diffeomorphic to the Cartesian product of the group $\mathrm{U}(n)$ with a convex open cone of a vector space of dimension $n(n+1)$ and thus that it is path-connected. Determine its fundamental group.

Exercise II.8. Prove that the symplectic form on $\mathbf{S}^2$ given in Examples II.1.7 can be written, in coordinates $(x,y,z)\in\mathbf{R}^3$,

$$\omega=x\,dy\wedge dz+y\,dz\wedge dx+z\,dx\wedge dy.$$

Endow the sphere $\mathbf{S}^2$ with the *cylindrical* "coordinates"

$$\begin{aligned}[\pi,-\pi[\times]-1,1[&\longrightarrow \mathbf{S}^2-\{N,S\}\\ (\theta,z)&\longmapsto(\sqrt{1-z^2}\cos\theta,\sqrt{1-z^2}\sin\theta,z).\end{aligned}$$

[7] This is actually a homeomorphism.
[8] Thus A and A^{-1} are similar too.

Check that the symplectic form can also be written

$$\omega = d\theta \wedge dz$$

(so that this is *the* area form on $\mathbf{S}^2$, the one that gives it total area 4π).

Exercise II.9. Consider the tangent bundle to the unit sphere $\mathbf{S}^n$ of $\mathbf{R}^{n+1}$ as the submanifold

$$\mathsf{T}\mathbf{S}^n = \left\{(q,p) \in \mathbf{R}^{n+1} \times \mathbf{R}^{n+1} \mid \|q\|^2 = 1 \text{ and } p \cdot q = 0\right\}.$$

Endow $\mathbf{R}^{n+1} \times \mathbf{R}^{n+1}$ with its symplectic form $\omega = \sum dp_i \wedge dq_i$. Prove that $\mathsf{T}\mathbf{S}^n$ is a symplectic submanifold. Using the Euclidean metric (already present in the very definition of the sphere), identify $\mathsf{T}\mathbf{S}^n$ with $\mathsf{T}^\star\mathbf{S}^n$. Compare the symplectic form above with the canonical symplectic form of the cotangent bundle.

Exercise II.10. Let $\mathcal{D}_n$ denote the space of all oriented affine lines in $\mathbf{R}^{n+1}$. Prove that $\mathcal{D}_n$ can be identified with the submanifold

$$\left\{(u,v) \in \mathbf{R}^{n+1} \times \mathbf{R}^{n+1} \mid \|u\| = 1 \text{ and } u \cdot v = 0\right\}.$$

Deduce that $\mathcal{D}_n$ can be endowed with the structure of a symplectic manifold.

Exercise II.11. What is the symplectic normal bundle of a Lagrangian submanifold?

***Exercise II.12* (The Liouville form is canonical).** A differential 1-form on a manifold V can be considered as a section of the cotangent bundle $\mathsf{T}^\star V$,

$$\eta : V \longrightarrow \mathsf{T}^\star V \qquad \eta(x) \in \mathsf{T}^\star_x V.$$

Denote by α the Liouville form. Prove that

$$\eta^\star \alpha = \eta.$$

Exercise II.13. We consider the symplectic form $\widetilde{\omega}$ on the orbit $\mathcal{H}_\lambda$ for $\lambda = (\mu_1, \mu_2)$ with $\mu_1 \neq \mu_2$ (so that the orbit is a complex projective line). Compute $\int_{\mathcal{H}_\lambda} \widetilde{\omega}$ in terms of μ_1 and μ_2.

Exercise II.14. Let Ω_n be the set of all nondegenerate skew-symmetric bilinear forms on $\mathbf{R}^{2n}$, and, for $\omega \in \Omega_n$, let $\mathcal{J}_c(\omega)$ be the set of almost complex structures calibrated by ω. The symplectic group is denoted by $\mathrm{Sp}(2n;\mathbf{R})$. Using the action of $\mathrm{GL}(n;\mathbf{R})$ on Ω_n by $(g,\omega) \mapsto \omega(g^{-1}\cdot, g^{-1}\cdot)$, we have proved in Exercise II.4 that

$$\Omega_n \cong \mathrm{GL}(2n,\mathbf{R})/\,\mathrm{Sp}(2n;\mathbf{R}).$$

Prove that

$$\mathcal{J}_c(\omega) \cong \mathrm{Sp}(2n;\mathbf{R})/\,\mathrm{U}(n).$$

Exercise II.15. Assume J is a (linear) calibrated complex structure on the symplectic vector space E. Show that the bilinear form $(x, y) \mapsto \omega(x, Jy) = (x, y)_J$ is an inner product on $\mathbf{R}^{2n}$, for which one has

$$\begin{cases} E^\perp = (JE)^\circ \\ E^\circ = (JE)^\perp \\ JE \cap E^\circ = 0, \end{cases}$$

$^\circ$ and $^\perp$ denoting orthogonality respectively for ω and $(\ ,\ )_J$.

Let L be any Lagrangian subspace of $\mathbf{R}^{2n}$. Check that $JL \in \Lambda_L$ (the space of all Lagrangian subspaces transversal to L) and that $JL = L^\perp$ (for $(\ ,\ )_J$). Deduce that there exists a basis $(x_1, \ldots, x_n, Jx_1, \ldots, Jx_n)$ of $\mathbf{R}^{2n}$ which is symplectic.

Exercise II.16. Let L be a Lagrangian subspace. If $J \in \mathcal{J}_c(\omega)$, $JL \in \Lambda_L$ and $(\ ,\ )_J$ defines a positive definite symmetric bilinear form on L.

(1) Conversely, given $L' \in L$ and an inner product g on L, construct an element of $\mathcal{J}_c(\omega)$: for a nonzero vector $x \in L$, consider its g-orthogonal $x^\perp \subset L$ as a subspace of E. Now $\left(x^\perp\right)^\circ$ is an $n+1$-dimensional subspace of E. Show that $\dim \left(x^\perp\right)^\circ \cap L' = 1$. Define Jx to be the unique vector of this line such that $\omega(x, Jx) = 1$ and check that this defines the required element of $\mathcal{J}_c(\omega)$.
(2) Deduce that there is a one-to-one correspondence between $\mathcal{J}_c(\omega)$ and $\Lambda_L \times \mathcal{Q}(L)$, where $\mathcal{Q}(L)$ is the set of all positive definite quadratic forms on L.
(3) Prove that Λ_L can be identified with the vector space of all symmetric $n \times n$ matrices.
(4) Deduce that $\mathcal{J}_c(\omega)$ is contractible.
(5) Consider $\mathcal{C}_n = \{(\omega, J) \mid J \text{ is calibrated by } \omega\}$. Show that $\mathcal{C}_n$ can be identified with the homogeneous space $\mathrm{GL}(2n, \mathbf{R})/\,\mathrm{U}(n)$. Show that the first projection $\mathcal{C}_n \to \Omega_n$ is a homotopy equivalence.

***Exercise II.17* (The Nijenhuis tensor).** Consider, on an almost complex manifold W, J, the "Nijenhuis tensor"

$$N(X, Y) = [JX, JY] - J[X, JY] - J[JX, Y] - [X, Y].$$

(1) Check that, if $X \mapsto [X, Y]$ is $\mathbf{C}$-linear (in the sense that it commutes with J), then $N \equiv 0$.
(2) Show that N is actually a tensor, that is: $N(X, Y)$ depends only on the values $X(x)$, $Y(x)$ and not really on the vector fields X and Y.
(3) Compute $N(X, JX)$ and deduce that, if W is a *surface*, N vanishes identically (on vectors) and everywhere (on W).

Exercise II.18. For any map $f : \mathbf{R}^{2n} \to \mathbf{C}$ and any vector field X on $\mathbf{R}^{2n}$, recall that $X \cdot f = X \cdot (f_1 + if_2) = X \cdot f_1 + iX \cdot f_2$ so that $f \mapsto X \cdot f$ is a complex linear map.

(1) Let $\mathbf{R}^{2n}$ be endowed with an almost complex structure J and suppose f is a *holomorphic* function, that is $df \circ J = i\,df$. Show that $df\,(N(X,Y)) = 0$ for all X, Y.
(2) Suppose there exist n holomorphic functions on $\mathbf{R}^{2n}$, which are independent at some point x. Show that N identically vanishes at x.
(3) Assume W is a complex manifold and J is its (integrable) complex structure. Show that the Nijenhuis tensor vanishes identically and everywhere on W.

Exercise II.19. We consider here the complex structure J and the symplectic form $\widetilde{\omega}$ on $\mathcal{H}_\lambda$ as defined in § II.1.d. Show that $\widetilde{\omega}\,(JX, JY) = \widetilde{\omega}\,(X,Y)$: the complex structure is an isometry of $\widetilde{\omega}$. Compute $\widetilde{\omega}\,(X, JY)$ in terms of the blocks in X and Y and in terms of the μ_j. Deduce that (if the μ_j are as above) $\widetilde{\omega}\,(X, JY)$ is a Riemannian metric on $\mathcal{H}_\lambda$. Thus J is calibrated by $\widetilde{\omega}$.

Exercise II.20. Calling d', d'' the Dolbeault operators defined by

$$d'f = \sum \frac{\partial f_j}{\partial dz_j} dz_j \qquad d''f = \sum \frac{\partial f_j}{\partial d\overline{z}_j} d\overline{z}_j,$$

consider the function $f : \mathbf{C}^n \to \mathbf{R}$ defined by $f(v) = \log\left(1 + \|v\|^2\right)$. Let ω be the 2-form $\omega = id'd''f$. Show that ω is a Kähler form on $\mathbf{C}^n$.

Exercise II.21. Let $\varphi_k : U_k \to \mathbf{C}^n$ $(0 \leqslant k \leqslant n)$ be the affine charts in $\mathbf{P}^n(\mathbf{C})$:

$$U_k = \{[x_0, \ldots, x_n] \mid x_k \neq 0\} \text{ and } \varphi_k\left([x_0, \ldots, x_n]\right) = \left(\frac{x_i}{x_k}\right)_{i \neq k} \in \mathbf{C}^n.$$

Let ω be the Kähler form on $\mathbf{C}^n$ defined in Exercise II.20. Put $\omega_k = \varphi_k^\star \omega$, so that

$$\omega_k = id'd'' \log \frac{\sum |x_i|^2}{|x_k|^2}.$$

Show that, on $U_k \cap U_l$, ω_k and ω_l coincide. Deduce that the ω_k $(0 \leqslant k \leqslant n)$ define a Kähler form on $\mathbf{P}^n(\mathbf{C})$. The associated Kähler metric is called the *Fubini–Study* metric.

Exercise II.22. Prove that the Kähler form of Exercise II.21 is given, in coordinates

$$x + iy \in \mathbf{C} \subset \mathbf{C} \cup \{\infty\} = \mathbf{P}^1(\mathbf{C}),$$

by the formula

$$\omega = \frac{dx \wedge dy}{(1 + x^2 + y^2)^2}.$$

Compute the area

$$\int_{\mathbf{P}^1(\mathbf{C})} \omega.$$

Using stereographic projection, prove that this ω agrees, up to a scalar factor, with the symplectic area form (also called ω) on $\mathbf{S}^2$, as described for instance in Exercise II.8.

Exercise II.23. Prove that the assignation $X \mapsto i_X\omega$ defines a surjective linear map $\mathcal{H}_{\text{loc}}(W) \to H^1(W;\mathbf{R})$. Deduce an exact sequence of real vector spaces:

$$0 \longrightarrow \mathcal{H}(W) \longrightarrow \mathcal{H}_{\text{loc}}(W) \longrightarrow H^1(W;\mathbf{R}) \longrightarrow 0.$$

Exercise II.24. Let α be a differential form and X, Y be two vector fields on W. Show that

$$\mathcal{L}_X i_Y \alpha - i_Y \mathcal{L}_X \alpha = i_{[X,Y]}\alpha.$$

Hint: one can check the formula when α is a 1-form(9), then by induction for decomposed k-forms $\alpha_1 \wedge \alpha_2$ where α_2 is a 1-form.

Exercise II.25. Let ω be a symplectic form and let X and Y be two locally Hamiltonian vector fields. Prove that

$$i_{[X,Y]}\omega = d i_X i_Y \omega$$

and deduce that the Poisson bracket of two locally Hamiltonian vector fields is (globally) Hamiltonian.

Exercise II.26. Let X be a skew-Hermitian matrix. Recall, from § II.3.c, that the fundamental vector field associated with $X \in \mathfrak{u}(n)$ on $\mathcal{H}$ (for the $\mathrm{U}(n)$-action by conjugation) is $\underline{X}_h = [X, h]$. Deduce that the function

$$\begin{aligned} f_X : \mathcal{H}_\lambda &\longrightarrow \mathbf{R} \\ h &\longmapsto \operatorname{tr}(iXh) \end{aligned}$$

is a Hamiltonian for the vector field $\underline{X}$.

Exercise II.27. For any nondegenerate (but not necessarily closed) 2-form ω on a manifold W, define Hamiltonian vector fields and a bracket $\{\,,\}$ on functions. Prove that ω is closed if and only if $\{\,,\}$ satisfies the Jacobi identity.

Exercise II.28. Let $\mathcal{Q}$ denote the vector space of all quadratic forms on $\mathbf{R}^{2n}$. Determine its dimension. Endow $\mathbf{R}^{2n}$ with the standard linear symplectic form $\omega = \sum dx_i \wedge dy_i$. Prove that $\mathcal{Q}$ is stable under the Poisson bracket and that the Lie algebra $(\mathcal{Q}, \{\ ,\ \})$ is isomorphic with the Lie algebra $\mathfrak{sp}(2n;\mathbf{R})$ of the symplectic group $\mathrm{Sp}(2n;\mathbf{R})$.

---

(9) For example, using local coordinates.

Exercise II.29. Let G be a Lie group. Prove that the bilinear form ω_ξ defined, for $\xi \in \mathfrak{g}^\star$, by the formula

$$\omega_\xi(X,Y) = \langle \xi, [X,Y] \rangle$$

defines a closed 2-form on the orbit of ξ in $\mathfrak{g}^\star$. Hint: look at the proof in the case where $\mathrm{G} = \mathrm{U}(n)$.

***Exercise II.30* (Geodesic flow on the round sphere).** In this exercise, we consider the unit sphere $\mathbf{S}^n$ in $\mathbf{R}^{n+1}$, its tangent bundle and the symplectic form defined in Exercise II.9, the notation of which we use. Consider the Hamiltonian

$$H(q,p) = \|p\|$$

which is smooth outside the zero section. Prove that, there,

$$\begin{aligned} X_H(q,p) &= \frac{1}{\|p\|}\sum_{i=1}^{n+1} p_i \frac{\partial}{\partial q_i} - \|p\| \sum_{i=1}^{n+1} q_i \frac{\partial}{\partial p_i} \\ &\in \mathsf{T}_{(q,p)}(\mathsf{T}\mathbf{S}^n) = \{(Q,P) \mid Q \cdot q = 0 \text{ and } p \cdot Q + P \cdot q = 0\}. \end{aligned}$$

The flow of X_H is called the *geodesic flow* on the sphere (although this is rather a flow on $\mathsf{T}\mathbf{S}^n$). Prove that the trajectory of a given point (q,p) projects (on $\mathbf{S}^n$) onto the great circle defined by the plane (q,p) (this is the geodesic through q with tangent vector p). Prove that H is the *periodic* Hamiltonian associated with the $\mathbf{S}^1$-action on the complement of the zero section by rotations

$$e^{i\theta} \cdot (q,p) = \left((\cos\theta) q + \frac{\sin\theta}{\|p\|} p, -(\|p\| \sin\theta) q + (\cos\theta) p \right).$$

Exercise II.31. Let W be a Kähler manifold of complex dimension n endowed with a *real structure*, namely an involution S which is *anti-holomorphic*, in the sense that, for any holomorphic function

$$f : U \longrightarrow \mathbf{C}$$

defined on an open subset of W, the function

$$\overline{f \circ S} : S(U) \longrightarrow \mathbf{C}$$

is a holomorphic function. Prove that the set $W_{\mathbf{R}}$ of fixed points of S (the *real part* of W) is a submanifold, all the connected components of which have dimension n..., but that $W_{\mathbf{R}}$ can be empty. Assume now that S satisfies $S^\star\omega = -\omega$. What can be said of the components of $W_{\mathbf{R}}$?

Work out the examples of the involution

$$S([z_0, \ldots, z_n]) = [\overline{z}_0, \ldots, \overline{z}_n]$$

taking $W = \mathbf{P}^n(\mathbf{C})$, respectively taking for W the quadric of homogeneous equation

$$\sum_{i=0}^{n} z_i^2 = 0.$$

Determine a Kähler manifold that contains the sphere S^n as a Lagrangian submanifold.

CHAPTER III

SYMPLECTIC AND HAMILTONIAN GROUP ACTIONS

In this chapter, we define symplectic and Hamiltonian actions. These are the main topics of this book, so that we spend some time on their properties. Hamiltonian actions of tori of maximal dimension are a special case of integrable systems. We prove (this is the Arnold–Liouville theorem), that they are also the local form of all integrable systems with compact level sets.

III.1. Hamiltonian group actions

III.1.a. Symplectic actions. We have already said that a G-action on the symplectic manifold (W, ω) is *symplectic* if any element g of G defines a diffeomorphism which preserves ω:

$$g^\star \omega = \omega.$$

Let us write now the infinitesimal version of this equality. Let $X \in \mathfrak{g}$ and let $\underline{X}$ be the associated fundamental vector field. Let g_t be the flow of $\underline{X}$. If we differentiate (with respect to the time t) the equation $g_t^\star \omega = \omega$, we get:

$$\mathcal{L}_{\underline{X}} \omega = \frac{d}{dt} g_t^\star \omega|_{t=0} = 0.$$

Thus the infinitesimal version of our equality is:

$$\forall X \in \mathfrak{g}, \quad \mathcal{L}_{\underline{X}} \omega = 0.$$

We use now both the Cartan formula

$$\mathcal{L}_X = d i_X + i_X d$$

and the fact that ω is closed, to get:

$$d i_{\underline{X}} \omega = 0.$$

Proposition III.1.1. *If the* G*-action preserves the symplectic form* ω*, then all the fundamental vector fields of the action are locally Hamiltonian.* □

III.1.b. Hamiltonian actions. Consider now the following diagram.

$$\begin{array}{ccccccc}
\mathcal{C}^\infty(W) & \overset{\widetilde{\mu}}{\longleftarrow} & \mathfrak{g} & & & & \\
\downarrow & & \downarrow & & & & \\
\mathcal{H}(W) & \longrightarrow & \mathcal{H}_{\mathrm{loc}}(W) & \longrightarrow & H^1(W;\mathbf{R}) & \longrightarrow & 0
\end{array}$$

One says that the symplectic G-action on W is *Hamiltonian* if there exists a morphism $\widetilde{\mu}$ of Lie algebras $\mathfrak{g} \to \mathcal{C}^\infty(W)$ making the diagram commute. It follows in particular that the fundamental vector fields of the action are then Hamiltonian.

Remarks III.1.2

(1) If the group is connected, we know (see Proposition I.4.5) that each of its elements is a product of exponentials

$$g = \exp X_1 \exp X_2 \cdots \exp X_n$$

and thus the existence of μ, which always implies the infinitesimal "symplecticity" of the action, implies that here the action is actually symplectic. It is thus useless to require that the action is symplectic: this will be a consequence of the existence of the map $\widetilde{\mu}$.

(2) Some authors, among them some of the best ones (for instance Souriau, who was one of the inventors of Hamiltonian actions[(1)] [**123**], see also [**96**]), call *Hamiltonian* an action for which there exists a map $\widetilde{\mu}$ making the diagram commute, without requiring that μ is a Lie algebra morphism. Suppose that $X \mapsto \widetilde{\mu}_X$ is only a map from $\mathfrak{g}$ to $\mathcal{C}^\infty(W)$. Then $\widetilde{\mu}_X$ is a function the symplectic gradient field of which is $\underline{X}$, in particular $\{\widetilde{\mu}_X, \widetilde{\mu}_Y\}$ is a Hamiltonian for $\underline{[X,Y]}$. Thus it is always true that $\{\widetilde{\mu}_X, \widetilde{\mu}_Y\} - \widetilde{\mu}_{[X,Y]}$ is (locally) constant on W. We are just assuming, in our definition, that the constant vanishes.

(3) To finish with remarks, note that if G is commutative and W compact (the case we are mainly interested in), any lift $\widetilde{\mu}$ is automatically a Lie algebra morphism: the Lie bracket in $\mathfrak{g}$ is trivial thus $\{\widetilde{\mu}_X, \widetilde{\mu}_Y\}$ is locally constant. The function $\widetilde{\mu}_X$ must have a critical point on W, which we assumed to be compact, thus there exists a point at which $\{\widetilde{\mu}_X, \widetilde{\mu}_Y\}$ vanishes, and so it is zero everywhere.

---

(1) and who created the terminology *application moment.*

The momentum mapping. Associated with $\widetilde{\mu}$ is the *momentum mapping*:

$$\begin{aligned} \mu : W &\longrightarrow \mathfrak{g}^\star = \operatorname{Hom}(\mathfrak{g}, \mathbf{R}) \\ x &\longmapsto (X \mapsto \widetilde{\mu}_X(x)). \end{aligned}$$

Then the fundamental vector field $\underline{X}$ is the Hamiltonian vector field of the function $x \mapsto \langle \mu(x), X \rangle$. The fact that $\widetilde{\mu}$ is a morphism of Lie algebras gives, for the momentum mapping μ:

Proposition III.1.3. *The momentum mapping $\mu : W \to \mathfrak{g}^\star$ is a Poisson mapping.*

Proof. We want to prove that, for any two maps $f, g : \mathfrak{g}^\star \to \mathbf{R}$ and any point x of the manifold W, one has

$$\{f \circ \mu, g \circ \mu\}(x) = \{f, g\}(\mu(x))$$

(using, as usual, the same notation for the Poisson bracket on W and that on the dual of the Lie algebra $\mathfrak{g}^\star$). By density of polynomials in $\mathcal{C}^\infty$ functions, it is enough to prove this for two polynomials f and g. Using the Leibniz rule, we can even assume that f and g are linear functions on $\mathfrak{g}^\star$, that is, $f = X \in \mathfrak{g}$, $g = Y \in \mathfrak{g}$.

For such an f, the Hamiltonian vector field of the composed map $f \circ \mu$ is defined by

$$\begin{aligned} \omega_x(Z, X_{f\circ\mu}) &= (df)_{\mu(x)} \mathsf{T}_x\mu(Z) \\ &= \langle \mathsf{T}_x\mu(Z), X \rangle, \end{aligned}$$

so that $X_{f\circ\mu} = \underline{X}$ and $X_{g\circ\mu} = \underline{Y}$. We have now

$$\begin{aligned} \{f \circ \mu, g \circ \mu\}(x) &= \omega_x(\underline{X}_x, \underline{Y}_x) \\ &= \langle \mathsf{T}_x\mu(\underline{X}_x), Y \rangle \\ &= \langle \underline{X}_{\mu(x)}, Y \rangle \end{aligned}$$

as above. On the other hand,

$$\{f, g\}(\mu(x)) = \langle \mu(x), [X, Y] \rangle = \langle \underline{X}_{\mu(x)}, Y \rangle$$

as well. □

The definition of a Hamiltonian action can be reformulated in terms of the map μ instead of $\widetilde{\mu}$. The condition that $\widetilde{\mu}_X$ is a Hamiltonian for the fundamental vector field $\underline{X}$ translates into a condition on the tangent mapping

$$\begin{aligned} \mathsf{T}_x\mu : \mathsf{T}_x W &\longrightarrow \mathfrak{g}^\star \\ Y &\longmapsto \mathsf{T}_x\mu(Y), \end{aligned}$$

namely

$$\forall X \in \mathfrak{g}, \quad \langle \mathsf{T}_x\mu(Y), X \rangle = -i_{\underline{X}}\omega(Y) = \omega(Y, \underline{X}).$$

Example III.1.4. If $H : W \to \mathbf{R}$ is any function and if the Hamiltonian vector field X_H is complete, its flow defines a Hamiltonian $\mathbf{R}$-action, the momentum mapping of which is H.

Example III.1.5. More generally, let $H_1, \dots, H_k$ be k commuting functions (in the sense that $\{H_i, H_j\} = 0$ for all i, j) on a symplectic manifold W. Assume that, for some reason (for instance because W is compact), the flows $\varphi_1, \dots, \varphi_k$ of their Hamiltonian vector fields are complete). Then, these flows determine a Hamiltonian $\mathbf{R}^k$-action with momentum mapping

$$\mu = (H_1, \dots, H_k) : W \longrightarrow \mathbf{R}^k.$$

This is true because the vector fields $X_{H_1}, \dots, X_{H_k}$ commute

$$[H_{H_i}, X_{H_j}] = X_{\{H_i, H_j\}} = 0,$$

so that the flows $\varphi_1, \dots, \varphi_k$ commute.

Tori. Suppose that W is compact and the group is a torus $\mathbf{T}$. Choose a basis $X_1, \dots, X_k$ of $\mathfrak{t}$. The existence of the momentum mapping is equivalent to that of a primitive $\widetilde{\mu}_{X_i}$ of $i_{X_i}\omega$ for any i. It is thus sufficient, in order to define $\widetilde{\mu}$, to write

$$\widetilde{\mu}_X = \sum_{i=1}^{k} \lambda_i \widetilde{\mu}_{X_i} \text{ for } X = \sum_{i=1}^{k} \lambda_i X_i.$$

Assume for instance that $H^1(W; \mathbf{R}) = 0$ (for example because W is simply connected). Then any symplectic $\mathbf{T}$-action is Hamiltonian. The momentum mapping μ is then well-defined up to the addition of a constant vector of $\mathfrak{t}^\star$.

Example III.1.6. The circle $\mathbf{S}^1$ acts on $\mathbf{C}^n$ by $u \cdot (z_1, \dots, z_n) = (uz_1, \dots, uz_n)$. The fundamental vector field associated with

$$\frac{\partial}{\partial \theta} \in \mathsf{T}_1\mathbf{S}^1 \text{ is } X_H = \sum_{j=1}^{n} \left(-y_j \frac{\partial}{\partial x_j} + x_j \frac{\partial}{\partial y_j} \right).$$

The momentum mapping for this action is the Hamiltonian

$$H = \frac{1}{2} \sum_{j=1}^{n} |z_j|^2 .$$

Example III.1.7. Similarly, the torus $\mathbf{T}^n = \{(t_1, \dots, t_n) \in \mathbf{C}^n \mid |t_i| = 1\}$ acts on $\mathbf{C}^n$ by $(t_1, \dots, t_n) \cdot (z_1, \dots, z_n) = (t_1 z_1, \dots, t_n z_n)$. This is a Hamiltonian action with momentum mapping

$$\mu(z_1, \dots, z_n) = \frac{1}{2}(|z_1|^2, \dots, |z_n|^2) + \text{constant} \in \mathbf{R}^n \cong (\mathfrak{t}^n)^\star.$$

One should notice that the image of μ is the "first quadrant", namely the subset of $\mathbf{R}^n$ defined by $(x_1 \geqslant 0, \ldots, x_n \geqslant 0)$, a fact which is not unrelated with the convex polyhedra we shall encounter in the next chapter.

Semisimple groups. If the Lie algebra $\mathfrak{g}$ is generated by the Lie brackets of its elements, namely if $\mathfrak{g} = [\mathfrak{g}, \mathfrak{g}]$, all the symplectic G-actions are Hamiltonian(2). It is not hard to see (Exercise II.25) that, if X and Y are two locally Hamiltonian vector fields, then the vector field $[X, Y]$ is *globally* Hamiltonian. But, if $\mathfrak{g}$ satisfies $\mathfrak{g} = [\mathfrak{g}, \mathfrak{g}]$, any element of $\mathfrak{g}$ may be written as a sum of brackets, and thus defines a fundamental vector field which is globally Hamiltonian. This is the case, for example, for the groups $\mathrm{G} = \mathrm{SO}(3)$ or $\mathrm{SU}(2)$ (see Exercise I.19).

Let us go back to the general case. The momentum μ was defined by a map $\widetilde{\mu} : \mathfrak{g} \to \mathcal{C}^\infty(W)$ such that the function $\widetilde{\mu}_X$ is a primitive of the 1-form $i_X\omega$. Hence, the momentum mapping is not *a priori* unique. One can add locally constant functions c_X such that

$$\widetilde{\mu}_{[X,Y]} + c_{[X,Y]} = \{\widetilde{\mu}_X + c_X, \widetilde{\mu}_Y + c_Y\} = \{\widetilde{\mu}_X, \widetilde{\mu}_Y\} = \widetilde{\mu}_{[X,Y]}$$

(that is c vanishes on $[\mathfrak{g}, \mathfrak{g}]$) to $\widetilde{\mu}_X$. Thus μ is defined up to the addition of a constant linear form in the annihilator $[\mathfrak{g}, \mathfrak{g}]^\circ$ of $[\mathfrak{g}, \mathfrak{g}]$. For instance, as we have already said, in the case where the group G is commutative, any constant will do. However, if G is semisimple, the momentum mapping, if it exists, is unique.

Notice that there is a simple procedure to construct examples of momentum mappings.

Proposition III.1.8. *Let μ be the momentum mapping for the Hamiltonian action of a Lie group* G *on a symplectic manifold W. Then, for any Lie subgroup* H *of* G, *the composition of μ and the canonical projection:*

$$W \xrightarrow{\ \mu\ } \mathfrak{g}^\star \longrightarrow \mathfrak{h}^\star$$

is a momentum mapping for the induced action. □

III.1.c. The coadjoint representation of a Lie group. The dual-of-a-Lie-algebra-coadjoint-action machinery can be used to produce examples.

The canonical momentum mapping. Let $\mathrm{G} \cdot \xi$ be a coadjoint orbit in $\mathfrak{g}^\star$. We thus have a Lie group G acting on a symplectic manifold $\mathrm{G} \cdot \xi$ and (obviously!) preserving the symplectic form. Everything is so canonical that the following proposition is almost tautological:

(2) Compact Lie groups the Lie algebra of which satisfy $\mathfrak{g} = [\mathfrak{g}, \mathfrak{g}]$ are said to be *semisimple*, see Exercise I.19.

Proposition III.1.9. *The inclusion* $\mathrm{G}\cdot\xi \subset \mathfrak{g}^\star$ *is a momentum mapping for the* G*-action on its orbit.*

Proof. We want to check that the inclusion $\mu : \mathrm{G}\cdot\xi \to \mathfrak{g}^\star$ satisfies the relation

$$\langle \mathsf{T}_\xi\mu(Y), X\rangle = \omega_\xi(Y, X)$$

for all tangent vectors Y to $\mathrm{G}\cdot\xi$. The tangent space to the orbit is generated by the fundamental vector fields, so that the relation we want to prove is just

$$\langle {}^{\mathfrak{g}^\star}\underline{X}_\xi, Y\rangle = \langle \xi, [Y, X]\rangle,$$

but this is exactly the relation defining the fundamental vector field on $\mathfrak{g}^\star$. □

This construction generates a lot of examples, using Proposition III.1.8. For instance we have seen (Remark II.3.5) that the type of the principal coadjoint orbits is the conjugacy class of any maximal torus. It is thus especially interesting to study the action of a fixed maximal torus $\mathbf{T} \subset \mathrm{G}$ on these orbits(3). We deduce a lot of examples of *torus* actions on symplectic manifolds... thus, even if one is only interested in commutative groups, this excursion into noncommutativity was not useless.

The case of $\mathrm{U}(n)$**.** Let us make the torus of *diagonal* unitary matrices act on the orbits $\mathcal{H}_\lambda$.

Proposition III.1.10. *The map* $\mathcal{H}_\lambda \to \mathbf{R}^n$*, which associates with any Hermitian matrix its diagonal entries, is a momentum mapping for the* $\mathbf{T}$*-action.*

Proof. According to Propositions III.1.8 and III.1.9, the momentum mapping is the composition

$$\mathcal{H}_\lambda \subset \mathcal{H} = \mathfrak{u}(n)^\star \longrightarrow \mathfrak{t}^\star,$$

for the map $\mathfrak{u}(n)^\star \to \mathfrak{t}^\star$ defined by the inclusion $\mathbf{T} \subset \mathrm{U}(n)$. Now, $\mathbf{T}$ is the subgroup of diagonal (in the canonical basis of $\mathbf{C}^n$) unitary matrices, hence $\mathfrak{t}^\star$ is isomorphic to $\mathbf{R}^n$ and the projection is indeed the mapping that maps a matrix to its diagonal entries. □

Example III.1.11. Let $\lambda = (\lambda_1, \lambda_2, \dots, \lambda_2)$. We have seen in § II.1.d that $\mathcal{H}_\lambda$ is a complex projective space $\mathbf{P}^{n-1}(\mathbf{C})$. We have thus exhibited a Hamiltonian $\mathbf{T}^n$-action on the complex projective space $\mathbf{P}^{n-1}(\mathbf{C})$.

(3) For instance, the fixed points will be the points in the orbit the stabilizer of which is precisely $\mathbf{T}$ and not one of its conjugates.

The case of SO(3). The Lie algebra $\mathfrak{so}(3)$ is the Lie algebra of 3×3 skew-symmetric matrices. Identify it with $\mathbf{R}^3$ in such a way that the Lie bracket becomes the vector cross product and with its dual $\mathfrak{so}(3)^\star$ with help of the canonical Euclidean structure of $\mathbf{R}^3$. The adjoint and coadjoint actions are then identified with the usual action of SO(3) on $\mathbf{R}^3$ by rotations (see, if necessary, Exercise I.16). The coadjoint orbits are thus the point 0 and the concentric spheres $\mathbf{S}^2$'s (with stabilizer SO(2)). We choose the third vector of the canonical basis, its orbit is

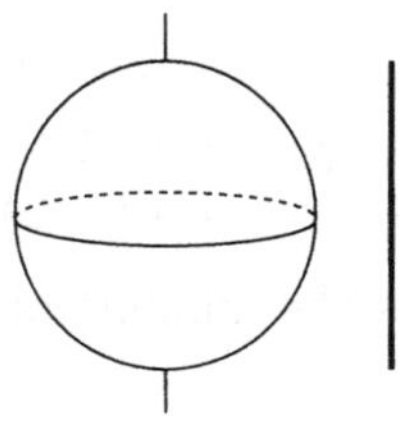

Figure 1

the unit sphere, the corresponding maximal "torus" is simply a circle that acts by rotations about the z-axis and the momentum mapping is just the Hamiltonian $(x, y, z) \mapsto z$. We merely copy here a figure that already appeared in Chapter I. Notice that the quotient map there is our momentum mapping here, a special case of a more general result (Exercise IV.12).

III.2. Properties of momentum mappings

III.2.a. Fixed points. A momentum mapping μ is a map from our manifold W into the vector space $\mathfrak{g}^\star$. We will use it to investigate the group action. Here is a simple example of what can be expected of such a situation.

Proposition III.2.1. *Let W be a compact symplectic manifold endowed with an action of $\mathbf{S}^1$. Assume the action is Hamiltonian. Then it has fixed points.*

Proof. A momentum mapping for an $\mathbf{S}^1$-action is simply a function

$$H : W \longrightarrow \mathbf{R}.$$

The manifold being compact, any function on it must have critical points. Let x be such a point, namely x is such that $(dH)_x = 0$, and hence, such that the Hamiltonian vector field X_H vanishes at x. But the latter is the fundamental vector field associated with the action. Thus, x is a fixed point. □

Think of the situation depicted in Figure 1, where the fixed points of the $\mathbf{S}^1$-action are the extrema of the momentum mapping. Notice that there are symplectic

$\mathbf{S}^1$-actions without fixed points. For instance, rotations on one factor $\mathbf{S}^1$ on the torus $\mathbf{S}^1 \times \mathbf{S}^1$, namely the formula

$$t \cdot (x, y) = (t \cdot x, y)$$

defines a symplectic action (which is not Hamiltonian).

III.2.b. The tangent map. The tangent map

$$\mathsf{T}_x\mu : \mathsf{T}_x W \longrightarrow \mathfrak{g}^\star$$

is, by definition, the transpose of

$$\begin{aligned} {}^t\mathsf{T}_x\mu : \mathfrak{g} &\longrightarrow \mathsf{T}_x^\star W \\ X &\longmapsto (i_{\underline{X}}\omega)_x. \end{aligned}$$

In particular, its image $\operatorname{Im} \mathsf{T}_x\mu$ is the annihilator in $\mathfrak{g}^\star$ of $\operatorname{Ker} {}^t\mathsf{T}_x\mu$, that is, the annihilator of

$$\left\{X \in \mathfrak{g} \mid (i_{\underline{X}}\omega)_x = 0\right\} = \left\{X \in \mathfrak{g} \mid \underline{X}_x = 0\right\} = \mathfrak{g}_x.$$

Proposition III.2.2. *The annihilator in $\mathfrak{g}^\star$ of $\mathfrak{g}_x$ is the subspace $\operatorname{Im} \mathsf{T}_x\mu$. Dually, the annihilator in $\mathfrak{g}$ of $\operatorname{Im} \mathsf{T}_x\mu \subset \mathfrak{g}^\star$ is $\mathfrak{g}_x$.* □

Thus the rank of $\mathsf{T}_x\mu$ is the dimension of the orbit of x. We deduce, in particular:

Proposition III.2.3. *The momentum mapping μ is a submersion at the point x if and only if the stabilizer G_x is discrete.*

Proof. This property is indeed equivalent to the nullity of $\mathfrak{g}_x$. □

Corollary III.2.4. *Let G be a commutative Lie group acting in an effective way on a symplectic manifold W. Assume the action is Hamiltonian with momentum mapping $\mu : W \to \mathfrak{g}^\star$. Then μ is a submersion on the open dense subset of principal orbits in W.*

Proof. We already know that the stabilizer of the principal orbits is the (discrete) subgroup $\{1\} \subset \mathrm{G}$. □

If G is not commutative, it may happen that the momentum mapping μ is nowhere submersive, even if the action is effective, as the next example shows.

Example III.2.5. Let the group SO(3) of rotations act on $\mathbf{S}^2 \times W$ by the usual (effective) action on $\mathbf{S}^2$ and by the trivial action on W. It preserves any "product" symplectic form $\omega_1 \oplus \omega_2$, and the image of the momentum mapping

$$\mu : \mathbf{S}^2 \times W \longrightarrow \mathfrak{so}(3)^\star = \mathbf{R}^3$$

is a sphere $\mathbf{S}^2$ (and, in particular, contains no open dense subset of $\mathfrak{so}(3)^\star$) so that μ is nowhere submersive.

Proposition III.2.6. *The kernel* $\operatorname{Ker} \mathsf{T}_x\mu$ *is the orthogonal (for* ω_x*) of the tangent space to the orbit through* x.

Proof. This is straightforward, $\mathsf{T}_x\mu(Y)$ is zero if and only if $\langle \mathsf{T}_x\mu(Y), X\rangle = 0$ for all vectors $X \in \mathfrak{g}$, that is, if and only if $\omega_x(\underline{X}_x, Y) = 0$ for all X, that is, if and only if Y is orthogonal to the subspace generated by the fundamental vector fields. □

III.2.c. Noether theorem. We consider now the "levels" of the momentum mapping[(4)] $\mu : W \to \mathfrak{g}^\star$ of a Hamiltonian G-action on the symplectic manifold (W, ω). The most classical form of Noether's theorem seems to be stated nowadays as follows:

Theorem III.2.7. *Let* H *be a function on* W *which is invariant under the* G*-action. Then* μ *is constant on the trajectories of the Hamiltonian vector field* X_H.

Proof. Indeed, if $\gamma(t)$ is a trajectory of X_H, one can write, for any $X \in \mathfrak{g}$:

$$\begin{aligned}\frac{d}{dt}\langle \mu\circ\gamma(t), X\rangle &= \langle \mathsf{T}_{\gamma(t)}\mu(X_H(\gamma(t))), X\rangle\\ &= \langle X_H(\gamma(t)), {}^t\mathsf{T}_{\gamma(t)}\mu(\underline{X})\rangle\\ &= \langle X_H(\gamma(t)), (i_X\omega)_{\gamma(t)}\rangle\\ &= \omega(\underline{X}, X_H)_{\gamma(t)}\\ &= -dH_{\gamma(t)}(\underline{X}).\end{aligned}$$

But the Hamiltonian H is invariant and $\underline{X}$ is a fundamental vector field of the action, so that

$$H(\exp(sX)\cdot\gamma(t)) = H(\gamma(t))$$

a relation which, when differentiated at $s = 0$, gives

$$dH(\underline{X}(\gamma(t))) = 0.$$

□

Notice that this theorem means that the Hamiltonian vector field X_H is tangent to the levels $\mu^{-1}(\xi)$ of the momentum mapping.

***Example III.2.8* (Periodic Hamiltonians).** Even the most trivial cases of the previous theorem are already interesting. Consider for instance the case where the group acting is $\mathbf{S}^1$ and the function is the very momentum mapping of this action. We get that the fundamental vector field is tangent to the hypersurfaces of constant "energy" H.

Such a function is called a *periodic Hamiltonian*. We see that all the regular levels of a periodic Hamiltonian are oriented submanifolds endowed with a fixed point free $\mathbf{S}^1$-action. For example in dimension 4, the regular levels are Seifert

(4) I shall call *level* ξ the inverse image $\mu^{-1}(\xi)$ of an element ξ, even if ξ is not a number.

manifolds. We shall see later how it is possible to use the constructions of Chapter I to investigate periodic Hamiltonians.

III.2.d. Equivariant momentum mappings. The momentum mapping μ maps the G-manifold W into another G-manifold $\mathfrak{g}^\star$. It is thus natural to ask whether this is an equivariant map.

Proposition III.2.9. *Let μ be the momentum mapping for the Hamiltonian action of the connected group* G *on the symplectic manifold W. Then μ is an equivariant map.*

Proof. We consider an element X of $\mathfrak{g}$ and we prove

$$\mathsf{T}_x\mu({}^W\underline{X}_x) = {}^{\mathfrak{g}^\star}\underline{X}_{\mu(x)},$$

that is, the infinitesimal version of the proposition (this will be enough, because the group is connected). Thanks to the relation defining the fundamental vector fields of the coadjoint action, we have, on the one hand

$$\begin{aligned}\langle {}^{\mathfrak{g}^\star}\underline{X}_{\mu(x)}, Y\rangle &= \langle \mu(x), [Y, X]\rangle \\ &= -\widetilde{\mu}_{[X,Y]}(x),\end{aligned}$$

and on the other hand

$$\begin{aligned}\langle \mathsf{T}_x\mu({}^W\underline{X}_x), Y\rangle &= \langle {}^W\underline{X}_x, {}^t\mathsf{T}_x\mu(Y)\rangle \\ &= \langle {}^W\underline{X}_x, (i_{{}^W\underline{Y}}\omega)_x\rangle \\ &= \omega_x({}^W\underline{Y}, {}^W\underline{X}) \\ &= -\left\{\widetilde{\mu}_X, \widetilde{\mu}_Y\right\}(x) \\ &= -\widetilde{\mu}_{[X,Y]}(x).\end{aligned}$$

Hence the two sides are equal. □

This endows Hamiltonian actions with very specific properties. For example, Proposition III.2.9 forces the inclusion of stabilizers $\mathrm{G}_x \subset \mathrm{G}_{\mu(x)}$. Here is an example:

Proposition III.2.10. *Let W be a symplectic manifold endowed with a nontrivial symplectic action of* $\mathrm{G} = \mathrm{SO}(3)$ *or* SU(2). *Then the stabilizer of the principal orbits is a* commutative *subgroup of* G.

Proof. We have seen in § III.1.b that, because G is semisimple, the action is Hamiltonian. Let $\mu : W \to \mathfrak{g}^\star$ be the (unique) momentum mapping for this action. Let x be a point in W. Then we have the inclusion $\mathrm{G}_x \subset \mathrm{G}_{\mu(x)}$ and the orbit stabilizers in $\mathfrak{g}^\star$ are either G or the circle $\mathrm{SO}(2) \subset \mathrm{SO}(3)$ or SU(2) (see § III.1.c). If G_x is not contained in SO(2), we have $\mathrm{G}_{\mu(x)} = \mathrm{G}$, or $\mu(x) = 0$. If this was the case for all the points of the principal orbits, μ would vanish on an open dense subset of W, thus everywhere. But the operation would then be trivial. □

In the same way, one can show more generally (using Remark II.3.5) that, if μ is a submersion at at least one point, then "the" stabilizer of the principal orbits is commutative and discrete.

III.2.e. Level sets. Let us return to the general case. For a regular value $\xi \in \mathfrak{g}^\star$ of μ, call $V_\xi = \mu^{-1}(\xi)$. As μ is equivariant, the subgroup $\mathrm{G}_\xi \subset \mathrm{G}$ keeps the set V_ξ invariant: if $x \in V_\xi$ and $g \in \mathrm{G}_\xi$, then

$$\mu(g \cdot x) = g \cdot \mu(x) = g \cdot \xi = \xi.$$

Look now at what happens to the symplectic form when we restrict it by the inclusion map $j_\xi : V_\xi \subset W$:

Lemma III.2.11. *The kernel of the pulled back 2-form $j_\xi^\star \omega$ at a point x is the subspace $\mathsf{T}_x(\mathrm{G}_\xi \cdot x)$. The rank of this 2-form is constant (namely, it does not depend on x) and*

$$\operatorname{rk} j_\xi^\star \omega = 2 \dim V_\xi + \dim(\mathrm{G} \cdot \xi) - \dim W.$$

Proof. We have, by definition,

$$\operatorname{Ker}(j_\xi^\star \omega)_x = \mathsf{T}_x V_\xi \cap (\mathsf{T}_x V_\xi)^\circ.$$

On the other hand, $\mathsf{T}_x V_\xi = \operatorname{Ker} \mathsf{T}_x \mu$ and we have (this is Proposition III.2.6) that

$$\operatorname{Ker} \mathsf{T}_x \mu = (\mathsf{T}_x(\mathrm{G} \cdot x))^\circ.$$

Thus

$$\operatorname{Ker}(j_\xi^\star \omega)_x = \operatorname{Ker} \mathsf{T}_x \mu \cap \mathsf{T}_x(\mathrm{G} \cdot x).$$

The tangent space $\mathsf{T}_x(\mathrm{G} \cdot x)$ to the orbit is generated by the fundamental vector fields ${}^W\underline{X}$. According to the proof of Proposition III.2.9, $\underline{X}_x$ is in $\operatorname{Ker} \mathsf{T}_x \mu$ if and only if the corresponding fundamental vector field in $\mathfrak{g}^\star$ vanishes at $\mu(x)$, that is, if and only if X belongs to the Lie algebra $\mathfrak{g}_{\mu(x)}$ of the stabilizer $\mathrm{G}_{\mu(x)}$. Thus $\operatorname{Ker}(j_\xi^\star \omega)_x$ is generated by the fundamental vector fields coming from $\mathfrak{g}_\xi$. Hence $\operatorname{Ker}(j_\xi^\star \omega)_x = \mathsf{T}_x(\mathrm{G}_\xi \cdot x)$, its dimension is

$$\dim \mathrm{G}_\xi - \dim \mathrm{G}_x = \dim \mathrm{G}_\xi$$

(because ξ is a regular value so that G_x is discrete). Its rank is

$$\begin{aligned} \operatorname{rk}(j_\xi^\star \omega)_x &= \dim V_\xi - (\dim \mathrm{G}_\xi - \dim \mathrm{G}_x) \\ &= \dim V_\xi \underbrace{- \dim \mathrm{G}_\xi + \dim \mathrm{G}}_{\dim \mathrm{G} \cdot \xi} - \dim W + \dim V_\xi \\ &= 2 \dim V_\xi - \dim W + \dim \mathrm{G} \cdot \xi \end{aligned}$$

and thus is constant on V_ξ. □

These results allow us to characterize the Hamiltonian actions of Abelian groups.

Proposition III.2.12. *Let* G *be a connected Lie group acting effectively on a symplectic manifold* W *with momentum mapping* μ. *The three following properties are equivalent.*

(1) *The orbits of* G *are isotropic.*
(2) *The momentum mapping* μ *is constant on the orbits.*
(3) *The group* G *is commutative.*

In this case, $\dim W \geqslant 2\dim \mathrm{G}$.

Proof. The orbit $\mathrm{G}\cdot x$ is isotropic if and only if

$$\mathsf{T}_x(\mathrm{G}\cdot x) \subset \mathsf{T}_x(\mathrm{G}\cdot x)^\circ = \operatorname{Ker} \mathsf{T}_x\mu.$$

In this case, for any $X, Y \in \mathfrak{g}$, the map

$$x \longmapsto -\omega_x(\underline{X}_x, \underline{Y}_x)$$

is identically zero. But we have seen (in Exercise II.25) that its differential is the 1-form $i_{[\underline{X},\underline{Y}]}\omega$. Thus $i_{[\underline{X},\underline{Y}]}\omega = 0$, from which it follows that $[\underline{X},\underline{Y}] = 0$ and, due to the fact that the fundamental vector field of the bracket is the bracket of fundamental vector fields (see § I.1.b), that $[X,Y] = 0$ for any $X, Y \in \mathrm{G}$, so that G is commutative.

Reciprocally, if G is commutative, $\mathrm{G}_\xi = \mathrm{G}$ for all ξ in $\mathfrak{g}^\star$ and thus $\operatorname{Ker} j_\xi^\star\omega$ is the whole tangent space to the orbit which is thus isotropic. □

The inequality in III.2.12 is easily generalized.

Proposition III.2.13. *Let* W *be a symplectic manifold endowed with a Hamiltonian action of a compact connected Lie group* G, *with momentum mapping* $\mu : W \to \mathfrak{g}^\star$. *Let* $\mathbf{T}$ *be a maximal torus in* G. *If* μ *is submersive in at least one point in* W, *then*

$$\dim \mathrm{G} + \dim T \leqslant \dim W.$$

Proof. It is sufficient to observe that the rank of $(j_\xi^\star\omega)_x$ is a nonnegative integer, and to compute:

$$\begin{aligned} \operatorname{rk} j_\xi^\star\omega &= 2\dim V_\xi - \dim W + \dim \mathrm{G}\cdot\xi \\ &= 2(\dim W - \dim \mathrm{G}) - \dim W + \dim \mathrm{G}\cdot\xi \\ &= \dim W - 2\dim \mathrm{G} + \dim \mathrm{G} - \dim \mathrm{G}_\xi. \end{aligned}$$

Hence $\dim W \geqslant \dim \mathrm{G} + \dim \mathrm{G}_\xi \geqslant \dim \mathrm{G} + \dim T$, thanks to Remark II.3.5. □

For example a manifold endowed with a symplectic SO(3)-action the momentum mapping of which is submersive somewhere has dimension at least 4.

The hypothesis "submersion at one point" is necessary: for example the SO(3)-action on the unit sphere $\mathbf{S}^2$ by isometries preserves the volume form and is thus symplectic.

III.2.f. Symplectic reduction, examples. The essence of what is called symplectic reduction is contained in a rather obvious algebraic lemma. Recall that we call *coisotropic* any vector subspace (*resp.* submanifold) of a symplectic vector space (*resp.* manifold) the symplectic orthogonal of which is *isotropic*: $F \subset E$ is coisotropic if and only if $F^\circ \subset F$.

***Lemma III.2.14* (Symplectic reduction).** *Let E be a vector space endowed with a (constant) symplectic form ω. Let F be a coisotropic subspace.*

(1) *The form ω induces a symplectic form on F/F°.*
(2) *If L is any* isotropic *subspace transversal to F, then the composite map $L \cap F \subset F \to F/F^\circ$ is the inclusion of an isotropic subspace.*

Proof. The subspace F° is, by definition, the kernel of the restriction of ω to F. Hence the first assertion. For the second one, we just have to check that $(L \cap F) \cap F^\circ = \{0\}$, but we have

$$(L \cap F) \cap F^\circ = L \cap F^\circ = (L^\circ + F)^\circ \subset (L + F)^\circ = \{0\}. \qquad \square$$

Straightaway, we have an application. Consider a torus $\mathbf{T}$ acting effectively on a compact symplectic manifold (W, ω) with momentum mapping $\mu : W \to \mathfrak{t}^\star$. We know that in this case there must exist regular values. Let ξ be such a value. For $x \in V_\xi$, we have

$$\mathsf{T}_x V_\xi = \operatorname{Ker} \mathsf{T}_x \mu = (\mathsf{T}_x(\mathbf{T} \cdot x))^\circ.$$

Moreover, the tangent space to the orbit, being generated by fundamental vector fields, is isotropic since $\mathbf{T}$ is commutative (Proposition III.2.12). Thus $\mathsf{T}_x V_\xi$ is coisotropic and its orthogonal is the tangent space to the orbit. Hence ω_x induces a nondegenerate skew-symmetric bilinear form on $\mathsf{T}_x V_\xi / \mathsf{T}_x(\mathbf{T} \cdot x)$.

The $\mathbf{T}$-action on W defines by restriction a $\mathbf{T}$-action on the level V_ξ, of which we know that all its stabilizers are finite, as $\mathsf{T}_x \mu$ is surjective thus $\mathfrak{t}^\star$ is the annihilator of $\mathfrak{t}_x$.

If moreover all these stabilizers are trivial, in other words if the $\mathbf{T}$-action on the regular level V_ξ is free, then the quotient $V_\xi / \mathbf{T}$ is a manifold B_ξ (of dimension $\dim W - 2 \dim \mathbf{T}$) and $\mathsf{T}_x V_\xi / \mathsf{T}_x(\mathbf{T} \cdot x)$ may be identified to its tangent space at the point $[x] = \mathbf{T} \cdot x$.

Proposition III.2.15. *Let the torus $\mathbf{T}$ act on the symplectic manifold W, with momentum mapping $\mu : W \to \mathfrak{t}^\star$. Assume it acts freely on the regular level V_ξ of μ. Then the orbit space V_ξ / T is a manifold naturally endowed with a symplectic form σ_ξ called the reduced symplectic form.*

Proof. The form defined by ω_x on $\mathsf{T}_x V_\xi / \mathsf{T}_x(\mathbf{T}\cdot x)$ does not depend on the point x we chose in the orbit $\mathbf{T}\cdot x$, thanks to the invariance of ω. The diagram

$$\begin{array}{ccc} V_\xi & \xhookrightarrow{j_\xi} & W \\ {\scriptstyle p_\xi}\big\downarrow & & \\ B_\xi & & \end{array}$$

defines a 2-form σ_ξ by $p_\xi^\star \sigma_\xi = j_\xi^\star \omega$. From which it follows that $p_\xi^\star d\sigma_\xi = 0$ and that σ_ξ is closed. □

Let G be a (not necessarily commutative) Lie group. Consider a Hamiltonian action with momentum mapping

$$\mu : W \longrightarrow \mathfrak{g}^\star.$$

Let ξ be a regular value of μ and G_ξ its stabilizer (for the coadjoint action). According to Proposition III.2.6, the 2-form induced by $j_\xi^\star \omega$ is nondegenerate on $\mu^{-1}(\xi)/\mathrm{G}_\xi$. So that we have:

Proposition III.2.16. *Let the Lie group* G *act on the symplectic manifold W, with momentum mapping $\mu : W \to \mathfrak{g}^\star$. Assume ξ is a regular value of μ the stabilizer G_ξ of which acts freely on the regular level $\mu^{-1}\xi$. Then the orbit space $\mu^{-1}(\xi)/\mathrm{G}_\xi$ is a manifold naturally endowed with a symplectic form σ_ξ.* □

Corollary III.2.17. *Assume* 0 *is a regular value of the momentum mapping μ and* G *acts freely on $\mu^{-1}(0)$. Then $\mu^{-1}(0)/\mathrm{G}$ is a manifold naturally endowed with a symplectic form.* □

***Example III.2.18* (Symplectic form on the complex projective space).** Consider the $\mathbf{S}^1$-action on $\mathbf{C}^n$ by multiplication $t\cdot(z_1,\ldots,z_n) = (tz_1,\ldots,tz_n)$. This is a Hamiltonian action with momentum mapping

$$\mu(z_1,\ldots,z_n) = \frac{1}{2}\sum_{i=1}^{n} |z_i|^2,$$

as follows, for instance, using Example III.1.7 and Proposition III.1.8, from the fact that the projection map $\mathbf{R}^n \to \mathbf{R}$ induced by the diagonal inclusion $\mathbf{S}^1 \to \mathbf{T}^n$ is the map

$$(x_1,\ldots,x_n) \longmapsto x_1 + \cdots + x_n.$$

All nonzero real numbers are regular values of μ. Choose any such value $a > 0$. The level $\mu^{-1}(a)$ is the sphere of radius $\sqrt{2a}$. The quotient by the $\mathbf{S}^1$-action is the projective space $\mathbf{P}^{n-1}(\mathbf{C})$. The symplectic reduction procedure provides a symplectic form (for each value of a) on this projective space.

Symplectic orbifolds. In general, the torus action may well have exceptional (that is, with finite stabilizer) orbits in regular levels. It is then no longer absolutely correct that the orbit space B_ξ is a symplectic *manifold.* We have already met this kind of problem in § I.3.c, where the *topological* structure of the quotient B_ξ happened to be enough for what we needed. Now we have to look carefully at what is preserved of the differentiable structure near the singular points in order to be able to speak of a symplectic form on the quotient.

The right notion here is that of orbifold, invented by Satake [**120**] under the name of V-manifold. The idea of a symplectic form on an orbifold was made precise by Weinstein in [**129**]. Since then, symplectic orbifolds (and the way the type of techniques explained in this book extend to them) have been extensively investigated by numerous authors, among which are Prato (unpublished) and Lerman and Tolman [**95**].

Locally, orbifolds are the open subsets of the quotient spaces $\mathbf{R}^n/\Gamma$, where Γ is a finite group equipped with a representation ρ in $\mathrm{GL}(n, \mathbf{R})$, the fixed point set of which has codimension at least 2. The local isomorphisms $U_1 \to U_2$ are the pairs (φ, γ), where $\gamma : \Gamma_1 \to \Gamma_2$ is a group homomorphism and $\varphi : \widetilde{U}_1 \to \widetilde{U}_2$ is a diffeomorphism of the saturated open subsets of $\mathbf{R}^n$, over U_1 and U_2, and all the diagrams

$$\begin{array}{ccc} \widetilde{U}_1 & \xrightarrow{\varphi} & \widetilde{U}_2 \\ {\scriptstyle \rho_1(g)}\uparrow & & \uparrow{\scriptstyle \rho_2(g)} \\ \widetilde{U}_1 & \xrightarrow[\varphi]{} & \widetilde{U}_2 \end{array}$$

are commutative.

We call *orbifold* a topological space modeled on the local structure just described. A *differential form* on an orbifold is of course given by the prescription, on each open local chart U, of a differential form (that is, a Γ-invariant differential form on $\widetilde{U}$). A *symplectic* form is a 2-form, which, in each local chart, is closed (this is a local notion!) and nondegenerate.

Example III.2.19. The cyclic group $\Gamma = \mathbf{Z}/m$ acts by rotations on $\mathbf{R}^2$. Then $\mathbf{R}^2/\Gamma$ is an orbifold. The surfaces B, quotients of Seifert manifolds by their $\mathbf{S}^1$-action, that we have discussed in § I.3.c are orbifolds. See also Exercises III.17 and III.18 and the orbifolds shown in Figure 6 (page 101).

Proposition III.2.20. *Let V_ξ be a regular level of the map $\mu : W \to \mathfrak{t}^\star$. The orbit space $B_\xi = V_\xi/\mathbf{T}$ is an orbifold, naturally endowed with a* reduced *symplectic form σ_ξ.*

Proof. The symplectic reduction lemma (Lemma III.2.14) asserts that, for any point x, the bilinear form ω_x induces a nondegenerate skew-symmetric bilinear form on the quotient space

$$\mathsf{T}_x V_\xi / \mathsf{T}_x(\mathbf{T}\cdot x) = E_x.$$

The latter is a slice at x for the $\mathbf{T}$-action on V_ξ. A neighborhood of $\mathbf{T}\cdot x$ in V_ξ has the form $\mathbf{T}\times_{\mathbf{T}_x} E_x$. Thus a neighborhood of x in B_ξ has the form $E_x/\mathbf{T}_x$. Now $\mathbf{T}_x$ is a finite group of isomorphisms of E_x, and we may assume that it preserves a complex structure (calibrated by the form induced from ω_x), so that the (co-)dimension of its fixed point set is even. If the codimension is positive, it is then greater than or equal to 2. □

Example III.2.21. Let H be a periodic Hamiltonian on a compact 4-dimensional symplectic manifold. Its regular levels are Seifert manifolds, with oriented topological surfaces as base spaces (one should check that symplectic orbifolds are oriented). We shall see in Chapter VIII that all the Seifert manifolds we have met in § I.3.c appear in this framework.

Poisson structures on quotient spaces. Assume that the group G acts on the symplectic manifold W and that the quotient W/G is a manifold. Of course, there is no reason why this quotient should be symplectic. However, if we assume that the G-action is Hamiltonian, there is something left from the symplectic structure, as asserted by the next proposition, a rather useful result that we shall have the opportunity to use in Chapter V.

Proposition III.2.22. *Let* G *be a Lie group acting on a symplectic manifold W with momentum mapping*

$$\mu : W \longrightarrow \mathfrak{g}^\star.$$

The Poisson bracket of W defines a Poisson bracket on W/G the symplectic leaves of which are in one-to-one correspondence, via μ, *with the coadjoint orbits of* G *in the image $\mu(W)\subset\mathfrak{g}^\star$.*

Proof. The Poisson bracket of two functions $f, g : W/\mathrm{G} \to \mathbf{R}$ is defined by

$$\{f,g\} = \{f\circ\pi, g\circ\pi\} \text{ where } \pi \text{ is the projection } W \longrightarrow W/\mathrm{G}.$$

Let $\xi\in\mathfrak{g}^\star$ and $\mathrm{G}\cdot\xi$ be its orbit. Endow $\mathrm{G}\cdot\xi$ with the opposite of its standard symplectic form and consider G acting on $W\times\mathrm{G}\cdot\xi$ diagonally, so that

$$\begin{aligned} J : W\times \mathrm{G}_\xi &\longrightarrow \mathfrak{g}^\star \\ (x,\eta) &\longmapsto \mu(x)-\eta \end{aligned}$$

is the momentum mapping of the action. Perform the symplectic reduction of $W\times\mathrm{G}\cdot\xi$ at the level 0 of J, getting a symplectic structure on $J^{-1}(0)/\mathrm{G}$. The mapping

$$J^{-1}(0)/\mathrm{G} = \{(x,\eta) \in W \times \mathrm{G}\cdot\xi \mid \mu(x) = \eta\}/\mathrm{G} \longrightarrow \mu^{-1}(\mathrm{G}\cdot\xi)/\mathrm{G}$$

is an isomorphism, and this puts a symplectic structure on the latter space. □

Example III.2.23. Look once again at the standard $\mathbf{S}^1$-action on $\mathbf{C}^{n+1}$. When restricted to $\mathbf{C}^{n+1} - \{0\}$, this is a free action. The quotient $\mathbf{C}^{n+1} - \{0\}/\mathbf{S}^1$ is a manifold, foliated by the projective spaces $\mu^{-1}(a)/\mathbf{S}^1$ $(a > 0)$. We know that all these leaves are symplectic manifolds (Example III.2.18 above). The proposition asserts that this foliation is the symplectic foliation associated with a Poisson structure on the quotient.

III.3. Torus actions and integrable systems

III.3.a. Integrable systems. Consider a Hamiltonian torus action on a symplectic manifold, with momentum mapping $\mu : W \to \mathbf{R}^k$. Assume the action is effective. We know that, in this case, $2k \leqslant \dim W = 2n$ (Proposition III.2.12). Consider the maximal case, namely the case where the dimension of the torus acting is half the dimension of the manifold ($k = n$). Look at the n component functions of the momentum mapping

$$\mu = (f_1, \ldots, f_n) : W \longrightarrow \mathbf{R}^n.$$

They are independent over the open dense subset of regular points of μ and they commute in the sense that

$$\{f_i, f_j\} = 0 \quad \forall i, j.$$

We will call, more generally, *completely integrable system* or, more simply, *integrable system*, a collection of n independent (in the sense above) commuting functions defined over a symplectic manifold of dimension $2n$.

Notice that, by definition, the momentum mapping of an effective Hamiltonian torus action on a symplectic manifold of the right dimension is an integrable system. This is why we will call *completely integrable* the Hamiltonian action of a torus on a symplectic manifold of dimension double (of that of the torus). Notice however that any nonconstant function on a surface is an integrable system—so that there are many completely integrable systems that do not come from integrable torus actions.

Why is this called a "system"? This is because one thinks of the differential (Hamiltonian) system defined by one of the Hamiltonian vector fields, X_{f_1}, say, namely

$$\dot{x}(t) = X_{f_1}(x(t)).$$

The other functions are then playing an auxiliary role. As $\{f_i, f_j\} = 0$, they will be constant along the trajectories of X_{f_1}—and thus said to be *first integrals* of the vector field X_{f_1}.

Remark III.3.1. Although it gives a symmetric role to all the f_i's, the present definition is certainly not the best possible. Very often, one is only interested in the subalgebra of $\mathcal{C}^\infty(W)$ generated by the f_i's and not in a specific set of generators. For a formally better definition, see, for instance, [**126**].

Example III.3.2 **(The simple pendulum).** The simple pendulum is the mechanical system constituted by a bob at the end of a rope the other end of which is fixed, in a constant gravitation field (Figure 2). It is considered that the bob stays in a vertical plane. Call θ the angle between the rope and the vertical (this is the direction of the gravitation field Γ). The differential equation describing the motion of the bob is $\ddot{\theta} = -\sin\theta$. The phase space is $\mathbf{S}^1 \times \mathbf{R}$ (a symplectic manifold as

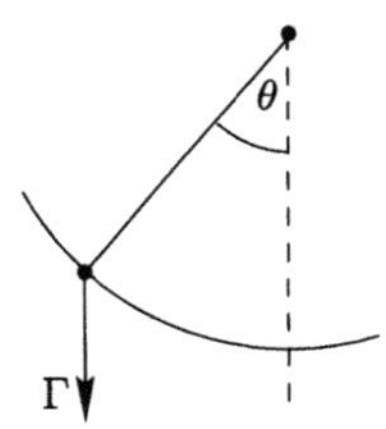

Figure 2. The simple pendulum

the cotangent bundle $\mathsf{T}^\star\mathbf{S}^1$). We use a real variable q (such that θ is q modulo 2π) and the dual variable $p = \dot{q}$. The differential equation is equivalent to the system

$$\dot{q} = p, \qquad \dot{p} = -\sin q.$$

The vector field $X(q,p) = (p, -\sin q)$ is the Hamiltonian vector field associated with the function $H = \frac{1}{2}p^2 - \cos q = \frac{1}{2}p^2 - \cos\theta$ (this is the total energy of the simple pendulum).

There are numerous beautiful examples of integrable systems (see, for example, [**15, 18**]). The aim of this section is to prove that, near a compact connected component of a regular level of an integrable system, there is a Hamiltonian torus action with the same levels. This is the Arnold–Liouville theorem.

III.3.b. The Arnold–Liouville theorem. We consider an integrable system, namely a symplectic manifold (W, ω) of dimension $2n$ and a map $f : W \to \mathbf{R}^n$ the components $f_1, \dots, f_n$ of which are independent and commute.

The local $\mathbf{R}^n$-action. Let $X_1, \ldots, X_n$ denote the Hamiltonian vector fields associated with the f_i's. They commute

$$[X_i, X_j] = X_{\{f_i, f_j\}} = 0$$

and define an $\mathbf{R}^n$-action by

$$t \cdot x = \varphi_n^{t_n} \circ \varphi_{n-1}^{t_{n-1}} \circ \cdots \circ \varphi_1^{t_1}(x)$$

where φ_i denotes the flow of X_i and $t = (t_1, \ldots, t_n) \in \mathbf{R}^n$ belongs to a neighborhood of 0, in such a way that $\varphi_i^{t_i}$ is defined. This action is locally free at any regular point, since the fundamental vector fields are independent.

Completeness. Assume now that the vector fields X_i are complete, namely that the flows φ_i^t are defined for all real numbers t. What we have now is a locally free $\mathbf{R}^n$-action on the open subset of regular points in W. Notice that this preserves the levels of f, because the Hamiltonian vector fields X_i are tangent to these levels.

The connected components of the regular levels of f are homogeneous spaces, quotients of $\mathbf{R}^n$ by discrete subgroups. The discrete subgroups of $\mathbf{R}^n$ are the lattices $\mathbf{Z}^k$ in the k-dimensional linear subspaces. The connected components of the regular levels must thus be of the form $\mathbf{R}^{n-k} \times \mathbf{T}^k$ for some k such that $0 \leqslant k \leqslant n$.

Liouville tori. The compact connected components of the regular levels are thus tori $\mathbf{T}^n$ and these tori are Lagrangian. We have thus proved the first assertion of Arnold–Liouville's theorem.

***Theorem III.3.3* (The Arnold–Liouville theorem [5]).** *Let $f_1, \ldots, f_n$ be n independent commuting functions on a symplectic manifold (W, ω) of dimension $2n$. Any compact connected component of a regular level of $(f_1, \ldots, f_n)$ is a Lagrangian torus. Such a torus has a neighborhood on which there are coordinates $(J_1, \ldots, J_n, \varphi_1, \ldots, \varphi_n)$, with*

$$\omega = \frac{1}{2\pi} \sum_{i=1}^{n} dJ_i \wedge d\varphi_i \text{ and } f_i = F_i(J_1, \ldots, J_n).$$

In these coordinates, the Hamiltonian system associated with f_j is

$$\begin{cases} \dot{J}_i = 0 \\ \dot{\varphi}_i = a_i(J_1, \ldots, J_n). \end{cases}$$

The coordinates J_i are the *action* coordinates, while the φ_i's are the *angle* coordinates. The rest of this chapter is devoted to the proof of this theorem *(cum commento)*.

Remark III.3.4. The Arnold–Liouville theorem is proved in [**5**] (an early 1970s book), where it is attributed to Liouville, who certainly did not even state it. Hence, probably following Arnold's theory (according to which a theorem is given the name of the last person who proves it), it is called the Arnold–Liouville theorem. The statement and a sketch of a proof were also given by Mineur [**110**], who committed what is called in French a *plagiat par anticipation* [**92**]. For the history of this theorem, see [**127**].

Torus action and action variables. The regular levels of an integrable system thus have a natural affine structure, defined by the integration of the Hamiltonian vector fields of the first integrals. Notice that, by definition, the solutions of the Hamiltonian system are *linear* in this affine structure.

What we aim to describe now is a transverse affine structure, determined by the *action* coordinates, that we are going to define. We fix a regular value q of the map f. We assume that the level $f^{-1}(q)$ has a compact connected component F_q, that we fix as well. We investigate the situation in a neighborhood of F_q. The regular values of f constitute an open subset of W, so that there is a neighborhood U of q in $\mathbf{R}^n$ which consists of regular values.

***Example III.3.5* (The simple pendulum, continuation).** The zeroes of the Hamiltonian vector field are obtained for $p = 0$ and $\sin q = 0$, thus the levels $H = \pm 1$ are singular and the other ones regular. Here U is any open subset of $\mathbf{R} - \{-1, 1\}$. As Figure 3 shows, the regular levels are indeed unions of tori (circles). Let us

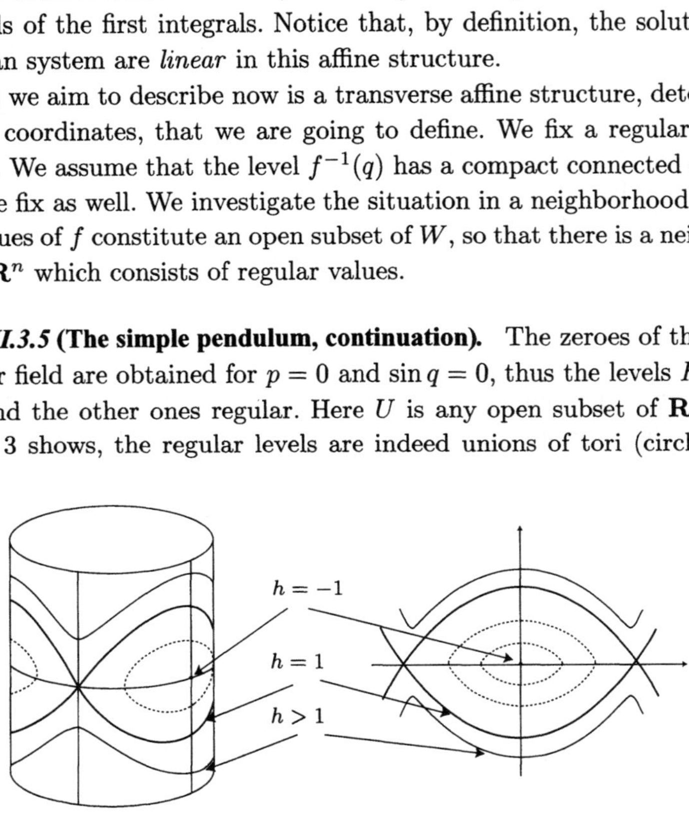

Figure 3. Energy levels for the simple pendulum

consider, for instance, $U =]-1, 1[$. Note that $H^{-1}(U)$ has two connected components. We restrict our attention to one of them. This is what the picture on the right shows. The Hamiltonian vector field has periodic orbits. The difference with a periodic Hamiltonian is that here the period depends on the orbit.

The submanifold (torus) F_q being compact, there is a relatively compact neighborhood V of F_q in W such that $f(\overline{V}) \subset U$. Let us call g the restriction of f to V. The mapping $g : V \to U$ is submersive, that we can assume to be proper. If U and V are chosen small enough, the fibration g can be trivialized (this is the Ehresmann fibration theorem [**46, 47**]). We thus get a diffeomorphism

$$\tau : V \longrightarrow U \times F_q$$

such that $\tau^{-1}(\{u\} \times F_q) = g^{-1}(u)$ (this is the fiber of u). Using τ, we can also construct sections of g. Let us fix a point p of F_q. In the model $U \times F_q$, we have an obvious section $u \mapsto (u, p)$. Look at its image σ under τ^{-1} (Figure 4):

$$\sigma(u) = \tau^{-1}(u, p).$$

This is indeed a section of g.

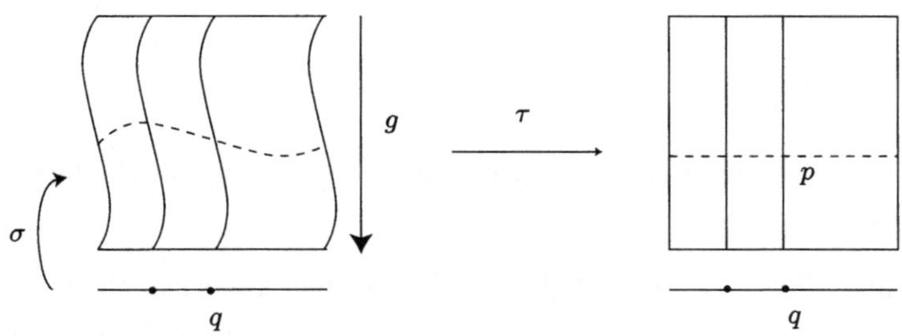

Figure 4

Using the section σ, we prove now that the tori $g^{-1}(u)$ deform continuously, in the sense that $g^{-1}(u)$ is the quotient of $\mathbf{R}^n$ by the stabilizer $P(\sigma(u))$ (I denote $P(p)$ the stabilizer of p, this is a lattice(5) in $\mathbf{R}^n$), a lattice that depends only on u (and depends smoothly on u). We want to apply the implicit function theorem to the equation

$$t \cdot \sigma(u) = \sigma(u).$$

Let thus t_0 be a point in $P(p) \subset \mathbf{R}^n$. Let A be a neighborhood of t_0 in $\mathbf{R}^n$ and B be a neighborhood of p in F_q, small enough so that

$$t \in A \text{ and } u \in U \Longrightarrow \tau(t \cdot \sigma(u)) \in \{u\} \times B.$$

The open subset B of F_q can be considered as an open subset of $\mathbf{R}^n$, since $F_q \cong \mathbf{R}^n/P(p)$. We have thus defined a map

$$\begin{aligned} \Theta : A \times U &\longrightarrow B \\ (t, u) &\longmapsto \tau(t \cdot \sigma(u)) - p \end{aligned}$$

(5) P is for period, [**57**].

(the minus sign is to be understood in the sense of the affine structure on the torus F_q). We have

$$\begin{aligned}\Theta(t,u)=0 &\Longleftrightarrow \tau(t\cdot\sigma(u))=p\\ &\Longleftrightarrow t\cdot\sigma(u)=\sigma(u)\\ &\Longleftrightarrow t\in P(\sigma(u)).\end{aligned}$$

The partial differential $(d_t\Theta)_{(t,u)}$ of Θ with respect to the variables t is the composition of the differential $(d\tau)_{t\cdot\sigma(u)}$ of the diffeomorphism τ and the differential of $t\mapsto t\cdot\sigma(u)$. The latter has rank n, because it maps the vectors $\dfrac{\partial}{\partial t_i}$ of the canonical basis onto the independent vectors $X_i(\sigma(u))$.

The partial differential $(d_t\Theta)_{(t,u)}$ being of rank n, the implicit function theorem gives a function $T:U\to\mathbf{R}^n$ (we may have to shrink the open subset U) such that

$$\begin{cases}(t,u)\in A\times U \text{ and } t=T(u)\Leftrightarrow t\cdot\sigma(u)=\sigma(u)\\ T(p)=t_0.\end{cases}$$

Making t_0 vary in a basis of the lattice $P(p)$, we get a basis of the lattice $P(\sigma(u))$, depending smoothly on u.

Let thus $(T_1(u),\ldots,T_n(u))$ (for $u\in U$) be such a basis of the lattice $P(\sigma(u))$ in $\mathbf{R}^n$. Its avatar $(Y_1(v),\ldots,Y_n(v))$ in T_vV is a basis of the tangent space to the fiber. In formulas, if $T_i(u)$ is written, in the canonical basis of $\mathbf{R}^n$,

$$T_i(u)=\sum_{i=1}^{n}T_{i,j}(u)\frac{\partial}{\partial t_i},$$

then

$$Y_i(v)=\sum_{j=1}^{n}T_{i,j}(f(v))X_j(v).$$

The vector fields $Y_1,\ldots,Y_n$ are periodic of period 1 by definition. Moreover, we have, denoting ψ_i^t the flow of Y_i,

$$\psi_i^t(v)=(tT_i(f(v)))\cdot v.$$

We deduce

$$\begin{aligned}\psi_i^t\circ\psi_j^{t'}(v)&=tT_i(f(\psi_j^{t'}(v)))\cdot\psi_j^{t'}(v)\\ &=tT_i(f(v))\cdot(t'T_j(f(v))\cdot v)\\ &=tt'T_i(f(v))T_j(f(v))\cdot v\end{aligned}$$

(using the fact that $f(\psi_j^{t'}(v))=f(v)$). Hence Y_i and Y_j commute. Hence we have defined an action of the torus $\mathbf{T}^n=\mathbf{R}^n/\mathbf{Z}^n$ on V.

Action coordinates. Action coordinates are defined by the integration of a 1-form on the trajectories of the vector fields Y_i. We need to construct firstly such a 1-form. We can assume that the open subset U is a ball, and thus that F_q is a retract of V. In particular, we have $H^2(V;\mathbf{R}) \cong H^2(F_q;\mathbf{R})$. The torus F_q is isotropic, so that the cohomology class $[\omega]$ of the symplectic form ω is zero in $H^2(F_q;\mathbf{R})$. Hence it is also zero in $H^2(V;\mathbf{R})$. The symplectic form is thus exact on V. Let λ be one of its primitives, that is, a 1-form such that $\omega = d\lambda$. We define

$$J_i(v) = \int_0^1 \lambda(Y_i) \circ \psi_i^t(v)dt,$$

that is, we integrate the function of t that is obtained computing λ, at the point $\psi_i^t(v)$, on the tangent vector Y_i. Putting $\gamma_i(t) = \psi_i^t(v)$, we define a loop parametrized by $[0,1]$, the orbit of v for the circle action corresponding to the vector field Y_i. The function $J_i(v)$ is the integral of the form λ on the loop γ_i, that is,

$$J_i(v) = \int_{\gamma_i} \lambda.$$

Proposition III.3.6. *The $\mathbf{T}^n$-action on V is Hamiltonian and its momentum mapping is*

$$J = (J_1, \dots, J_n) : V \longrightarrow \mathbf{R}^n.$$

Proof. We apply a somewhat more general lemma (Lemma III.3.8 below). The only thing we need to check is the fact that any of the fundamental vector fields Y_i satisfies $i_{Y_i}\mathfrak{L}_{Y_i} = 0$. Using the Cartan formula, we get

$$\begin{aligned} i_{Y_i}\mathfrak{L}_{Y_i}\omega &= i_{Y_i} d i_{Y_i}\omega \\ &= i_{Y_i} d\left(\sum T_{i,j} df_j\right) \\ &= i_{Y_i}\left(\sum dT_{i,j} \wedge df_j\right). \end{aligned}$$

But Y_i is tangent to the fibers of f and $T_{i,J}$ is constant on these fibers, hence Y_i is in the kernel of $dT_{i,j}$ and in that of df_j as well. □

***Example III.3.7* (The simple pendulum (continuation)).** On $\mathbf{S}^1 \times \mathbf{R}$, the 1-form $\lambda = pdq$ is a primitive of the symplectic form (this is, up to sign, the Liouville form). Let h be a regular value of the Hamiltonian H, say $h \in]-1,1[$ and let W_h be a component of the submanifold $H^{-1}(h)$. Of course, the integral

$$\int_{W_h} pdq$$

is the area of the domain bounded by W_h (this is one of the dotted curves of Figure 3). This is "the" action coordinate in this case.

Lemma III.3.8. *Let* (W, ω) *be an* exact *symplectic manifold. Let* λ *be a* 1*-form on* W *such that* $\omega = d\lambda$*. Let* $\mathbf{S}^1$ *act on* W *with fundamental vector field* Y*. Assume that* $i_Y \mathcal{L}_Y \omega = 0$*. Then the* $\mathbf{S}^1$*-action is Hamiltonian, with Hamiltonian*

$$J(x) = \int_{\mathbf{S}^1 \cdot x} \lambda.$$

Proof. The proof is direct. We compute $(dJ)_x(Z)$ for Z in $\mathsf{T}_x W$ and we prove that

$$(\mathsf{T}_x J(Z) = \omega(Z, Y(x)).$$

Let thus x be a point of W and Z be a tangent vector to W at this point. First, we extend Z to an invariant vector field along the orbit of x by putting

$$Z(t \cdot x) = \mathsf{T}_x t(Z).$$

Then, we extend Z to a vector field on W, still denoted Z. We write

$$J(x) = \int_0^1 \lambda_{t \cdot x}(Y(t \cdot x)) dt$$

(in which expression, t denotes both an element of $\mathbf{S}^1 = \mathbf{R}/\mathbf{Z}$ and the corresponding element in $[0, 1]$). We differentiate. We have to compute, for a fixed t in $\mathbf{S}^1$, the differential of the function

$$x \longmapsto \lambda_{t \cdot x}(Y(t \cdot x)),$$

evaluated on Z. This is

$$(\mathcal{L}_Z \lambda)_{t \cdot x}(Y(t \cdot x)) + \lambda_{t \cdot x}((\mathcal{L}_Z Y)(t \cdot x)).$$

The second term of this sum is zero, since $\mathcal{L}_Z Y = -\mathcal{L}_Y Z$ and $(\mathcal{L}_Y Z)(t \cdot x)$ depends only on Z along the orbit of x; the vector field Z being invariant along the orbit, we have $\mathcal{L}_Y Z(t \cdot x) = 0$.

Let us compute the first term. We have

$$(d\lambda)(Z, Y) = Z \cdot \lambda(Y) - Y \cdot \lambda(Z) - \lambda([Z, Y]).$$

We have said that, at the points $t \cdot x$, the Lie bracket $[Z, Y]$ vanishes. Hence the last term vanishes and we have

$$(\mathcal{L}_Z \lambda)_{t \cdot x}(Y(t \cdot x)) = (d\lambda)_{t \cdot x}(Z(t \cdot x), Y(t \cdot x)) + (\mathcal{L}_Y \lambda)_{t \cdot x}(Z(t \cdot x)).$$

To end the computation of $\mathsf{T}_x J(Z)$, we have to integrate these two terms on $[0, 1]$. Let us start with the latter. We differentiate the function

$$t \longmapsto \lambda_{t \cdot x}(Z(t \cdot x))$$

and get

$$\frac{d}{dt}(\lambda_{t \cdot x}(Z(t \cdot x))) = (\mathcal{L}_Y \lambda)_{t \cdot x}(Z(t \cdot x)) + \lambda_{t \cdot x}((\mathcal{L}_Y Z)(t \cdot x)) = (\mathcal{L}_Y \lambda)_{t \cdot x}(Z(t \cdot x)).$$

We thus get, by integration,

$$\int_0^1 (\mathcal{L}_Y\lambda)_{t\cdot x}(Z(t\cdot x))dt = \int_0^1 \frac{d}{dt}\left(\lambda_{t\cdot x}(Z(t\cdot x))\right)dt = 0.$$

Up to now, we have used the fact that the 2-form ω is exact and that it was possible to extend Z as a vector field which is invariant along the orbit of x. We need now to use the properties of the fundamental vector field Y. Since Y and Z are invariant along the orbit of x, the derivative of the function

$$t \longmapsto \omega_{t\cdot x}(Z(t\cdot x), Y(t\cdot x))$$

with respect to t is

$$\left(\frac{d}{dt}\omega_{t\cdot x}\right)(Z(t\cdot x), Y(t\cdot x)) + \omega_{t\cdot x}\left((\mathcal{L}_Y Z)_{t\cdot x}, Y(t\cdot x)\right) + \omega_{t\cdot x}\left(Z(t\cdot x), (\mathcal{L}_Y Y)(t\cdot x)\right).$$

The second term is zero by the assumption we have made on Y. The two others are obviously zero. The derivative with respect to t is zero, hence the function itself does not depend on t. We get eventually

$$\int_0^1 \omega_{t\cdot x}(Z(t\cdot x), Y(t\cdot x))dt = \omega_x(Z, Y(x)),$$

namely

$$\mathsf{T}_x J(Z) = \omega_x(Z, Y(x))$$

and this is what we wanted to prove. □

Notice that the map J has rank n, hence the torus action is effective. We have indeed

$$\mathsf{T}_x J(Z) = (\omega_x(Z, Y_1), \ldots, \omega_x(Z, Y_n))$$

and this linear mapping has rank n because the vectors Y_i are independent and the form ω is nondegenerate.

Angle coordinates. What we have got up to now is a symplectic fibration

$$J : V \longrightarrow U,$$

the fibers of which are Lagrangian tori, and a $\mathbf{T}^n$-action with momentum mapping J. We can use, as above, a section σ of the fibration and a diffeomorphism τ with $U \times F_q$. As F_q is a torus $\mathbf{R}^n/\mathbf{Z}^n$, we have coordinates $(\theta_1, \ldots, \theta_n)$ on F_q and thus also coordinates $(J_1, \ldots, J_n, \theta_1, \ldots, \theta_n)$ on V. Notice that the θ_i's, being angles, belong to $\mathbf{R}/2\pi\mathbf{Z}$, while our torus is $\mathbf{R}^n/\mathbf{Z}^n$ (hence the normalization in the formula for the symplectic form).

In these coordinates, the symplectic form is written

$$\omega = \frac{1}{2\pi}\sum dJ_i \wedge d\theta_i + \sum a_{i,j} dJ_i \wedge dJ_j$$

(there are no terms $d\theta_i \wedge d\theta_j$ because the fibers are Lagrangian).

Put $\eta = \sum a_{i,j} dJ_i \wedge dJ_j$. The $a_{i,j}$'s might depend on the θ_k's, but this is not the case: the form ω being closed, η is closed as well, in other words, η does not depend on the θ_k's. In a more intrinsic way, $\eta = \sigma^\star\omega$ is closed as ω is. Thus we have a closed form on U. We have already assumed (and used the fact) that U is contractible. Hence η is exact, $\eta = dA$ for some 1-form A on U that we write

$$A = \sum A_i dJ_i.$$

The A_i's are functions on U. We use them to modify the section σ, putting

$$\varphi_i = \theta_i - A_i$$

(the new "angles") or

$$s(J_1, \ldots, J_n) = (J_1, \ldots, J_n, \theta_1 - A_1, \ldots, \theta_n - A_n),$$

in short

$$s(J_1, \ldots, J_n) = \sigma(J_1, \ldots, J_n) - (A_1, \ldots, A_n).$$

Eventually,

$$s^\star\omega = \sigma^\star\omega - dA = 0$$

thus

$$\omega = \frac{1}{2\pi} \sum dJ_i \wedge d\varphi_i.$$

Notice that J can be modified by the addition of a constant, φ by the addition of a function of J.

Remark III.3.9. The symplectic volume of our neighborhood V is simply the (Lebesgue) measure of U.

The Hamiltonian system in the action-angle coordinates. Let H be any function in the algebra generated by the f_j's. The Hamiltonian system $\dot{x} = X_H(x)$ has a specially simple form in the action-angle coordinates. Firstly, the Hamiltonian H is in the algebra generated by the first integrals and thus is a function of $J_1, \ldots, J_n$. We put

$$H(x) = h(J_1, \ldots, J_n).$$

Then, the coordinates $(J_1, \ldots, J_n, \varphi_1, \ldots, \varphi_n)$ are symplectic (up to a factor 2π, thus the Hamiltonian system is written, in these coordinates,

$$\dot{J}_i = -\frac{\partial h}{\partial \varphi_i} = 0 \text{ and } \dot{\varphi}_i = \frac{\partial h}{\partial J_i} = a_i(J_1, \ldots, J_n).$$

In this form, it is easy to check that the solutions are indeed angles φ_i depending linearly on the time on the levels of J. It is clear that such a system can be solved by quadratures, hence the terminology "integrable system".

I will give examples of action-angle coordinates in Chapter V.

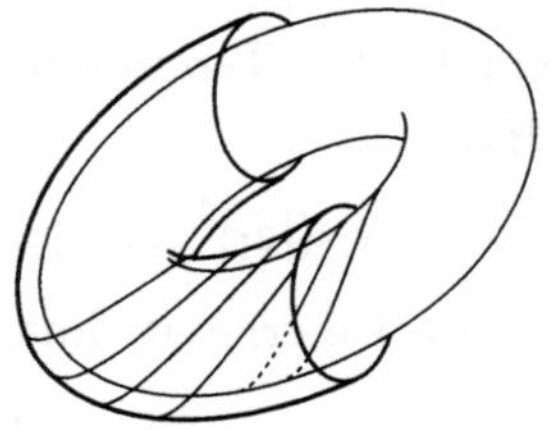

Figure 5

Exercises

Exercise III.1. Let G be a Lie group acting on a manifold V. Define a G-action on the cotangent bundle $\mathsf{T}^\star V$ by

$$\widetilde{g}(x,\varphi) = (g(x), \varphi \circ \mathsf{T}_{g(x)} g^{-1})$$

where $x \in V$, $\varphi \in \mathsf{T}^\star_x V$ is a linear form $\mathsf{T}_x W \to \mathbf{R}$. Check that this is indeed a (left) action. Show that the Liouville form α is invariant(6) ($\widetilde{g}^\star \alpha = \alpha$) and conclude that the action is Hamiltonian, with $\widetilde{\mu}_X = -i_X \alpha$.

Exercise III.2. Which surfaces admit symplectic $\mathbf{S}^1$-actions (use the results of § I.3.a)? Prove that the only surface admitting a Hamiltonian $\mathbf{S}^1$-action is the 2-sphere.

Exercise III.3. Let W_1 and W_2 be two symplectic manifolds acted on by the same Lie group G. Assume that the two actions are Hamiltonian with momentum mappings

$$\mu_1 : W_1 \longrightarrow \mathfrak{g}^\star, \quad \mu_2 : W_2 \longrightarrow \mathfrak{g}^\star.$$

Prove that the diagonal action of G on $W_1 \times W_2$ is Hamiltonian with momentum mapping

$$\begin{aligned} \mu : W_1 \times W_2 &\longrightarrow \mathfrak{g}^\star \\ (x_1, x_2) &\longmapsto \mu_1(x_1) + \mu_2(x_2). \end{aligned}$$

Exercise III.4. Let λ_1 and λ_2 be two distinct real numbers and let $\lambda = (\lambda_1, \lambda_2, \ldots, \lambda_2)$, so that

$$\begin{aligned} \Psi_\lambda : \mathcal{H}_\lambda &\longrightarrow \mathbf{P}^{n-1}(\mathbf{C}) \\ A &\longmapsto \ell, \text{ the } \lambda_1\text{-eigenline} \end{aligned}$$

is a diffeomorphism. Prove that the composed map

$$\mu \circ \Psi_\lambda^{-1} : \mathbf{P}^{n-1}(\mathbf{C}) \longrightarrow \mathcal{H}_\lambda \longrightarrow \mathfrak{t}^\star = \mathbf{R}^n,$$

(6) This example can be found in Arnold's book [**5**], which I take this opportunity to recommend.

that is, the momentum mapping for the $\mathbf{T}$-action on $\mathbf{P}^{n-1}(\mathbf{C})$ endowed with the canonical symplectic form of $\mathcal{H}_\lambda$, is

$$[z_1, \ldots, z_n] \longmapsto (\lambda_1 - \lambda_2)(|z_1|^2, \ldots, |z_n|^2) + (\lambda_2, \ldots, \lambda_2)$$

where $(z_1, \ldots, z_n)$ is of unit length. Describe its image in $\mathbf{R}^n$.

Exercise III.5. Let W be a symplectic manifold endowed with a Hamiltonian G-action, with momentum mapping μ. Let x be a point in W. We know that μ sends the orbit $\mathrm{G} \cdot x$ into the coadjoint orbit $\mathrm{G} \cdot \mu(x) \subset \mathfrak{g}^\star$. Let ω be the symplectic form of the orbit $\mathrm{G} \cdot \mu(x)$. What can be said of the 2-form $\mu^\star\omega$ on $\mathrm{G} \cdot x$?

Exercise III.6. Show that the equality in Lemma III.2.11 remains true if one merely assumes that V_ξ is a submanifold the tangent space of which is $\operatorname{Ker} \mathsf{T}_x\mu$ (instead of requiring that ξ be a regular value). Hint: show that in this case $\dim \mathrm{G}_x = \dim \mathrm{G} - \dim W + \dim V_\xi$.

Exercise III.7. Consider the situation investigated in Proposition III.2.22. Assume that ξ is a regular value of μ and that G acts freely on $\mu^{-1}(\mathrm{G} \cdot \xi)$. Prove that

$$\mathsf{T}_x(\mu^{-1}(\mathrm{G} \cdot \xi)) = \{v \in \mathsf{T}_x W \mid \exists X \in \mathfrak{g} \text{ such that } \mathsf{T}_x\mu(v) = \operatorname{ad}^\star_X \mu(x)\},$$

that

$$\dim \mu^{-1}(\mathrm{G} \cdot \xi)/\mathrm{G} = \dim W + \dim \mathrm{G} \cdot \xi - 2 \dim \mathrm{G},$$

and that the (reduced) symplectic form is given by

$$\omega_x(v, v') = \omega(v, v') - \langle \mu(x), [X, X'] \rangle$$

(X and v being defined by the relation above).

Exercise III.8. Endow $\mathbf{C}^n$ with its canonical symplectic form. The unitary group $\mathrm{U}(n)$ acts on $\mathbf{C}^n$ (by definition). Prove that the formula

$$\widetilde{\mu}_X(z) = \frac{i}{2} {}^t\bar{z} X z, \qquad z \in \mathbf{C}^n, \quad X \in \mathfrak{u}(n)$$

defines a momentum mapping

$$\mu : \mathbf{C}^n \longrightarrow \mathfrak{u}(n)^\star$$

for this action. Deduce that the (induced) action of $\mathrm{U}(n)$ on $\mathbf{P}^{n-1}(\mathbf{C})$ is Hamiltonian and write a formula for its momentum mapping. Identifying $\mathfrak{u}(n)$ with $\mathfrak{u}(n)^\star$ by the map

$$A \longmapsto \left(X \longmapsto \frac{i}{2} \operatorname{tr}({}^t\bar{A} X)\right),$$

check that the inclusion $\mathrm{SU}(n) \subset \mathrm{U}(n)$ gives, at the (dual of) Lie algebra level

$$\begin{aligned} \mathfrak{u}(n)^\star &\longrightarrow \mathfrak{su}(n)^\star \\ A &\longmapsto A - \left(\frac{1}{n} \operatorname{tr} A\right) \mathrm{Id}. \end{aligned}$$

Deduce that the momentum mapping for the action of $\mathrm{SU}(n)$ on $\mathbf{C}^n$ is

$$z \longmapsto z^t z - \frac{1}{n}\operatorname{tr}(z^t z\,\mathrm{Id}).$$

What is the momentum mapping

$$\mu : \mathbf{P}^{n-1}(\mathbf{C}) \longrightarrow \mathfrak{su}(n)?$$

Prove that the composition of μ with the stereographic projection

$$\begin{array}{ccccc} \mathbf{S}^2 & \longrightarrow & \mathbf{C}\cup\{\infty\} = \mathbf{P}^1(\mathbf{C}) \xrightarrow{\ \mu\ } & \mathfrak{su}(2) & = \mathbf{C}\times\mathbf{R} \\ (x,y,z) & \longmapsto & \dfrac{x+iy}{1-z} & \begin{pmatrix} a & b \\ \bar{b} & -a \end{pmatrix} & \mapsto (b,a) \end{array}$$

is just the inclusion $(x,y,z) \mapsto \dfrac{1}{2}(x,y,z)$.

Exercise III.9. Consider the natural action of $\mathrm{SO}(3)$ on $\mathbf{S}^2 \subset \mathbf{R}^3$. Recall that the momentum mapping is just the inclusion. Determine the momentum mapping for the diagonal action of $\mathrm{SO}(3)$ on $\mathbf{S}^2 \times \cdots \times \mathbf{S}^2$.

Using the fact that $\mathbf{S}^2 \cong \mathbf{P}^1(\mathbf{C})$, check that the quotient

$$\underbrace{\mathbf{S}^2 \times \cdots \times \mathbf{S}^2}_{n \text{ times}} / \mathfrak{S}_n$$

by the permutation of factors is diffeomorphic to $\mathbf{P}^n(\mathbf{C})$ (hint: $\mathbf{C}^n/\mathfrak{S}_n$ is the affine space of monic polynomials of degree n by $(z_1,\dots,z_n) \mapsto (z-z_1)\cdots(z-z_n)$, you just have to compactify both spaces). Deduce a Hamiltonian $\mathrm{SO}(3)$-action on the projective space $\mathbf{P}^n(\mathbf{C})$.

Exercise III.10. The group $\mathrm{SU}(2)$ acts on $\mathbf{C}^2$, hence it acts on the vector space (of dimension $n+1$) of homogeneous polynomials of degree n in two variables by

$$(A\cdot P)(x,y) = P((x,y)\cdot A).$$

Check that the action of $\mathrm{SU}(2)$ on $\mathbf{P}^n(\mathbf{C})$ defined this way induces an *effective* action of $\mathrm{SO}(3)$... that coincides with the action defined in Exercise III.9.

***Exercise III.11* (Symplectic area of the sphere).** Consider, on $\mathbf{P}^n(\mathbf{C})$, the reduced symplectic form σ_a at the level a (namely, coming from the sphere of radius $\sqrt{2a}$) as in Example III.2.18. Consider a projective line $\mathbf{P}^1(\mathbf{C}) \subset \mathbf{P}^n(\mathbf{C})$ (or let n be 1). Check that

$$\int_{\mathbf{P}^1(\mathbf{C})} \sigma_a = 2\pi a.$$

***Exercise III.12* (Reduction in stages).** Let $\mathrm{G} = \mathrm{G}_1 \times \mathrm{G}_2$ act in a Hamiltonian way on the symplectic manifold W. Let $\mu_i : W \to \mathfrak{g}_i^\star$ be the components of the momentum mapping

$$\mu : W \longrightarrow \mathfrak{g}_1^\star \oplus \mathfrak{g}_2^\star.$$

Check that μ_2 is invariant under the G_1-action. Let $V_1 = \mu_1^{-1}(0)$. Assume that G_1 acts freely on V_1 so that $W_1 = V_1/G_1$ is a symplectic manifold. Prove that G_2 acts on W_1 with momentum mapping $\widetilde{\mu}_2 : W_1 \to \mathfrak{g}_2^\star$ making the diagram

$$\begin{array}{ccccc} V_1 & \longrightarrow & W & \xrightarrow{\mu_2} & \mathfrak{g}_2^\star \\ \downarrow & & & \nearrow_{\widetilde{\mu}_2} & \\ W_1 & & & & \end{array}$$

commute. Assume that G acts freely on $\mu^{-1}(0,0)$. Prove that G_2 acts freely on $\widetilde{\mu}_2^{-1}(0)$ and that there is a natural symplectic diffeomorphism

$$\widetilde{\mu}_2^{-1}(0)/G_2 \longrightarrow \mu^{-1}(0,0)/G.$$

Exercise III.13. Let W be the complement of the zero section in the tangent bundle $\mathsf{T}S^n$. Consider the $\mathbf{S}^1$-action of the geodesic flow (of the round metric) on $\mathsf{T}S^n$ (Exercise II.30). Let a be a positive real number. Prove that the manifold obtained by symplectic reduction of the level a is diffeomorphic with the Grassmannian $\widetilde{G}_2(\mathbf{R}^{n+1})$ of oriented 2-planes in $\mathbf{R}^{n+1}$ (that we have already met in Exercise I.21 and which is thus a symplectic manifold). Identify the reduced surface and its symplectic form when $n = 2$.

Exercise III.14. The isometry group $SO(n+1)$ acts on $\mathbf{R}^{n+1}$ and (diagonally) on $\mathbf{C}^{n+1}$ by

$$A \cdot (x + iy) = A \cdot x + iA \cdot y.$$

Identifying $\mathfrak{so}(n+1)^\star$ with the vector space $\bigwedge^2 \mathbf{R}^{n+1}$ (as we did for $\mathfrak{so}(4)^\star$ in Exercise I.20), prove that the momentum mapping of this action is

$$x + iy \longmapsto x \wedge y.$$

Deduce a Hamiltonian action of $SO(n+1)$ on $\mathbf{P}^n(\mathbf{C})$ with momentum mapping

$$[x + iy] \xrightarrow{\ \mu\ } \frac{x \wedge y}{\|x\|^2 + \|y\|^2}.$$

Prove that the function $\|\mu\|^2$ is maximal along the submanifold of $\mathbf{P}^n(\mathbf{C})$ of equation

$$\sum z_i^2 = 0$$

and that this quadric hypersurface is diffeomorphic with the Grassmannian $\widetilde{G}_2(\mathbf{R}^{n+1})$ (already(7) met in Exercises I.21 and III.13). Check that the complement of this quadric in $\mathbf{P}^n(\mathbf{C})$ is diffeomorphic with an open disc bundle(8) in the tangent bundle $\mathsf{T}\mathbf{P}^n(\mathbf{R})$.

(7) This is another way to see this Grassmannian as a symplectic (actually Kähler) manifold. See Exercise III.15.

(8) This is an example of a Lagrangian barrier, see [**23**].

Exercise III.15. Consider now the various descriptions of the Grassmannian $\widetilde{G}_2(\mathbf{R}^{n+2})$ as a symplectic manifold of dimension $2n$ that we have met:

- This is a coadjoint orbit in $\mathfrak{so}(n+2)^\star$ (Exercise I.21).
- This is a symplectic reduction, starting from $\mathsf{T}\mathbf{S}^{n+1}$ (Exercise III.13).
- This is a quadric in $\mathbf{P}^{n+1}(\mathbf{C})$ (Exercises I.21 and III.14).

Compare the symplectic forms got in these ways (notice that the two first descriptions depend on a parameter).

***Exercise III.16* (Complex Grassmannians).** The unitary group $\mathrm{U}(k)$ acts on the product $\mathbf{C}^n \times \cdots \times \mathbf{C}^n$ (k times) by

$$A \cdot (v_1, \ldots, v_k) = A^t(v_1, \ldots, v_k)$$

(multiplication of the $k \times k$-matrix A by the matrix the rows of which are the row vectors v_i). Notice that the case where $k = 1$ is that of the standard $\mathbf{S}^1$-action by multiplication on $\mathbf{C}^n$. Prove that this is a Hamiltonian action with momentum mapping

$$\mu(v_1, \ldots, v_k) = \frac{1}{2}\left(\langle v_i, v_j \rangle\right)_{1 \leqslant i,j \leqslant k} \in \mathfrak{u}(n)^\star$$

(the dual $\mathfrak{u}(n)^\star$ being identified, as usual, with the space $\mathcal{H}$ of Hermitian matrices). Use this to endow the complex Grassmannian $G_k(\mathbf{C}^n)$ of k-planes in $\mathbf{C}^n$ with (reduced) symplectic forms.

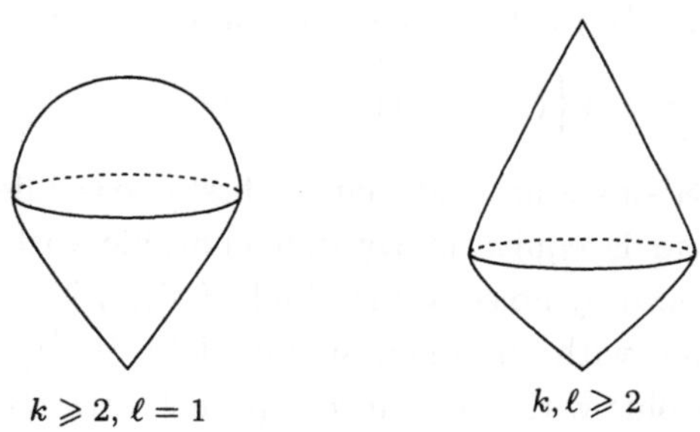

Figure 6. Symplectic teardrops

***Exercise III.17* (A symplectic teardrop).** Let $\mathbf{S}^1$ act on $\mathbf{C}^2$ by

$$u \cdot (x, y) = (u^k x, u^\ell y)$$

for two relatively prime positive integers k and ℓ. Check that this is a Hamiltonian action with momentum mapping

$$\mu(x, y) = \frac{1}{2}(k\,|x|^2 + \ell\,|y|^2).$$

Let a be any positive real number. What are the stabilizers of the points in the regular level $\mu^{-1}(a)$? Prove that the symplectic reduction is a symplectic orbifold (at least if k or $\ell \geqslant 2$).

***Exercise III.18* (Weighted projective spaces).** Let $k_1, \dots, k_{n+1}$ be relatively prime positive integers. Prove that the reduced spaces of the $\mathbf{S}^1$-action on $\mathbf{C}^{n+1}$ by

$$u \cdot (z_1, \dots, z_{n+1}) = (u^{k_1} z_1, \dots, u^{k_{n+1}} z_{n+1})$$

are compact symplectic orbifolds, the *weighted projective spaces.* Prove that the standard action of the torus $\mathbf{T}^{n+1}$ on $\mathbf{C}^{n+1}$ defines a Hamiltonian action on the weighted projective space.

Let $n = 2$, $(k_1, k_2, k_3) = (1, 2, 3)$. Determine the image of the momentum mapping for the $\mathbf{T}^3$-action above.

***Exercise III.19* (Symplectic cutting).** Let H be a periodic Hamiltonian on a symplectic manifold W. Consider the $\mathbf{S}^1$-action on the product $W \times \mathbf{C}$ by

$$u \cdot (x, z) = (u \cdot x, \overline{u} z)$$

and its momentum mapping

$$\widetilde{H}(x, z) = H(x) - \frac{1}{2} |z|^2 .$$

Assume the $\mathbf{S}^1$-action on the level $H^{-1}(t)$ is free. Prove that t is then a regular value of $\widetilde{H}$ and that $\widetilde{H}^{-1}(t)$ is the disjoint union of the two submanifolds

$$\{(x, 0) \mid H(x) = t\} \sqcup \left\{(x, z) \mid H(x) > t \text{ and } z = u \cdot \sqrt{2(H(x) - t)}\right\}.$$

Notice that both are $\mathbf{S}^1$-invariant and that the first one is diffeomorphic to $H^{-1}(t)$. Check that the second is equivariantly diffeomorphic with $H^{-1}(]t, +\infty[) \times \mathbf{S}^1$. Deduce that the reduced symplectic manifold $\widetilde{H}^{-1}(t)/\mathbf{S}^1$ contains an open submanifold diffeomorphic with the open submanifold $H^{-1}(]t, +\infty[)$ and that the complement is isomorphic with the reduced space $H^{-1}(t)/\mathbf{S}^1$.

We have thus *cut* the symplectic manifold W at the level t of H and "closed" it by the addition of a symplectic submanifold isomorphic with $H^{-1}(t)/\mathbf{S}^1$. This simple and elegant construction is due to Lerman [**93**], it has beautiful applications, see for instance [**94**].

Example. Assume $W = \mathbf{C}^n$ and $H(x) = \frac{1}{2} \|x\|^2$ is the Hamiltonian of the standard $\mathbf{S}^1$-action. let t be any positive real number. Prove that the symplectic cut of $\mathbf{C}^n$ at t is the manifold obtained by replacing a ball of radius $\sqrt{2t}$ in $\mathbf{C}^n$ by a complex projective space $\mathbf{P}^{n-1}(\mathbf{C})$. Let p be a point in a symplectic manifold. According to Darboux's theorem, a neighborhood of p contains a ball on which the symplectic form is the standard form on $\mathbf{C}^n$. Deduce that the symplectic cutting can be performed at any point of any symplectic manifold. This is the *symplectic blow up* of McDuff [**105**]. See also [**106**, § 6.2].

***Exercise III.20* (The harmonic oscillator).** Consider, on $\mathbf{R}^2$ endowed with its standard symplectic form $\omega = dp \wedge dq$, the Hamiltonian

$$H(p,q) = \frac{1}{2}p^2 + \frac{1}{2}aq^2$$

for some real number $a > 0$. When is this a periodic Hamiltonian? In the general case, determine the regular levels and the action-angle coordinates.

Exercise III.21. Consider, on $\mathbf{R}^2$ endowed with its standard symplectic form $dp \wedge dq$, the Hamiltonians

$$H(q,p) = \frac{1}{2}(p^2 + q^2) \text{ and } K(q,p) = H(q,p)^2.$$

The first one is periodic, associated with the standard $\mathbf{S}^1$-action on $\mathbf{R}^2$ by rotations. As any function on a symplectic manifold of dimension 2, K can be considered as an integrable system. What are its Liouville tori? Determine the action-angle coordinates for K on an (as big as possible) open subspace of $\mathbf{R}^2$. Prove that the flow of X_K is

$$\varphi^t(z) = e^{i|z|^2 t} \cdot z.$$

What is the period of φ^t on the orbit of z? Prove that the closure of the subgroup generated by φ^t in the group of diffeomorphisms of the plane is not compact.

***Exercise III.22* (The spherical pendulum).** The bob of a pendulum moves on the sphere centered at the fixed point, at the other end of the rope. The *spherical pendulum* is the Hamiltonian system associated with the Hamiltonian

$$H = \frac{1}{2}\|p\|^2 - \Gamma \cdot q$$

where p and q are vectors of $\mathbf{R}^3$ and Γ is the "vertical" and constant gravitation field. Prove that the phase space is

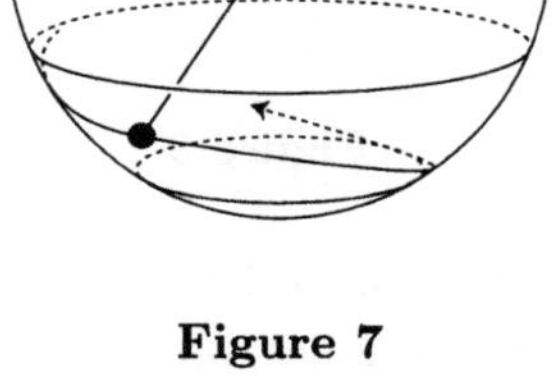

Figure 7

$$\mathsf{TS}^2 = \left\{(q,p) \in \mathbf{R}^3 \times \mathbf{R}^3 \mid \|q\|^2 = 1 \text{ and } q \cdot p = 0\right\},$$

the tangent bundle to the sphere $\mathbf{S}^2$, endowed with the restriction of the standard symplectic form of $\mathbf{R}^3 \times \mathbf{R}^3$, which is still symplectic (Exercise II.9). Check that the Hamiltonian vector field associated with H is $X_H = (Y, X)$ with $Y = p$ and $X = \Gamma - (q \cdot \Gamma + \|p\|^2)q$, so that the Hamiltonian system is

$$\begin{cases} \dot{q} = p \\ \dot{p} = \Gamma - (q \cdot \Gamma + \|p\|^2)q. \end{cases}$$

Prove that the system is invariant under the group of rotations around the vertical

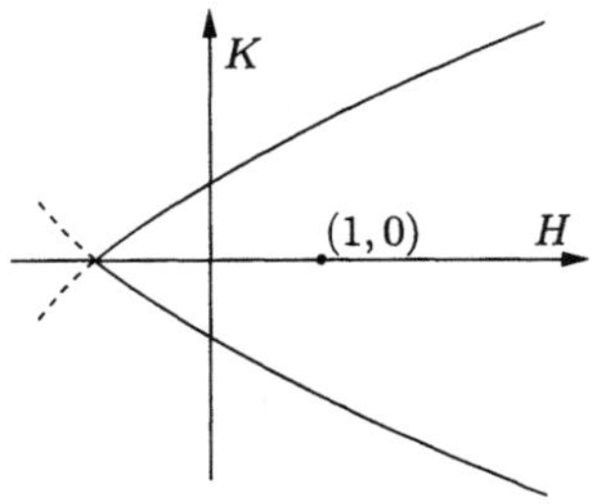

Figure 8

axis. What is the momentum mapping of this $\mathbf{S}^1$-action? Deduce that the function K defined by

$$K(q,p) = (q \wedge p) \cdot \Gamma$$

is a first integral, that commutes with H... and that the spherical pendulum is an integrable system. Determine the regular values (prove that the critical values are as shown in Figure 8). Prove that the regular levels of this system are tori (that is, compact and connected) and that, during its motion, the bob oscillates between two parallel circles, as shown in Figure 9.

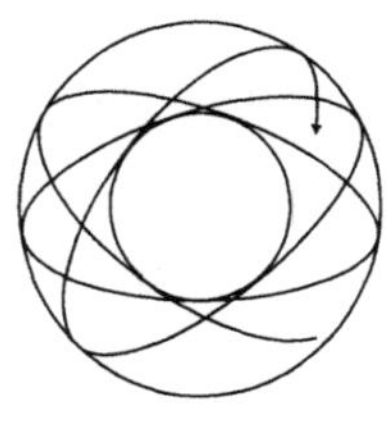

Figure 9

It can be shown, see [**42**] (and also [**18, 19**]), that there are no *global* (meaning defined over the whole set of regular values) action angle coordinates in this example.

CHAPTER IV

MORSE THEORY FOR HAMILTONIANS

This chapter is the heart of the book. Quite a few spectacular theorems will be proved. The main ones are

- the famous convexity theorem of Atiyah [**7**] and Guillemin–Sternberg [**63**] which asserts that the image of a compact connected symplectic manifold under the momentum mapping of a Hamiltonian torus action is a convex polyhedron (this is Theorem IV.4.3)
- the uniqueness theorem of Delzant [**39**], according to which, when the torus acting is half the dimension of the manifold, the polyhedron determines the manifold (this is Theorem IV.4.20).

I will also prove refinements (the Kähler version of the convexity theorem, due to Atiyah) and beautiful applications (the Schur–Horn, Toeplitz–Hausdorff and Kushnirenko theorems stated in the introduction).

In the case of the Hamiltonian action of a torus, the fixed points of the action are the critical points of a *function*. This is why the study of Hamiltonian actions is in some sense easy: one can use Morse theory. We shall not avoid it here, so we shall begin by proving that the functions under consideration have very convenient properties from that point of view (Frankel's theorem [**51**], here Theorem IV.2.3, from which we shall derive, as Atiyah does it, the basic result (Corollary IV.3.2) announced in the introduction, then the convexity theorem itself and some of its applications).

In this chapter, we thus consider a symplectic manifold endowed with the Hamiltonian action of a *torus* $\mathbf{T}$, with momentum mapping $\mu : W \to \mathfrak{t}^\star$.

IV.1. Critical points of almost periodic Hamiltonians

IV.1.a. Almost periodic Hamiltonians. We know that the fixed points of $\mathbf{T}$ correspond to certain critical points of μ. It would be more convenient to have a single function (with values in $\mathbf{R}$) allowing us to study the whole $\mathbf{T}$-action. It is not very difficult to find: choose any $X \in \mathfrak{t}$ which generates $\mathbf{T}$ in the sense that the one parameter subgroup $\exp(tX)$ is dense in $\mathbf{T}$. The associated Hamiltonian

$$W \xrightarrow{\ \mu\ } \mathfrak{t}^\star \longrightarrow (\mathbf{R}X)^\star$$

has the fixed points of $\mathbf{T}$ as critical points.

We shall say that a function $H : W \to \mathbf{R}$ is an *almost periodic* Hamiltonian if the flow of its Hamiltonian vector field X_H generates a subgroup of the group of all diffeomorphisms of W the closure of which is a torus.

Remark IV.1.1. The closure of the subgroup generated by any vector is a connected *commutative* subgroup; it is thus equivalent to requiring that the subgroup generated by the flow of X_H is compact. This is *not* a mild assumption. It makes a difference, for instance, between a Hamiltonian H and its square H^2 (see for instance Exercise III.21).

***Example IV.1.2* (The complex projective space).** Consider the projective space $\mathbf{P}^n(\mathbf{C})$, endowed with its standard symplectic form (Example III.2.18). The torus

$$\mathbf{T}^n = \left\{ (t_0, \ldots, t_n) \in \mathbf{T}^{n+1} \mid \prod_{i=0}^{n} t_i = 1 \right\}$$

acts on $\mathbf{P}^n(\mathbf{C})$ by

$$(t_0, \ldots, t_n) \cdot [z_0, \ldots, z_n] = [t_0 z_0, \ldots, t_n z_n]$$

(brackets denote homogeneous coordinates). The fixed points of this action are the $n+1$ points $[0, \ldots, 1, 0, \ldots, 0]$. The action is, of course, Hamiltonian (as that of $\mathbf{T}^{n+1}$ on $\mathbf{C}^{n+1}$ was, see Example III.1.7) and is even a special case of the action described in Example III.1.11. The momentum mapping is deduced from that in

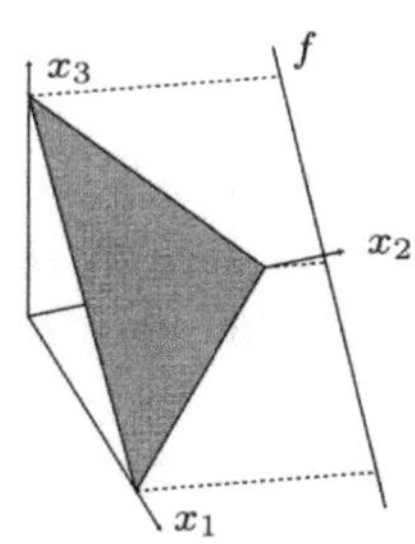

Figure 1

Example III.1.7, this is the map

$$\mu : [z_0, \ldots, z_n] \longmapsto \left(\frac{1}{2} \frac{|z_0|^2}{\sum |z_i^2|}, \ldots, \frac{1}{2} \frac{|z_n|^2}{\sum |z_i^2|} \right)$$

which takes its values in the hyperplane $\sum_{i=0}^n x_i = \frac{1}{2}$. The image of μ (in the case $n = 2$) is shown in Figure 1. Fix real numbers $a_0, \ldots, a_n$. The function (projection)

$$\begin{aligned} f : \mathbf{P}^n(\mathbf{C}) &\longrightarrow \mathbf{R} \\ [z_0, \ldots, z_n] &\longmapsto \frac{1}{2} \sum_{i=0}^{n} \frac{a_i \left|z_i\right|^2}{\sum_{i=0}^{n} \left|z_i\right|^2} \end{aligned}$$

is an almost periodic Hamiltonian.

IV.1.b. Critical points.

Let us now investigate the fixed points of an almost periodic Hamiltonian H. Call $\mathbf{T}$ the torus "generated by H". The critical points of H are the zeros of X_H, or the fixed points of $\mathbf{T}$. We have seen in Corollary I.2.3 that the set Z of fixed points is a submanifold in W.

The symplectic form ω is preserved by $\mathbf{T}$ and by definition[(1)]. With the help of an invariant Riemannian metric, one derives an invariant calibrated almost complex structure J and an invariant Hermitian metric (see § II.2).

Let $z \in Z$. Then the torus $\mathbf{T}$ acts on the complex vector space $\mathsf{T}_z W$ preserving J and the Hermitian form, that is to say, as a subgroup of $\mathrm{U}(n)$. Notice first that all these transformations are diagonalizable: the exponential map $\mathfrak{t} \to \mathbf{T}$ is onto and the elements of $\mathfrak{t}$, sitting in $\mathfrak{u}(n)$ (the skew-Hermitian matrices) cannot avoid being diagonalizable. As $\mathbf{T}$ is commutative, there is a basis of $\mathsf{T}_z W$ in which all the elements of $\mathbf{T}$ are diagonal. We can thus write:

$$\mathsf{T}_z W = V_0 \oplus V_1 \oplus \cdots \oplus V_k$$

where V_0 is the subspace of fixed points of $\mathbf{T}$, in other words $V_0 = \mathsf{T}_z Z \subset \mathsf{T}_z W$, and each V_j is $\mathbf{T}$-invariant.

Let us look at how $\exp X_H$ acts on this decomposition. On each complex subspace V_j, it acts as multiplication by some scalar $\exp(i\lambda_j)$ for a real number λ_j, since $X_H \in \mathfrak{t} \subset \mathfrak{u}(n)$. Moreover, λ_j is nonzero if $j \neq 0$, because X_H generates the whole torus $\mathbf{T}$. Notice, by the way:

Proposition IV.1.3. *Each component of the set Z of zeros of the almost periodic vector field X_H is a symplectic submanifold.*

Proof. As we have just seen, this is an almost complex submanifold for a structure calibrated by ω. □

At each fixed point z, the above gives the second derivative of H:

---

(1) This is one more *zeugma*.

Proposition IV.1.4. *The second derivative of H at the critical point z, is, in the above notation, the quadratic form*

$$\frac{1}{2}\sum_{j=1}^{k}\lambda_j\,|v_j|^2.$$

Proof. If $v_0^1,\ldots,v_0^r,v_1,\ldots,v_k$ are local coordinates corresponding to the decomposition above and if $v_j=q_j+ip_j$, it also follows that

$$X_H=\sum_{j=1}^{k}\lambda_j\left(q_j\frac{\partial}{\partial p_j}-p_j\frac{\partial}{\partial q_j}\right),$$

in other words that

$$dH=\sum_{j=1}^{k}\lambda_j(p_jdp_j+q_jdq_j).$$

In these coordinates,

$$H=\frac{1}{2}\sum_{j=1}^{k}\lambda_j(p_j^2+q_j^2)+o(|v|^2).$$ □

Notice also that the second derivative is nondegenerate in the direction transverse to the fixed submanifold. This leads us to the next notion.

IV.2. Morse functions (in the sense of Bott)

IV.2.a. Definitions. A function f on a manifold W is called a *Morse function* (in the sense of Bott) if its critical set is a submanifold of W, and if its second derivative is a nondegenerate quadratic form in the transverse directions.

Example IV.2.1. If the critical points are isolated, this means that the second derivative is nondegenerate and f is a Morse function (in the sense of Morse!)

We will use a Morse lemma "with parameter":

Proposition IV.2.2. *There exist local coordinates (x,y) in the neighborhood of the point z of the critical submanifold Z in which*

1. *the submanifold Z is described by $y=0$,*
2. *the function f may be written*

$$f(x,y)=f(z)+q_x(y)$$

where q_x is a quadratic form, nondegenerate in the y variables (transverse to Z).

Proof. All the constructions in the classical proofs of Morse lemma (see [**108**] for example) can be parametrized without any difficulty. □

The index of the second derivative (number of "negative squares") is locally constant along the critical submanifold. Its value on a connected component Z_j will be called the *index* of the critical submanifold Z_j and denoted by $\lambda(Z_j)$.

Thanks to the Morse lemma, one can construct subbundles (each isomorphic to each other) of rank $\lambda(Z_j)$ of the normal bundle to Z_j on which the second derivative is negative. Such a subbundle is called "the" *negative normal bundle.*

What we have said may be summarized as:

***Theorem IV.2.3* (Frankel [51]).** *Let H be an almost periodic Hamiltonian on a symplectic manifold (W, ω). Then H is a Morse function, all the critical submanifolds of which have an even index.* □

***Example IV.2.4* (The complex projective space (continuation)).** We consider again the projective space with a Morse function f deduced from the momentum mapping of the $\mathbf{T}^n$-action, in two different cases (shown in Figure 2).

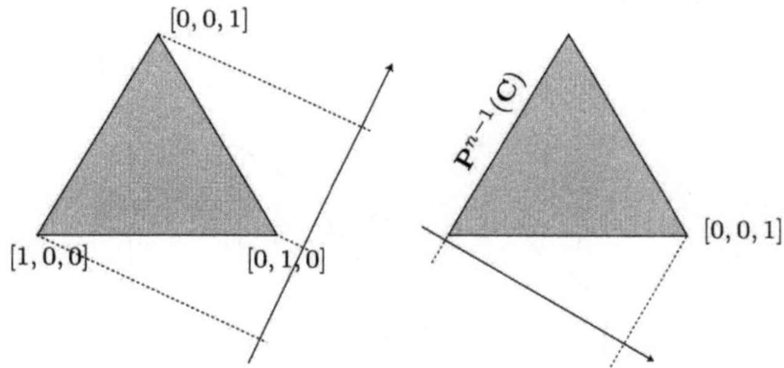

Figure 2

- Choose distinct values of the real numbers a_i, ordered as

$$a_0 < a_1 < \cdots < a_n.$$

The function f is then a Morse function. The point $[1, 0, \ldots, 0]$ is the absolute minimum, the critical point $[0, \ldots, 0, 1, 0, \ldots, 0]$ (with the 1 at the k-th place) is a critical point of index $2k$, corresponding to the k-th critical value, a_k. The function f is a perfect[(2)] Morse function.

- On the contrary, consider the case where $a_0 = \cdots = a_{n-1} = 0$ and $a_n = 1$, so that

$$f([z_0, \ldots, z_n]) = \frac{1}{2} \frac{|z_n|^2}{\sum |z_i^2|}.$$

[(2)]so perfect that this is *the* example of a Morse function given by Milnor in [**108**]

There are only two critical values. The minimum, 0, is obtained on the submanifold $z_n = 0$, a projective hyperplane, critical submanifold of index 0. The maximum, $\frac{1}{2}$, is obtained at the point $[0, \ldots, 0, 1]$.

IV.2.b. Reconstruction. In the case of isolated critical points, one can *reconstruct* the manifold with the help of the critical points and their indices (see [**108**]). When he introduced these more general Morse functions, Bott proved that that was also the case for them. Let, for $a \in \mathbf{R}$, $W_a = \{x \in W \mid f(x) \leqslant a\}$ and $V_a = \{x \in W \mid f(x) = a\}$ the level a. The two main results of the theory are the two next theorems.

Theorem IV.2.5. *If W is compact and if the interval $[a, b] \subset \mathbf{R}$ contains no critical value of f, then W_b is diffeomorphic to W_a.*

Hence the topology does not change if we are not going through a critical level. Here is what happens when we cross one:

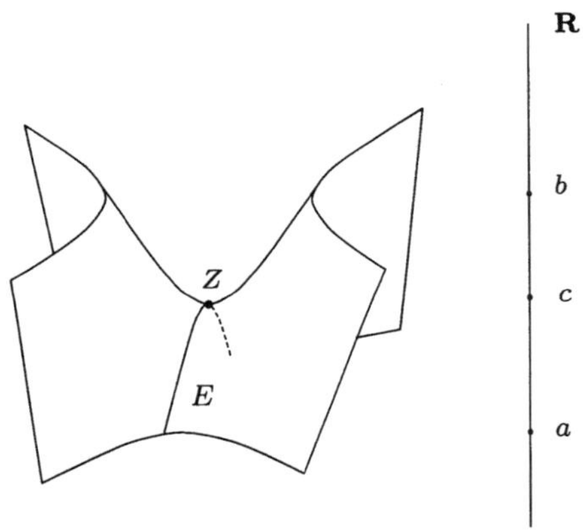

Figure 3

Theorem IV.2.6. *If $c \in [a, b]$ is the unique critical value of f in this interval, the homotopy type of W_b is described by the addition to W_a of the negative normal bundle of the critical submanifold at level c.*

See Figure 3, in which the critical submanifold is a point (the point Z), and its negative normal bundle an interval (the curve E).

Sketch of a proof of Theorem IV.2.5. See [**108**] for details. Choose a Riemannian metric on W, thus getting a gradient vector field $\operatorname{grad} f$, which does not vanish on $f^{-1}[a, b]$. Modify it with the help of a differentiable function ρ on W which takes the value $1/\|\operatorname{grad} f\|^2$ on $f^{-1}[a,b]$ and which vanishes outside a neighborhood, as follows:

$$X = \rho \operatorname{grad} f.$$

Calling φ_t the flow of X, it is easily checked that φ_{b-a} sends W_b onto W_a. □

Proof of Theorem IV.2.6. Let Z be the critical submanifold corresponding to c and $\mathcal{U}_\varepsilon$ be a small enough tubular neighborhood of Z in W. We describe now the homotopy type of the pair

$$\left(W_c, \mathring{W}_c\right) \cap \mathcal{U}_\varepsilon.$$

Let N be the normal bundle to Z in W and E be the negative normal bundle along Z. Identify $\mathcal{U}_\varepsilon$ with a part of N, thus getting a projection $p : \mathcal{U}_\varepsilon \to Z$. Call A_ε the image of E:

$$A_\varepsilon = \left\{ v \in \mathcal{U}_\varepsilon \mid v \in E_{p(v)} \right\}$$

and A_ε^- the complement of Z in A_ε (see Figure 4).

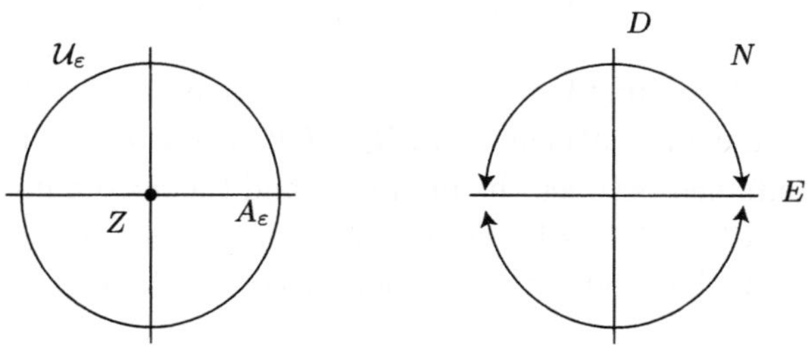

Figure 4

Choosing a supplement bundle D of E in N, it is now very easy to write down a retraction $N - D \to E$, and it is not more difficult to prove the following lemma.

Lemma IV.2.7. *For ε small enough, there exists a retraction (by deformation)*

$$\left(W_c, \mathring{W}_c\right) \cap \mathcal{U}_\varepsilon \longrightarrow (A_\varepsilon, A_\varepsilon^-).$$ □

This ends the proof. □

IV.3. Connectedness of the fibers of the momentum mapping

Let us begin by stating and proving the key result of [**7**].

Connectedness of the levels.

Theorem IV.3.1. *Let f be a nondegenerate function (in the sense of Bott) which has no critical submanifold of index 1 or $n-1$. Then it has a unique local minimum and a unique local maximum. Moreover, all of its nonempty levels are connected.*

Proof. We use Theorem IV.2.6. The homotopy type of W_a can change only by crossing a critical level, in which case it changes by adding to W_a the negative normal bundle of the critical submanifold. If the critical submanifold has index 0, that is if this is a local minimum, we add a connected component. To connect all pieces later on, as W is connected, we must go through a new critical level, for which the sphere bundle of the negative bundle must be connected... but this is impossible except if the index of the submanifold is 1. Thus it is seen that there can be only one local minimum, and, applying the result to $-f$, only one local maximum

Moreover, W_a is connected, and, for the same reasons,

$$\overline{W - W_a} = \{x \mid f(x) \geqslant a\}$$

is connected as well. Assume $V_c = f^{-1}(c)$ is a nonconnected level. Any component of V_c thus defines a nontrivial element in $H_{n-1}(W_c)$. But this group is zero: indeed, if c is strictly contained between the minimum and the maximum of f, the critical submanifolds of critical levels lower than c all have negative normal bundles of dimension $\leqslant n-2$ which cannot create nonzero elements of H_{n-1}. □

The case of almost periodic Hamiltonians. One easily deduces the next corollary from Theorems IV.2.3 and IV.3.1.

Corollary IV.3.2. *Let (W,ω) be a compact connected symplectic manifold and let $H : W \to \mathbf{R}$ be an almost periodic Hamiltonian. All the levels $H^{-1}(t)$ of H are empty or connected.* □

Remark IV.3.3. This use of Morse theory and this connectedness result are the core of the topic treated in this book. Morse theory will also be used in Chapter VIII. It also plays a role in more difficult problems, for instance in the results analogous to Theorem IV.4.3 for noncommutative groups, which for a relatively long period were not completely solved. When Kirwan succeeded, in [**86**], this was with the help of the function $\|\mu\|^2$, which is rather degenerate, but has nevertheless enough of the properties of Morse functions to be handled and give results [**85**]. There are now alternative ways of proving the convexity theorems, for instance [**34, 2, 94**].

IV.4. Application to convexity theorems

IV.4.a. Ginzburg's proof of the Toeplitz–Hausdorff theorem. Recall its statement, already given in the introduction.

Theorem IV.4.1. *Let A be an $n \times n$ complex matrix, and let $\lambda \in \mathbf{R}^n$. The image of the mapping*

$$\begin{aligned} f_A : \mathcal{H}_\lambda &\longrightarrow \mathbf{C} \\ X &\longmapsto \operatorname{tr}(AX) \end{aligned}$$

is a convex subset of $\mathbf{C}$.

We begin by proving a direct application of Corollary IV.3.2.

Lemma IV.4.2. *Let (W, ω) be a symplectic manifold. Let f_1 and f_2 be two functions, the Hamiltonian vector fields of which generate a relatively compact subgroup in the diffeomorphism group of W. The image of $f_1 \times f_2$ is a convex subset of $\mathbf{R}^2$.*

Proof of the lemma. Let $H : W \to \mathbf{R}$ be the composition of $f_1 \times f_2$ with any linear projection $\mathbf{R}^2 \to \mathbf{R}$. The Hamiltonian vector field of H generates a relatively compact subgroup in the group of diffeomorphisms of W, thus H is almost periodic. Its fibers are connected... but all these fibers, when H varies, are all the intersections of the image of $f_1 \times f_2$ with the straight lines in $\mathbf{R}^2$. Hence this image meets any straight line in an interval, thus it is convex. □

Proof of the theorem. Apply the lemma to

$$W = \mathcal{H}_\lambda, \quad f_1 = \operatorname{Re} f_A, \quad f_2 = \operatorname{Im} f_A.$$

The only thing we need to check is that the Hamiltonian vector fields of f_1 and f_2 generate a relatively compact subgroup in the diffeomorphism group of $\mathcal{H}_\lambda$. This is true because these are fundamental vector fields for the action of $\mathrm{U}(n)$ on $\mathcal{H}_\lambda$, the subgroup they generate is then a subgroup of the compact group $\mathrm{U}(n)$: indeed if one writes $A = U + iV$ with U and V Hermitian matrices, then $f_1(h) = \operatorname{tr}(Uh)$, $f_2(h) = \operatorname{tr}(Vh)$, and f_1 is a Hamiltonian for the vector field $i\underline{U}$, f_2 a Hamiltonian for the vector field $i\underline{V}$, as we have seen in Exercise II.26. □

IV.4.b. Convexity of the image of the momentum mapping. Imitating the proof of Lemma IV.4.2, we get Atiyah's proof of the Atiyah [**7**] and Guillemin–Sternberg [**63**] convexity theorem.

Theorem IV.4.3. *Let (W, ω) be a compact connected symplectic manifold. Assume that $f_1, \ldots, f_n$ are n functions on W, the flows of the Hamiltonian vector fields of which generate a subgroup of the diffeomorphism group of W the closure of which is a torus. Then the image of W by the mapping $f = (f_1, \ldots, f_n)$ is a convex subset of $\mathbf{R}^n$. Moreover if $Z_1, \ldots, Z_N$ denote the connected components of the set* *of common critical points of the f_j's, then $f(Z_j)$ is a point c_j and $f(W)$ is the convex hull of the points c_j.*

Remark IV.4.4. As we have noticed (in Remark IV.1.1), the fact that X_{f_i} generates a compact subgroup is very sensitive to the actual function f_i used. This is also, even more obviously, the case for the conclusion of the theorem (imagine what happens to the polyhedron if you replace one of the f_i's by its square for instance).

In particular, if μ is the momentum mapping for a Hamiltonian action of the torus $\mathbf{T}^n$ on W, the choice of a basis of $\mathbf{T}^n$ identifies $\mathfrak{t}^\star$ to $\mathbf{R}^n$ and the components of μ to n functions $f_1, \dots, f_n$ satisfying the assumption of the theorem.

Corollary IV.4.5. *Let (W, ω) be a compact connected symplectic manifold endowed with a Hamiltonian torus action with momentum mapping μ. The image of μ is a convex polyhedron in $\mathfrak{t}^\star$, the convex hull of the images of the fixed points of the torus action.*

***Example IV.4.6* (The complex projective space (continuation)).** The image of $\mathbf{P}^{n-1}(\mathbf{C})$ under the momentum mapping (shown in Figure 1) is the convex hull of the n points $\frac{1}{2}(0, \dots, 0, 1, 0, \dots, 0)$, images of the fixed points $[0, \dots, 0, 1, 0, \dots, 0]$ of the $\mathbf{T}^n$-action.

Proof of the theorem. We prove the following statements (in which n is the number of component functions of f):

(A_n) The level set $f^{-1}(t)$ is empty or connected for all $t \in \mathbf{R}^n$.

(B_n) The image $f(W)$ is convex.

(C_n) If $Z_1, \dots, Z_N$ are the connected components of the set of common critical points of the f_i's, then $f(Z_j)$ is a point c_j and $f(W)$ is the convex hull of the points c_j.

Having noticed that (B_1) is obvious, we first prove that (A_n) implies (B_{n+1}). For this, consider the diagram

$$\begin{array}{ccc} W & \xrightarrow{(f_1, \dots, f_{n+1})} & \mathbf{R}^{n+1} \\ & & \Big\downarrow \pi \\ & & \mathbf{R}^n \end{array}$$

where π is any linear projection. The assumption applies to $g = \pi \circ f$, since any connected closed subgroup of a torus is a torus. As in the proof of Lemma IV.4.2, the subsets

$$f(W) \cap \pi^{-1}(t) = f(g^{-1}(t))$$

describe, when π and t vary, the intersections of $f(W)$ with all the straight lines in $\mathbf{R}^{n+1}$. By Property (A_n), $g^{-1}(t)$ is empty or connected, likewise its image by f, thus $f(W)$ is convex.

Let us now prove that (B_n) implies (C_n). Denote by $X_1, \ldots, X_n$ the Hamiltonian vector fields associated with $f_1, \ldots, f_n$. Let $\mathbf{T}$ be the torus they generate. The Z_j's are the connected components of the set of fixed points of $\mathbf{T}$. On such a component, the X_i's vanish, thus the f_i's are constant, so that $f(Z_j) = c_j$ is indeed a point of $\mathbf{R}^n$. Let now φ be a linear combination $\varphi = \sum \lambda_i f_i$, the real numbers λ_i's being chosen so generic that the vector field $X = \sum \lambda_i X_i$ generates a dense subgroup of $\mathbf{T}$. Then the Z_j's are the components of the set of critical points of φ. In particular, the maximum of φ occurs on Z. Hence the restriction of the linear form $\sum \lambda_i \xi_i$ on $\mathbf{R}^n$ to $f(W)$ reaches its maximum at one of the c_j's. As this is true for almost any choice of the λ_i's, the image of W is contained in the convex hull of the points c_j. We already know that it is convex. It contains c_j by definition, so we have proven (C_n).

Let us now show by induction on n that (A_n) holds. Statement (A_1) is exactly Corollary IV.3.2. We now assume, by induction assumption, that (A_n) is true. Let

$$f_1, \ldots, f_{n+1} : W \longrightarrow \mathbf{R}$$

be $n+1$ functions satisfying the assumption and let $\xi = (\xi_1, \ldots, \xi_{n+1})$ be a point in $\mathbf{R}^{n+1}$. We want to show that

$$f^{-1}(\xi) = f_1^{-1}(\xi_1) \cap \cdots \cap f_{n+1}^{-1}(\xi_{n+1})$$

is empty or connected. If f has no regular value, then one of the df_i is a linear combination of the others, we may forget it and apply (A_n), so that we are done. Otherwise, the set of regular values is dense. By continuity, we may thus assume that ξ is a regular value of f. Then $N = f_1^{-1}(\xi_1) \cap \cdots \cap f_n^{-1}(\xi_n)$ is a submanifold of W. By (A_n), we know that this submanifold is connected.

Let us now check that $f_{n+1}|_N$ satisfies the assumptions of Theorem IV.3.1. Consider its critical points on N. These are the points of W where (on W) df_{n+1} is a linear combination of the df_i's,

$$df_{n+1}(x) = \sum_{i=1}^{n} \lambda_i(x) df_i(x).$$

Lemma IV.4.7. *The λ_i's are locally constant on the set of critical points of f_{n+1} in N.*

Proof. Let x be such a critical point. We have

$$X_{n+1}(x) = \sum_{i=1}^{n} \lambda_i(x) X_i(x).$$

Hence the Lie algebra of the stabilizer of x contains the line generated by $X_{n+1} - \sum \lambda_i X_i$. Let X be any vector in this stabilizer. By definition of N, we know that the vector fields $X_1, \ldots, X_n$ are independent at x, so that X is a multiple of $X_{n+1} - \sum \mu_i X_i$ for some scalars μ_i's. Hence

$$(X_{n+1} - \sum_{i=1}^{n} \mu_i X_i)(x) = 0,$$

hence $\mu_i = \lambda_i(x)$ and the Lie algebra of the stabilizer of x is the line generated by $X_{n+1} - \sum \lambda_i(x) X_i$. Now the manifold W is compact, so that there are only finitely many orbit types and hence finitely many stabilizers. The λ_i's must thus be locally constant. □

Let $x \in N$ be such a critical point. The λ_i's are constants near x. Put $F = f_{n+1} - \sum \lambda_i f_i$ and let Z be the component of critical points of F (on W) which contains x. The function F is an almost periodic Hamiltonian on W as well, to which we cannot refrain from applying Frankel's theorem (Theorem IV.2.3).

Lemma IV.4.8. *The submanifolds Z and N are transversal at x.*

Proof. We must show that $df_1(x), \ldots, df_n(x)$ are still independent when restricted to Z. Let $X_1, \ldots, X_n$ be the Hamiltonian vector fields associated to $f_1, \ldots, f_n$. They are tangent to Z because the Poisson brackets $\{f_i, F\}$ vanish, but Z is symplectic, thus for all $(\alpha_1, \ldots, \alpha_n)$, there exists $Y \in T_x Z$ such that $\omega(\sum \alpha_i X_i, Y) \neq 0$, which may be equally well written $\sum \alpha_i df_i(Y) \neq 0$. □

Now we know that $Z \cap N$ is a critical submanifold of $F|_N$. Moreover, the transversality of Z and N implies that the normal bundle of $Z \cap N$ in N is the restriction to $Z \cap N$ of the normal bundle of Z in W. Thus, $Z \cap N$ is a critical submanifold of even index. Now, when restricted to N, $F = f_{n+1} - \sum \lambda_i \xi_i$, (and thus f_{n+1}) has the same property. We may then apply Theorem IV.3.1 to $f_{n+1}|_N$, from which it follows that

$$(f_{n+1}|_N)^{-1}(\xi_{n+1}) = N \cap f^{-1}(\xi_{n+1}) = f^{-1}(\xi)$$

is connected, hence (A_{n+1}) holds. And this is the end of the proof. □

Remark IV.4.9. We deduce that the vector fields $X_1, \ldots, X_n$ have common zeros (in other words the torus action does have fixed points): it follows from the theorem that Z is not empty. Moreover, there may be a lot of those fixed points, as the following corollary shows.

Corollary IV.4.10. *Let $\mathbf{T}$ be an n-dimensional torus acting in an effective and Hamiltonian way on a compact symplectic manifold (W, ω). The action has at least $n+1$ fixed points.*

Proof. We have noticed in Corollary III.2.4 that, for an effective action, there exists at least one point at which the momentum mapping μ is submersive, and thus open. Hence the image $\mu(W)$ is a convex polyhedron with nonempty interior in $\mathbf{R}^n$, it must thus have at least $n+1$ vertices, which cannot avoid being the images of at least as many fixed points. □

This corollary generalizes the remark we have made for $n = 1$ at the beginning of § III.2.

IV.4.c. Application: a theorem of Schur on Hermitian matrices. We shall apply the convexity theorem to the action by conjugation of the torus of diagonal unitary matrices on Hermitian matrices. Thanks to § III.1.c, we already know that the momentum mapping of this action maps a Hermitian matrix h to its diagonal entries $(h_{1,1}, \ldots, h_{n,n})$, and the convexity theorem says:

***Corollary IV.4.11* ([121], and also [88, 7, 63]).** *Let h be a Hermitian matrix with spectrum $\lambda = (\lambda_1, \ldots, \lambda_n) \in \mathbf{R}^n$ and let $S = \left\{(\lambda_{\sigma(1)}, \ldots, \lambda_{\sigma(n)}) \mid \sigma \in \mathfrak{S}_n\right\}$. Then the diagonal of the matrix h is in the convex hull $\widehat{S}$ of S. Conversely, any point of $\widehat{S}$ is the diagonal of a Hermitian matrix with spectrum λ.*

Proof. The only thing we have to check is that the points in S are the fixed points of the action of $\mathbf{T}$ on $\mathcal{H}_\lambda$. It is thus enough to notice that an element h of $\mathcal{H}_\lambda$ is fixed by $\mathbf{T}$ if and only if it is diagonal. □

We have $h_{1,1} + \cdots + h_{n,n} = \operatorname{tr} h = \lambda_1 + \cdots + \lambda_n$ constant on $\mathcal{H}_\lambda$, in particular, its image is included in a hyperplane and the action is not effective.

Figure 5 shows some examples, in the case $n = 3$. In the first picture, the three

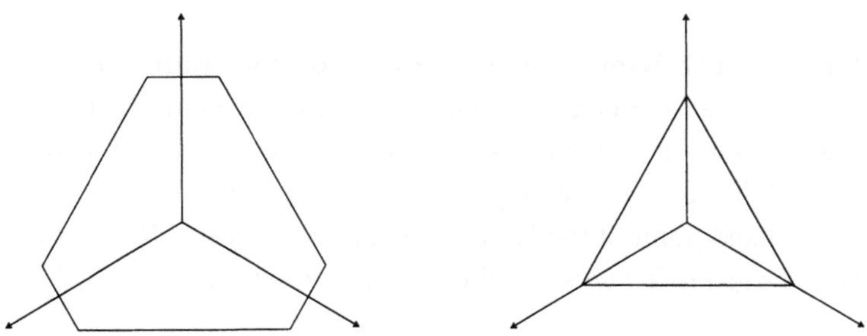

Figure 5

eigenvalues are distinct, $\mathcal{H}_\lambda$ is the manifold of complete flags in $\mathbf{C}^3$ (a symplectic manifold of dimension 6) and its image is a hexagon. In the second one, there are only two distinct eigenvalues, $\mathcal{H}_\lambda$ is a complex projective plane and the image a triangle (the very same triangle, indeed, that appears in Figure 1). Figure 6 shows the image of a principal SU(4)-orbit (this is the manifold of complete flags in $\mathbf{C}^4$, so that it has dimension 12). This polyhedron, the faces of which are rectangles and hexagons, is a *permutahedron*(3).

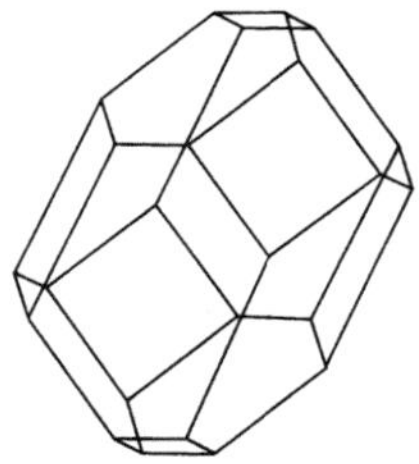

Figure 6

Permutahedra can be understood as geometric representations of the symmetric group. The permutahedron P_n is also the image by the momentum mapping $\mathcal{H}_{1,2,\ldots,n+1} \to \mathbf{R}^{n+1}$ of a principal coadjoint orbit of $\mathrm{SU}(n+1)$.

More generally, let $\mathcal{O}$ be a principal coadjoint orbit in the dual $\mathfrak{g}^\star$ of the Lie algebra of a compact Lie group G. Recall from Remark II.3.5 that $\mathcal{O}$ is the homogeneous space $\mathrm{G}/\mathbf{T}$ for some maximal torus $\mathbf{T}$ of G. Now, the fixed points of the $\mathbf{T}$-action on $\mathcal{O}$ correspond to elements in $N(\mathbf{T})/\mathbf{T} = \mathrm{W(G)}$, the *Weyl* group of G (here, $N(\mathbf{T})$ denotes the normalizer of $\mathbf{T}$ in G). The vertices of the image of the momentum mapping

$$\mu : \mathcal{O} \subset \mathfrak{g}^\star \longrightarrow \mathfrak{t}^\star$$

look like the Weyl group of G. For more information on images of orbits, see [**62**].

IV.4.d. Orbits, stabilizers and the image of the momentum mapping.

The polyhedra we get as images of momentum mappings are rather special ones. Applying the equivariant version of the Darboux Theorem, we shall now investigate them and especially look at the "equations" of their faces.

Let z be a fixed point of the Hamiltonian action of torus $\mathbf{T}^n$ on the symplectic manifold W. The group $\mathbf{T}^n$ acts in a linear way on the tangent space $\mathsf{T}_z W$ as above,

$$\mathsf{T}_z W = V_1 \oplus \cdots \oplus V_m.$$

Here, each V_i is a complex line on which $\mathbf{T}^n$ acts by multiplication

$$t \cdot v_i = t^{\alpha_i} v_i \text{ where } t = (t_1, \ldots, t_n) \in T^n = \mathbf{S}^1 \times \cdots \times \mathbf{S}^1 \subset \mathbf{C}^n$$

(3) The *permutahedron* P_n is defined as the convex hull of the points $(\sigma(1), \ldots, \sigma(n))$ when σ describes the symmetric group $\mathfrak{S}_n$. See Exercise IV.11.

for some multi-exponent $\alpha_i = (a_i^1, \ldots, a_i^n) \in \mathbf{Z}^n$. We can also consider α_i as the element of $\mathfrak{t}^\star$ defined by

$$\langle \alpha_i, e_j \rangle = a_j^i$$

where $(e_1, \ldots, e_n)$ is the basis of $\mathfrak{t}$ defined by the above decomposition. To say the same thing in a more intrinsic way, the $\mathbf{T}$-action on V_i is defined by its *character* $\alpha_i : \mathbf{T} \to \mathbf{S}^1$; differentiating and transposing, we get a map $\mathbf{R} \to \mathfrak{t}^\star$, and α_i is just the image of 1, the *weight* of the representation.

Proposition IV.4.12. *Let z be a fixed point in the symplectic manifold W and let $p = \mu(z) \in \mathfrak{t}^\star$ be its image under the momentum mapping of the torus action. There exists a neighborhood $\mathcal{U}$ of z in W and a neighborhood $\mathcal{V}$ of p in $\mathfrak{t}^\star$ such that*

$$\mu(\mathcal{U}) = \mathcal{V} \cap C_p(\alpha_1, \ldots, \alpha_n)$$

where $C_p(\alpha_1, \ldots, \alpha_n)$ denotes the convex cone with vertex p generated by the vectors $(\alpha_1, \ldots, \alpha_n)$ in $\mathfrak{t}^\star$.

Proof. We first apply the Darboux Theorem in its equivariant form (this is Remark II.1.13) near the fixed point z. The exponential of any invariant Riemannian metric

$$\exp : \mathsf{T}_z W \longrightarrow W$$

conjugates the $\mathbf{T}$-action on W and the linear action on $\mathsf{T}_z W$ described above. Moreover, it sends a neighborhood $\mathcal{U}_0$ of 0 in $\mathsf{T}_z W$ onto a neighborhood $\mathcal{U}$ of z in W. If $\mathcal{U}_0$ is small enough, one finds a $\mathbf{T}$-equivariant map

$$\psi : (\mathcal{U}_0, 0) \longmapsto (\mathsf{T}_z W, 0)$$

such that $\psi^\star \exp^\star \omega = \omega_0$. We now have two momentum mappings:

$$\mu' : \mathsf{T}_z W \supset \mathcal{U}_0 \xrightarrow{\exp} \mathcal{U} \xrightarrow{\mu} \mathfrak{t}^\star$$

on the one hand, and

$$\mu_0 : \mathsf{T}_z W \longrightarrow \mathfrak{t}^\star$$

(associated with the linear action) on the other hand. They can only differ by a constant vector, we may thus fix $\mu_0(0) = \mu'(0) = \mu(z) = p$ in order that they agree everywhere. Of course we have

$$\mu_0(v_1, \ldots, v_n) = p + \frac{1}{2}\Big(\sum_{i=1}^{m} a_1^i |v_i|^2, \ldots, \sum_{i=1}^{m} a_n^i |v_i|^2\Big)$$

and hence the image of $\mathcal{U}$ by μ (as that of $\mathcal{U}_0$ by μ_0) is the cone

$$p + \sum_{i=1}^{m} u_i \alpha_i \ (u_i \geqslant 0). \qquad \square$$

Corollary IV.4.13. *The hyperplanes which delimit the polyhedron P in $\mathfrak{t}^\star$ have integral equations.* $\square$

This statement may be understood, either in coordinates, or, in a more intrinsic way, as follows. The equation of any hyperplane in $\mathfrak{t}^\star$ may be written as $\langle X, \xi\rangle = a$ for some $X \in \mathfrak{t}$ and $a \in \mathbf{R}$; and in our case we can assume that X generates a *circle* S_X in $\mathbf{T}$, namely, that $\exp X = 1$.

Choose one of these hyperplanes, say H, and such an "integral" vector X. The vector field $\underline{X}$ is periodic and is the Hamiltonian vector field of the function

$$x \longmapsto \langle X, \mu(x)\rangle = \widetilde{\mu}_X(x).$$

Let $F \subset H$ be the face of P which is contained in H. Consider $\mu^{-1}(F)$. By the very definition, $\widetilde{\mu}_X$ is constant on $\mu^{-1}(F)$, that is to say that the circle S_X fixes all points of $\mu^{-1}(F)$, which is thus included in the union of the fixed submanifolds of S_X, and hence is a *symplectic submanifold* of W.

As the $\mathbf{T}$-action on W is effective, $\mathbf{T}/S_X$ is a torus with an effective action on $\mu^{-1}(F)$. The open set consisting of principal orbits corresponds to those points x of $\mu^{-1}(F)$ for which $\operatorname{Im} \mathsf{T}_x\mu = H$ (this is always a subspace of H). In this case,

$$(\operatorname{Im} \mathsf{T}_x\mu)^\circ = \mathfrak{s}_X$$

is indeed the Lie algebra of the stabilizer. In particular

$$\dim \mu^{-1}(F) = \dim H - 1.$$

The momentum mapping of the $\mathbf{T}/S_X$-action on $\mu^{-1}(F)$ is obtained from

– the exact sequence of groups

$$1 \longrightarrow S_X \longrightarrow T \longrightarrow T/S_X \longrightarrow 1$$

– which may be differentiated, giving the exact sequence of Lie algebras

$$0 \longrightarrow \mathfrak{s}_X \longrightarrow \mathfrak{t} \longrightarrow \mathfrak{t}/\mathfrak{s}_X \longrightarrow 0$$

– and then transposed to the exact sequence of vector spaces

$$0 \longrightarrow (\mathfrak{t}/\mathfrak{s}_X)^\star \longrightarrow \mathfrak{g}^\star \longrightarrow \mathfrak{s}_X^\star \longrightarrow 0$$

which exhibits $(\mathfrak{t}/\mathfrak{s}_X)^\star$ as a hyperplane of $\mathfrak{g}^\star$ (parallel to H!). The momentum mapping μ, which takes its values in this hyperplane (up to translation), is the momentum mapping for the action of the quotient group.

IV.4.e. Effective actions and Delzant's uniqueness theorem. We concentrate now on effective completely integrable actions, that is actions of half-dimensional tori.

Corollary IV.4.14. *Let W be a compact connected symplectic manifold endowed with an effective completely integrable action. For any point $z \in W$, the rank of the momentum mapping μ at z is the dimension of the face of $\mu(W)$ to which $\mu(z)$ belongs.*

Proof. Let H be the (neutral component of the) stabilizer of z. Let k be its dimension. We look at the action of H on W as a subgroup of $\mathbf{T}$. Now z is a fixed point. We can choose a basis $(X_1, \ldots, X_k)$ of $\mathfrak{h}$, completed as a basis $(X_1, \ldots, X_n)$ of $\mathfrak{t}$ (here $2n = \dim W$), so that the projection $\pi : \mathfrak{t}^\star \to \mathfrak{h}^\star$ is just

$$(\xi_1, \ldots, \xi_n) \longmapsto (\xi_1, \ldots, \xi_k)$$

and the momentum mapping μ_{H} is $\pi \circ \mu$. We apply Proposition IV.4.12 to the fixed point z of H and get

$$\mu_{\mathrm{H}}(\mathcal{U}) = \mathcal{V} \cap C_p(\alpha_1, \ldots, \alpha_k).$$

Now, as H is the (neutral component of the) stabilizer of z, the weights $\alpha_1, \ldots, \alpha_k$ generate (a rank-k sublattice of) the lattice $\mathbf{Z}^k$, so that the cone C_p has p as its unique vertex and the dimension of the face of $\mu(W)$ to which p belongs is $n-k$. □

Remark IV.4.15. As we assume the action to be effective, the interior of the polyhedron P under consideration is nonempty (μ being a submersion at some point). As we also assume that the torus acting is of maximal dimension (half the dimension of the manifold), the points with trivial stabilizer in W correspond to the points in the interior of the polyhedron. According to Proposition III.2.3, the interior points of the polyhedron are then(4) the regular values of the momentum mapping.

This proof of the corollary above gives a more precise result.

Proposition IV.4.16. *Let P be the image polyhedron of the momentum mapping μ of an effective completely integrable action. Let F be any k-dimensional face of P. Then $\mu^{-1}(F)$ is a symplectic submanifold of dimension $2k$ and*

$$\mu|_{\mu^{-1}(\mathring{F})} : \mu^{-1}(\mathring{F}) \longrightarrow \mathring{F}$$

is a fibration with fiber a torus $\mathbf{T}^k$. □

In this statement, $\mathring{F}$ denotes the relative interior of F, namely its interior in the affine space it spans.

Still considering the effective completely integrable action of a torus $\mathbf{T}$, let now z be a fixed point of the whole group $\mathbf{T}$ and let $p = \mu(z)$. We know that z is an isolated fixed point, hence the weights of the linearized action of $\mathbf{T}$ on

$$\mathsf{T}_z W = V_1 \oplus \cdots \oplus V_n$$

form a *basis* of $\mathfrak{t}^\star$. Hence, at each vertex of the polyhedron P, we see a polyhedral cone generated by n independent vectors. Such a polyhedron is said to be

(4) This is not the case when the dimension of the torus is smaller. See, for instance, Exercise IV.10.

simple. Recall that the action is also *effective*, so that the weights must form an integral basis of the lattice $\mathbf{Z}^n \in \mathfrak{t}^\star$. Such a simple polyhedron is called *primitive* (see Figure 7). We thus have proved the following proposition.

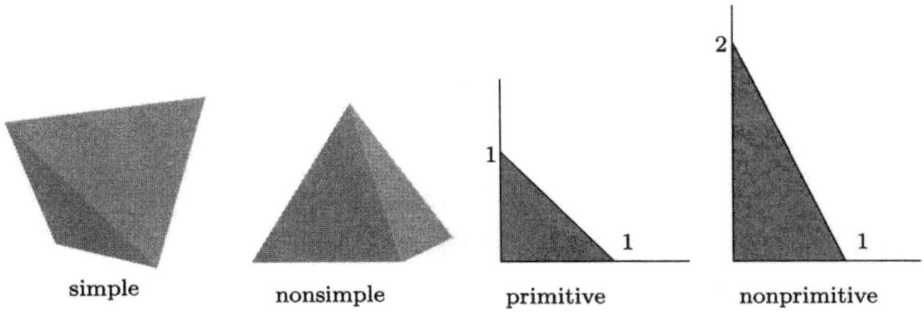

Figure 7

***Proposition IV.4.17* ([39]).** *The polyhedron image of the momentum mapping of an effective completely integrable action on a compact connected symplectic manifold is primitive.* □

A theorem of Delzant asserts the converse: all primitive polyhedra are indeed images of such momentum mappings[(5)]. We will check that this is true in Chapter VII.

Remarks IV.4.18

(1) If the action is not effective, the polyhedron will not be primitive. See Exercise IV.3.
(2) There is of course no claim that the polyhedron itself should be integral. The reader has certainly understood, for instance, that multiplying the symplectic form by λ will result in a dilation (by λ) on the polyhedron.

***Remark IV.4.19* (Global action-angle coordinates).** Let $\mu : W \to \mathbf{R}^n$ be the momentum mapping for an effective completely integrable action. We know (this is assertion (A_n) in the proof of Theorem IV.4.3) that all the levels of μ are connected. Moreover, we have seen (this is Remark IV.4.15) that the points in the interior of the polyhedron $P = \mu(W)$ are regular values of μ. Let t be a point of the interior $\mathring{P}$. Its inverse image $\mu^{-1}(t)$ is a $\mathbf{T}^n$-orbit and a Liouville torus (according to the terminology of § III.3.b). Also

$$\mu : \mu^{-1}(\mathring{P}) \longrightarrow \mathring{P}$$

(5) This is why primitive polyhedra are sometimes (see for instance [**61**, **31**]) called *Delzant polyhedra*.

is a submersion and, in particular (Ehresmann's theorem as in § III.3.b, plus the fact that $\mathring{P}$ is contractile), the open subset $\mu^{-1}(\mathring{P})$ is diffeomorphic to $\mathring{P} \times \mathbf{T}^n$.

Still because the interior $\mathring{P}$ is contractile, the symplectic form ω is exact on the inverse image $\mu^{-1}(\mathring{P})$: we have

$$[\omega] \in H^2(\mu^{-1}(\mathring{P}); \mathbf{R}) \cong H^2(\mathbf{T}^n; \mathbf{R}),$$

but the fibers $\mathbf{T}^n$ are Lagrangian, so that $[\omega] = 0$. Hence, following the proof of the Arnold–Liouville Theorem (Theorem III.3.3), the components of the momentum mapping μ are already action coordinates and we deduce angle coordinates using any section. Hence, we have seen that there are *global* action-angle coordinates[(6)] on $\mu^{-1}(\mathring{P})$.

Let us now come to the main result of this subsection, namely, Delzant's uniqueness theorem, which asserts that the polyhedron determines the symplectic manifold. Precisely:

***Theorem IV.4.20* (Delzant [39]).** *Let (W, ω), (W', ω') be two compact connected symplectic manifolds of dimension $2n$. Assume that both are endowed with an effective completely integrable action and that the two momentum mappings*

$$\mu : W \longrightarrow \mathbf{R}^n, \quad \mu' : W' \longrightarrow \mathbf{R}^n$$

have the same image. Then there exists an equivariant (symplectic) diffeomorphism

$$\varphi : W \longrightarrow W' \text{ such that } \mu' \circ \varphi = \mu.$$

The only reason why I have put the word "symplectic" in parentheses is that I will not prove that part of the statement, proving only that the equivariant diffeomorphism type of the symplectic manifold is determined by the polyhedron. The proof I give here is the original proof of [**39**].

Let $p \in \mu(W)$ be a point of the polyhedron P and let F be the face the relative interior of which p belongs to. From Proposition IV.4.16, we know that $\mu^{-1}(F)$ is a symplectic submanifold acted on by a half-dimensional torus $\mathbf{T}^k$. Moreover, its normal bundle is trivial. Using action-angle coordinates (see Theorem III.3.3), we get:

Proposition IV.4.21. *Let $\widetilde{U}$ be a small open ball centered at p, such that $u = \widetilde{U} \cap F$ is a relatively open ball in F. Then a neighborhood of $\mu^{-1}(U)$ is diffeomorphic to*

$$\mathbf{T}^k \times U \times B^{2n-2k}(\varepsilon) \subset \mathbf{T}^t \times U \times \mathbf{C}^{n-k}$$

---

(6) This is not always the case that there exist global action-angle coordinates over the set of regular values of an integrable system. See, for instance, the references about the spherical pendulum given in Exercise III.22.

by a diffeomorphism

– *which pulls back the symplectic form*

$$\sum_{j=1}^{k} d\theta_j \wedge dJ_j + \sum_{j=1}^{n-k} dx_j \wedge dy_j$$

to the symplectic form ω *and*

– *which is equivariant with respect to the* $\mathbf{T}^n = \mathbf{T}^k \times \mathbf{T}^{n-k}$*-action on* $\mathbf{T}^k \times U \times \mathbf{C}^{n-k}$ *by*

$$\begin{aligned}(e^{i\varphi_1}, \dots, e^{i\varphi_k}, u_1, \dots, u_{n-k}) \cdot \left(e^{i\theta_1}, \dots, e^{i\theta_k}, J_1, \dots, J_k, z_1, \dots, z_{n-k}\right) \\ = \left(e^{i\theta_1+\varphi_1}, \dots, e^{i\theta_k+\varphi_k}, J_1, \dots, J_k, u_1 z_1, \dots, u_{n-k} z_{n-k}\right).\end{aligned}$$

□

This gives a semi-local normal form for our manifold with $\mathbf{T}^n$-action around the pre-image of any point of P. Notice that the momentum mapping is

$$\mu(e^{i\theta_1}, \dots, e^{i\theta_k}, J_1, \dots, J_k, z_1, \dots, z_{n-k}) = p + \left(J_1, \dots, J_k, \frac{1}{2}|z_1|^2, \dots, \frac{1}{2}|z_{n-k}|^2\right).$$

The next lemma will allow us to glue these data together.

Lemma IV.4.22. *Let* φ *be a diffeomorphism of* $\mathbf{T}^k \times U \times \mathbf{S}^{2n-2k-1}(\varepsilon)$ *such that*

– φ *is* $\mathbf{T}^n$*-equivariant,*

– $\mu \circ \varphi = \mu$.

Then φ *can be extended to a diffeomorphism* $\widetilde{\varphi}$ *of* $\mathbf{T}^k \times U \times B^{2n-2k}(\varepsilon)$ *having the same properties.*

Proof. The diffeomorphism φ has the form

$$\varphi(\theta, J, z_1, \dots, z_{n-k}) = (\theta + \theta_0, J, \psi_1, \dots, \psi_{n-k})$$

with $|\psi_j|^2 = |z_j|^2$ and

$$\psi_j(\theta, J, u_1 z_1, \dots, u_{n-k} z_{n-k}) = u_j \psi_j(\theta, J, z_1, \dots, z_{n-k}).$$

We look at the function ψ_1. We have, for $(x_1, \dots, x_{n-k})$ real,

$$|\psi_1(\theta, J, x_1, \dots, x_{n-k})|^2 = |x_1|^2$$

and

$$\psi_1(\theta, J, -x_1, \dots, x_{n-k}) = -\psi_1(\theta, J, x_1, \dots, x_{n-k})$$

so that the function f defined by

$$x_1 f(\theta, J, x_1, \dots, x_{n-k}) = \psi_1(\theta, J, x_1, \dots, x_{n-k})$$

is an *even* smooth function of each of its entries x_j's, from the sphere $\mathbf{S}^{n-k-1}(\varepsilon)$ to the circle $\mathbf{S}^1$. According to a classical result of Whitney, we deduce the existence of a smooth function g, defined on the simplex

$$\sum_{j=1}^{n-k} X_j = 1, \quad X_1 \geqslant 0, \ldots, X_{n-k} \geqslant 0,$$

taking its values in $\mathbf{S}^1$, and such that

$$\begin{aligned} f(\theta, J, x_1, \ldots, x_{n-k}) &= x_1 g(\theta, J, x_1^2, \ldots, x_{n-k}^2) \\ &= x_1 \exp(ih(\theta, J, x_1^2, \ldots, x_{n-k}^2)). \end{aligned}$$

Now, for any (complex) $(z_1, \ldots, z_{n-k})$ in the sphere, we have

$$\psi_1(\theta, J, z_1, \ldots, z_{n-k}) = z_1 \exp(ih(\theta, J, z_1, \ldots, z_{n-k}))$$

a formula which allows us to extend ψ_1 (and analogously, the other ψ_j's). □

Proof of the (equivariant diffeomorphism assertion of the) theorem

The proof of Delzant's uniqueness theorem now proceeds as follows. We cover the polyhedron P by open subsets

$$P = \bigcup_{i,F} U_{i,F} \text{ where the } F \text{ are the faces,}$$

such that

- $U_{i,F} \cap U_{j,F'} = \varnothing$ for $F \neq F'$,
- $U_{i,F} \cap F$ is relatively closed in the relative interior $\overset{\circ}{F}$ and we have, over $U_{i,F}$, coordinates as in Proposition IV.4.21.
- $U_{i,F} \cap (\cup_{j>i} U_{j,F})$ contains the image of the $\mathbf{T}^k \times (U_{i,F} \cap F) \times \mathbf{S}^{2n-2k-1}(\varepsilon)$ contained in the neighborhood of $\mu^{-1}(U_{i,F} \cap F$ used in Lemma IV.4.22.

The mentioned proposition gives an equivariant diffeomorphism

$$\mu^{-1}(U_{i,F}) \longrightarrow \mu'^{-1}(U_{i,F})$$

compatible with the momentum mappings, while the mentioned lemma allows us to glue these diffeomorphisms by stages. □

IV.4.f. The Kähler case. Assume now that we have a (compact) torus acting in a Hamiltonian way on a compact Kähler manifold W. In addition to being symplectic, our manifold is now endowed with a Riemannian metric and a complex structure, all this being compatible, as we have said in § II.2.b. We require the torus action to preserve also the Riemannian metric, namely we ask that the torus act by isometries. Because of all the compatibility assumptions, this implies that the complex structure also is preserved.

Hence our torus acts by automorphisms of the *complex* manifold. This implies that the action extends to a complex torus action, either because we know that the group of automorphisms of a complex manifold is a complex Lie group, or, more concretely, as follows. We consider $\mathbf{T}^n$ as the compact torus in the complex torus

$$(\mathbf{C}^\star)^n = \{(e^{u_1}t_1, \ldots, e^{u_n}t_n) \mid (t_1, \ldots, t_n) \in \mathbf{T}^n \text{ and } u_i \in \mathbf{R}\} \cong \mathbf{R}^n \times \mathbf{T}^n.$$

Let $\underline{X}_j$ be the fundamental vector field associated with the action of the j-th circle $\{(1, \ldots, 1, t_j, 1, \ldots, 1\}$ in $\mathbf{T}^n$, consider the flow φ^j of the vector field $J\underline{X}_j$ and define

$$(e^{u_1}t_1, \ldots, e^{u_n}t_n) \cdot x = (\varphi^1_{u_1} \circ \cdots \circ \varphi^n_{u_n})((t_1, \ldots, t_n) \cdot x)$$

(this is well-defined because we have assumed the manifold to be compact). The Lie

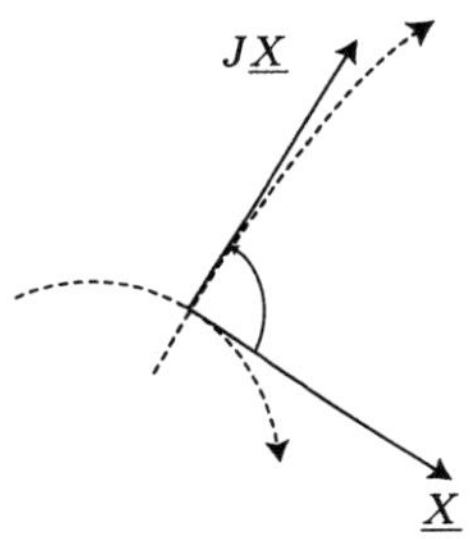

Figure 8

algebra of $\mathbf{T}^n$ is a copy $\mathfrak{t}^n$ of $\mathbf{R}^n$. That of $(\mathbf{C}^\star)^n$ is $\mathbf{C}^n$, decomposed as $\mathbf{R}^n \times \mathfrak{t}^n$, so that the fundamental vector field associated with $(Y, X) \in \mathbf{R}^n \times \mathbf{R}^n$ is $\underline{X} + J\underline{Y}$. In this situation, Atiyah proves, in [**7**], a refinement of his convexity theorem IV.4.3.

Theorem IV.4.23. *Let W be a compact connected Kähler manifold endowed with the Hamiltonian action of a compact torus $\mathbf{T}^n$, with momentum mapping*

$$\mu : W \longrightarrow \mathbf{R}^n.$$

Let $V \subset W$ be an orbit of the complexified action of the complex torus $(\mathbf{C}^\star)^n$ and let $\overline{V}$ be its closure. Let $Z_1, \ldots, Z_k$ be the components of the set of fixed points of $\mathbf{T}^n$ in W which intersect $\overline{V}$. Then $\mu(\overline{V})$ is the convex hull of the points $c_i = \mu(Z_i)$ in $\mathbf{R}^n$.

Proof. We can (and thus will) assume that the components of μ are independent functions on V (otherwise, we consider an independent subset). We are thus assuming that $\dim V = 2n = 2 \dim \mathbf{T}^n$ and that $\mathbf{T}^n$ (or a quotient of $\mathbf{T}^n$ by a finite group) acts freely on V, or that the restriction of μ to V is submersive.

We consider, as in the proof of Theorem IV.4.3, the components f_i of μ and the Morse–Bott function

$$f = \sum_{i=1}^{n} \lambda_i f_i.$$

Notice that the gradient of f (with respect to the Riemannian metric which is part of the Kähler structure) is

$$\operatorname{grad} f = JX_f$$

and its gradient flow φ^t is the action of the corresponding one-parameter subgroup in $(\mathbf{C}^\star)^n$.

Let now x be any point of the orbit V. We look at the limit points of $\varphi^t(x)$ for $t \to \infty$. Notice that, at a limit point, the vector field must vanish, so that the limit points are critical points.

Lemma IV.4.24. *The limit* $\lim_{t\to\infty}(\varphi^t(x))$ *exists and is the supremum of* f *on* V.

Proof of the lemma. We look at the function

$$\begin{aligned} \mathbf{R} &\longrightarrow \mathbf{R} \\ t &\longmapsto f(\varphi^t(x)), \end{aligned}$$

clearly an increasing function. As this is also a bounded function, it has a limit for $t \to +\infty$, which is equal to $f(x_\infty)$, for x_∞ a limit point. Now f is constant on the component, say Z, of its critical points to which x_∞ belongs. Moreover, Z is an invariant submanifold for the $(\mathbf{C}^\star)^n$-action. Hence,

$$\text{for any } z \in (\mathbf{C}^\star)^n, \quad \lim_{t\to+\infty} f(\varphi_t(z\cdot x)) = f(z\cdot x_\infty) = f(x_\infty),$$

so that, V being an orbit, $f(x_\infty)$ does not depend on anything and we have

$$f(x_\infty) = \sup_{x\in V} f(x). \qquad \square$$

Back to the proof of the theorem, we choose the λ_i's so that the function f is generic and the torus it generates is the whole $\mathbf{T}^n$. Then the critical submanifold Z above is one of the Z_j's. We thus have

$$\sup_{x\in V} f(x) = \sup_{j=1,\dots,k} f(Z_j)$$

and we have proved that $\mu(\overline{V})$ is *contained* in the convex hull P of $c_1, \dots, c_k$.

Let us prove the reverse inclusion. Let x be a point of V. We want to find a neighborhood of $\mu(x)$ in P which is contained in the image $\mu(V)$. Let δ be the distance of $\mu(x)$ to the boundary of P. Recall that we are assuming μ to be submersive on V, so that $\mu(x)$ must be an interior point of P and δ a positive

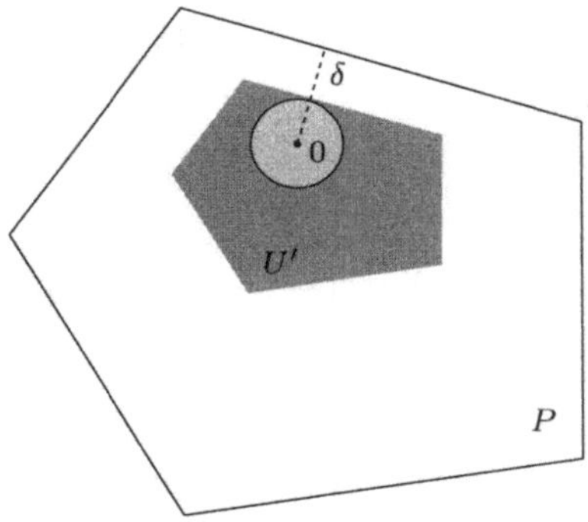

Figure 9

number. By adding the constant vector $-\mu(x)$ to μ, we can assume that $\mu(x)$ is the origin of $\mathbf{R}^n$. Consider the function

$$\begin{aligned} \sigma : \mathbf{S}^{n-1} &\longrightarrow \mathbf{R} \\ \lambda = (\lambda_1, \ldots, \lambda_n) &\longmapsto \sup_{\xi \in P} \left(\sum \lambda_i \xi_i \right) \end{aligned}$$

(the *support* function of P). Our positive number δ is now

$$\delta = \inf_{\lambda \in S^{n-1}} \sigma(\lambda).$$

Consider now the function $f_\lambda = \sum \lambda_i f_i$, so that

$$\begin{aligned} \sup_{x \in V} f_\lambda(x) &= \sup_{j=1,\ldots,k} f(Z_j) \\ &= \sup_{j=1,\ldots,k} \sum_{j=1}^{n} \lambda_i c_{i,j} \quad \text{(obvious notation)} \\ &= \sigma(\lambda). \end{aligned}$$

We look now at the trajectory of the gradient flow φ_λ^t of f_λ through our point x. For $t = 0$, we are at x and $f_\lambda(x) = 0$. Then the function increases along the trajectory until it reaches $\sigma(\lambda)$. Hence, our positive number δ is the image of a unique point $\varphi_\lambda^{t_\lambda}(x)$ of the trajectory. This gives us a number t_λ and $\lambda \mapsto t_\lambda$ is a continuous function of λ. Look now at the vectors Y in $\mathbf{R}^n$ that have the form $Y = r\lambda$ with $\lambda \in S^{n-1}$ and $r \leqslant t_\lambda/2$. This is a neighborhood U of 0 and the image of the composed mapping

$$\begin{array}{ccccc} U & \longrightarrow & V & \xrightarrow{\ \mu\ } & \mathbf{R}^n \\ Y & \longmapsto & \exp(Y)\cdot x & \longmapsto & \mu(\exp(Y)\cdot x) \end{array}$$

is a neighborhood U' of 0 in $\mu(V)$.

For any point $\xi \in \partial U'$, there is a λ (with $\|\lambda\| = 1$) such that $\sum \lambda_i \xi_i = \delta/2$ and in particular $\|\xi\| \geqslant \delta/2$, so that the ball centered at 0 with radius $\delta/2$ is contained in U'. We have thus proved that the interior of P is contained in the image $\mu(V)$. Eventually, Theorem IV.4.23 is proved. □

IV.4.g. Application: a theorem of Kushnirenko on monomial equations. We prove the theorem of Kushnirenko, stated in the introduction of this book, as a consequence of Theorem IV.4.23. This proof, written in [**8**], is due to Arnold and Khovanski.

Consider a finite set $S \subset \mathbf{Z}^n$ of multi-exponents and the system of n equations and n unknowns:

$$\sum_{\alpha \in S} c_\alpha^j z^\alpha = 0$$

where $1 \leqslant j \leqslant n$, the parameters c_α^j are complex and the unknown z sits in the complex torus $(\mathbf{C}^\star)^n$.

Theorem IV.4.25. *The number of solutions of*

$$\sum_{\alpha \in S} c_\alpha^j z^\alpha = 0$$

for a general enough choice of the coefficients c_α^j is

$$N(S) = n! \operatorname{Vol}(\widehat{S})$$

where $\widehat{S}$ denotes the convex hull of S in $\mathbf{R}^n$.

Proof. Let m be the cardinality of S and $\alpha_0, \ldots, \alpha_{m-1}$ be its elements. We consider the action of the *complex* torus $(\mathbf{C}^\star)^n$ on $\mathbf{C}^m$ by

$$z \cdot (u_0, \ldots, u_{m-1}) = (z^{\alpha_0} u_0, \ldots, z^{\alpha_{m-1}} u_{m-1}).$$

The orbit of the point $(1, \ldots, 1)$ is a faithful orbit (the orbit map is injective) if and only if the elements of S generate the lattice $\mathbf{Z}^n$.

In order to use Theorem IV.4.23, we need a torus action on a *compact* Kähler manifold, so that we will rather look at the same torus acting, by the same formula, on $\mathbf{P}^{m-1}(\mathbf{C})$ and at the orbit V_S of the point $[1, \ldots, 1]$. In order that V_S be a faithful orbit, we need that the set

$$\{\alpha - \beta \mid \alpha, \beta \in S\}$$

generate the lattice $\mathbf{Z}^n$. We will assume, to begin with, that this is true. Our number of solutions is then exactly the number of intersection points of V_S (an open complex variety of complex dimension n) with the codimension-n projective subspace $\mathbf{P}(E)$ defined by the linear equations

$$\sum_{\alpha \in S} c_\alpha^j z_\alpha = 0 \quad j = 1, \ldots, n.$$

We assume the coefficients to be generic, so that the intersection is transversal. Now the closure of V_S is an algebraic subvariety of $\mathbf{P}^{m-1}(\mathbf{C})$ (I will not discuss here the question of whether this is (or is not) a smooth submanifold of the projective space, see § VII.2.c), we have

$$\dim_{\mathbf{C}}(\overline{V}_S - V_S) < n,$$

so that the "boundary" does not intersect the generic projective space $\mathbf{P}(E)$.

Let ω be the standard[(7)] Kähler form on $\mathbf{P}^{m-1}(\mathbf{C})$, so that the cohomology class dual to $\mathbf{P}(E)$ is $[\omega]^n$ and that our number of intersection points is

$$N(S) = \int_{V_S} \omega^n = n! \operatorname{Vol}(V_S)$$

(recall from Remark II.1.8 that $\omega^n/n!$ is the Riemannian volume element). What the refined theorem IV.4.23 tells us is that the restriction of the momentum mapping

$$V_S \longrightarrow \overset{\circ}{P}$$

is a fibration with fiber $\mathbf{T}^n$, so that

$$\operatorname{Vol}(V_S) = \operatorname{Vol}(\mathbf{T}^n) \operatorname{Vol}\left(\overset{\circ}{P}\right) = \operatorname{Vol}(\widehat{S})$$

since $\mathbf{T}^n$ has volume 1 (see Remark III.3.9). Eventually, we have got the expected relation... under the assumption that the $\alpha - \beta$ generate $\mathbf{Z}^n$. If this is not true, this may be because

- they generate a lattice of rank $< n$, in this case both sides of the equality we want to prove vanish:
  - the convex hull $\widehat{S}$ is contained in a hyperplane, thus has volume zero, while
  - the orbit is too small and does not intersect the codimension-n subspace when the c_α^j's are general enough, hence $N(S) = 0$;
- or because they generate a lattice Λ of rank n, so that $\Gamma = \mathbf{Z}^n/\Lambda$ is a finite group. We can work with the torus $\mathbf{R}^n/\Lambda$ instead of our original $\mathbf{T}^n = \mathbf{R}^n/\mathbf{Z}^n$ and multiply both sides of the equation by the order of Γ. □

Remark IV.4.26. The closure $\overline{V}_S$ is the *toric variety* associated with the polyhedron $\widehat{S}$ (see Chapter VII).

---

(7) the one that gives volume 1 to a complex line $\mathbf{P}^1(\mathbf{C}) \subset \mathbf{P}^{m-1}(\mathbf{C})$

IV.5. Appendix: compact symplectic SU(2)-manifolds of dimension 4

In this appendix, we give a classification of the compact symplectic manifolds of dimension 4 endowed with a symplectic action of SU(2). The classification is, in essence, due to Iglesias [**70**], who actually considered the group SO(3), but the case of SU(2) is a little easier and a little less technical, that is why we have chosen it (the SO(3)-case will be presented in Exercise IV.18). As we have done in the case of Delzant's theorem (here Theorem IV.4.20), we shall not worry about the symplectic forms, but we shall concentrate instead on the manifolds with action.

Recall that SU(2) is the group of matrices

$$\mathrm{SU}(2) = \left\{ \begin{pmatrix} x & -\overline{y} \\ y & \overline{x} \end{pmatrix} \mid |x|^2 + |y|^2 = 1 \right\}$$

from which description it is easy to recognize that the manifold SU(2) is diffeomorphic to $\mathbf{S}^3$.

IV.5.a. A list of examples. We start from the standard linear action of SU(2) on $\mathbf{C}^2$ and from the induced action on $\mathbf{P}^1(\mathbf{C})$.

The complex projective space. We can of course compactify $\mathbf{C}^2$ into the projective plane $\mathbf{P}^2(\mathbf{C})$ by adding to it a projective line $\mathbf{P}^1(\mathbf{C})$ at infinity. The SU(2)-action extends. It can also be described as the action of the subgroup

$$\mathrm{SU}(2) \subset \mathrm{U}(2) \subset \mathrm{U}(3) \subset \mathrm{GL}(3; \mathbf{C})$$

on $\mathbf{P}^2(\mathbf{C})$. This is of course a symplectic action and thus a Hamiltonian action(8) (SU(2) is semi-simple) and it is effective. Notice that the origin of $\mathbf{C}^2$, alias the point $[0, 0, 1] \in \mathbf{P}^2(\mathbf{C})$, is a fixed point.

SU(2)*-actions on bundles over* $\mathbf{P}^1(\mathbf{C})$. Still starting from the linear action on $\mathbf{C}^2$, we can also cross $\mathbf{C}^2$ by a copy of $\mathbf{C}$ on which SU(2) acts trivially and twist everything by an $\mathbf{S}^1$-action. We look at the quotient $\mathbf{S}^3 \times_{\mathbf{S}^1} \mathbf{C}$, where

$$u \cdot ((a, b)z) = ((ua, ub), u^{-m}z) \text{ for some } m \in \mathbf{Z}.$$

The space itself is a complex line bundle over $\mathbf{P}^1(\mathbf{C})$, *via* the projection mapping

$$[(a, b), z] \longmapsto [a, b].$$

The SU(2)-action descends to the quotient so that we have a space E_{-m} with an SU(2)-action. This action is effective if and only if m is odd. If

$$(g \cdot (a, b), z) = ((ua, ub), u^{-m}z) \text{ for some } u \in \mathbf{S}^1,$$

(8) The momentum mapping can be deduced from Exercise III.8.

then (a,b) must be an eigenvector of g. If $g = \pm\,\mathrm{Id}$, all the vectors are eigenvectors and

$$(-\,\mathrm{Id}\cdot(a,b),z) = ((-a,-b)),z) \text{ if and only if } m \text{ is even.}$$

Note that if this is the case, we get an effective action of $\mathrm{SU}(2)/\left\{\pm\,\mathrm{Id}\right\} = \mathrm{SO}(3)$. Obviously, the points of the zero section ($z=0$) have stabilizer $\mathbf{S}^1$ (and the zero section is an orbit) while the other points have stabilizer $\mathbf{Z}/m$ (which is thus the principal stabilizer). The manifold E_{-m} can also be considered as a complex submanifold of a product $\mathbf{P}^1(\mathbf{C}) \times \mathbf{C}^N$ ($N = 2^m$). The mapping

$$\begin{aligned} \mathbf{S}^3 \times \mathbf{C} &\longrightarrow \mathbf{P}^1(\mathbf{C}) \times (\mathbf{C}^2)^{\otimes m} \\ (v,z) &\longmapsto ([v], z(v\otimes\cdots\otimes v)) \end{aligned}$$

defines an embedding

$$\mathbf{S}^3 \times_{\mathbf{S}^1} \mathbf{C} = E_{-m} \hookrightarrow \mathbf{P}^1(\mathbf{C}) \times (\mathbf{C}^2)^{\otimes m}$$

and our action of $\mathrm{SU}(2)$ is just the restriction of the diagonal action

$$\begin{aligned} \mathrm{SU}(2) \times (\mathbf{P}^1(\mathbf{C}) \times (\mathbf{C}^2)^{\otimes m}) &\longrightarrow \mathbf{P}^1(\mathbf{C}) \times (\mathbf{C}^2)^{\otimes m} \\ (A,\ell,w) &\longmapsto (A\cdot\ell, A^{\otimes m}\cdot w). \end{aligned}$$

The symplectic form on $\mathbf{C}^2$ being $\mathrm{SU}(2)$-invariant, the usual product Kähler form will be invariant as well.

To make E_{-m} a compact manifold, we glue to it a copy of E_m by the map

$$[(a,b),z] \longmapsto [(a,b),z^{-1}] \text{ for } z \neq 0.$$

The resulting compact $\mathrm{SU}(2)$-manifold W_m is the quotient

$$W_m = \mathbf{S}^3 \times_{\mathbf{S}^1} \mathbf{P}^1(\mathbf{C})$$

under the $\mathbf{S}^1$-action

$$u\cdot((a,b)[x,y]) = ((ua,ub),[x,u^m y]).$$

This is a bundle over $\mathbf{P}^1(\mathbf{C})$, the fibers themselves projective lines[(9)]. The bundle has two sections, "at zero" and "at infinity", both are orbits with stabilizer $\mathbf{S}^1$. All the other orbits have stabilizer $\mathbf{Z}/m$.

The embedding of E_{-m} into $\mathbf{P}^1(\mathbf{C}) \times (\mathbf{C}^2)^{\otimes m}$ extends to an embedding

$$\begin{aligned} W_m &\hookrightarrow \mathbf{P}^1(\mathbf{C}) \times \mathbf{P}((\mathbf{C}^2)^{\otimes m} \oplus \mathbf{C}) \\ (v,[x,y]) &\longmapsto ([v],[x(v\otimes\cdots\otimes v),y]). \end{aligned}$$

[(9)]This is the bundle $\mathbf{P}(\mathcal{O}(m)\oplus\mathbf{1})$, a complex manifold called a Hirzebruch surface, see Exercises IV.4 and IV.17.

As above, the standard Kähler form is SU(2)-invariant and thus defines an invariant symplectic form on W_m. The SU(2)-action cannot refrain from being Hamiltonian. We will not need it explicitly, but it is easy to write the momentum mapping for this action (this is Exercise IV.16).

Which gives us a list of compact symplectic manifolds of dimension 4 endowed with an effective Hamiltonian action: $\mathbf{P}^2(\mathbf{C})$ and the Hirzebruch surfaces W_m (for odd m). Notice that (this is proved in Exercise IV.17) all the W_m's for m odd are diffeomorphic with $\widetilde{\mathbf{P}}^2(\mathbf{C})$, the projective plane blown up at a point. However the SU(2)-actions are different (have different principal stabilizers, for instance).

IV.5.b. Classification. We assume now that W is a compact symplectic manifold of dimension 4 endowed with an effective symplectic action of SU(2). We will assume that the momentum mapping

$$\mu : W \longrightarrow \mathfrak{su}(2)^\star$$

is *submersive* at at least one point in W. We investigate the orbit types in W.

Principal orbits. From Proposition III.2.10, we know that the principal stabilizer is both discrete and a closed subgroup of the maximal torus

$$\mathbf{S}^1 = \left\{ \begin{pmatrix} z & 0 \\ 0 & \bar{z} \end{pmatrix} \mid z \in \mathbf{S}^1 \right\}.$$

Hence this is a cyclic group $\mathbf{Z}/m$. We deduce that principal orbits have dimension 3.

Exceptional orbits. The exceptional orbits have dimension 3 as well. Let x be a point in such an orbit. We know that

$$\operatorname{Ker} \mathsf{T}_x\mu = (\mathsf{T}_x(\mathrm{SU}(2) \cdot x))^\circ,$$

so that $\dim \operatorname{Ker} \mathsf{T}_x\mu = 1$ and μ is submersive at x. Moreover, the image of an orbit of dimension 3 has dimension 2 and is contained in a coadjoint orbit in $\mathfrak{su}(2)^\star$. This is thus a sphere $\mathbf{S}^2 \subset \mathfrak{su}(2)^\star$; hence the stabilizer of the orbit is a subgroup of the stabilizer of this point in the sphere. This is thus also a cyclic group, $\mathbf{Z}/q$, say. According to the slice theorem, a neighborhood of an orbit with stabilizer $\mathbf{Z}/q$ has the form $\mathrm{SU}(2) \times_{\mathbf{Z}/q} \mathbf{R}$. The cyclic group $\mathbf{Z}/q$ acts on $\mathbf{R}$, either trivially or (only if q is even)by $\pm\,\mathrm{Id}$. Notice that, in the case of a nontrivial action, the manifold would not be orientable, so that the $\mathbf{Z}/q$-action on $\mathbf{R}$ must be trivial. All the orbits in such a neighborhood have stabilizer $\mathbf{Z}/q$, therefore $q = m$ and there are no exceptional orbits.

Singular orbits. On the contrary, there are singular orbits. Consider a circle subgroup $\mathbf{S}^1 \subset \mathrm{SU}(2)$. The corresponding Hamiltonian H must have critical points on the compact manifold W. Let x be such a point. This is a fixed point of the $\mathbf{S}^1$-action and thus its stabilizer (for the SU(2)-action) contains our $\mathbf{S}^1$, so that the orbit of x is singular. The stabilizer of a point x is a subgroup of the stabilizer of $\mu(x)$ for the coadjoint action, and this is a circle unless $\mu(x) = 0$. Moreover, if this is the case, the orbit of x is isotropic, because

$$\mathsf{T}_x(\mathrm{SU}(2)\cdot x) \subset \operatorname{Ker} \mathsf{T}_x\mu = \{\mathsf{T}_x(\mathrm{SU}(2)\cdot x)\}^{\circ}.$$

Either the stabilizer of x is the whole group SU(2), x is a fixed point, or it is a closed subgroup of dimension 1, so that the orbit of x has dimension 2 and is Lagrangian.

The stabilizer of a singular orbit which is not a fixed point is a closed dimension-1 subgroup H of SU(2), that is, a subgroup covering SO(2) or O(2) by the restriction of the covering map

$$\varphi : \mathrm{SU}(2) \longrightarrow \mathrm{SO}(3).$$

The singular orbit has a neighborhood (once again the slice theorem) of the form $\mathrm{SU}(2) \times_{\mathrm{H}} \mathbf{R}^2$, where H acts on $\mathbf{R}^2$

- as O(2) is $\varphi(\mathrm{H}) = \mathrm{O}(2)$,
- as $\mathbf{S}^1$, either by $u \cdot z = uz$ or by $u \cdot z = u^2 z$.

Except in the case of the standard $\mathbf{S}^1$-action, this gives an open subset of W on which the SU(2)-action is not effective, in contradiction with the assumptions made (see, if necessary, Exercise I.7).

The dimension-2 singular orbits are thus spheres. Moreover, such a sphere cannot be Lagrangian. Otherwise, it would have a neighborhood both of the form $\mathrm{SU}(2) \times_{\mathbf{S}^1} \mathbf{R}^2$, with $\mathbf{S}^1$ acting on $\mathbf{R}^2$ by rotations (the slice theorem, plus effectiveness) and of the form $\mathsf{T}^\star \mathbf{S}^2$ (the Darboux theorem, see § II.1.c). But these two bundles over $\mathbf{S}^2$ are not diffeomorphic (as manifolds), the former has Euler class ± 1 while the Euler class of the latter is ± 2.

Notice eventually that a singular orbit which is not mapped to 0 by μ is a sphere and is mapped to a symplectic sphere $\mathbf{S}^2 \subset \mathfrak{su}(2)^\star$ by the immersive Poisson mapping μ ($\mathsf{T}_x\mu$ is injective when restricted to the orbit) and hence is a symplectic submanifold of W.

We summarize these remarks in a proposition.

Proposition IV.5.1. *A singular orbit of an effective symplectic* SU(2)*-action on a 4-dimensional compact manifold with somewhere submersive momentum mapping* μ *is*

- *either an isolated fixed point, mapped to* 0 *by* μ,
- *or a symplectic sphere mapped on a nontrivial sphere in* $\mathfrak{su}(2)^\star$. □

Remark IV.5.2. Notice that we have proved that there is no Lagrangian orbit. In the case of the group SO(3), such orbits can appear. For example if we make it act as a subgroup (the real part) of SU(3) ⊂ U(3) on $\mathbf{P}^2(\mathbf{C})$, the real part $\mathbf{P}^2(\mathbf{R})$

is a Lagrangian orbit [**70**]. This is the small difference which makes the investigation of SU(2)-actions a little easier than that of SO(3)-actions. See, however, Exercise IV.18.

To deduce a classification theorem, we look at the "square of the momentum mapping" function.

Lemma IV.5.3. *The function* $f = \frac{1}{2}\|\mu\|^2 : W \to [0,+\infty[$ *is a quotient map for the* SU(2)*-action on* W.

Proof. We know, thanks to the slice theorem and to the study of orbits and their stabilizers above that the quotient will be a manifold with boundary, of dimension 1, the ends corresponding to the singular orbits. If we assume W to be connected, this must be a compact interval and there are exactly two singular orbits.

The function f descends to the quotient as

$$g : W/\operatorname{SU}(2) \longrightarrow [0,+\infty[,$$

a smooth function from one interval to another. Computing its tangent mapping

$$\mathsf{T}_x f(Y) = \mu(x)\cdot \mathsf{T}_x\mu(Y),$$

we see that, if x is a fixed point with $\mu(x) \neq 0$, then

$$Y \in \operatorname{Ker} \mathsf{T}_x f \Leftrightarrow \mathsf{T}_x\mu(Y) \in \mu(x)^\perp = \mathsf{T}_{\mu(x)}\mathbf{S}^2$$

where $\mathbf{S}^2$ is the coadjoint orbit of $\mu(x)$. Hence $\mathsf{T}_x f$ vanishes exactly when x is on a singular orbit, and g is strictly monotone, which shows that f may be identified with the quotient mapping. □

Remark IV.5.4. The map

$$\begin{aligned} \mathfrak{su}(2)^\star &\longrightarrow [0,+\infty[\\ A &\longmapsto \tfrac{1}{2}\|A\|^2 \end{aligned}$$

is a quotient map itself (see Exercise I.16).

Let $[a,b] = f(W)$ and let $c \in\,]a,b[$. Then the maximum $f^{-1}(b)$ is a symplectic sphere, $f^{-1}([c,b])$ is a tubular neighborhood of this sphere, given by the slice theorem and so this is an $E(\pm m)$ used in our list above ($|m|$, odd, is the order of the principal stabilizer). There are only two possibilities:

- Either $a > 0$, then $f^{-1}([a,c])$ is an $E(n)$ as well, necessarily $|m| = |n|$ and actually $n = -m$ in order to be able to glue, so that W is a Hirzebruch surface W_m.
- Or $a = 0$, then $f^{-1}(a)$ is an isolated fixed point near which the SU(2)-action is the standard linear action on $\mathbf{C}^2$. In particular, $|m| = 1$ and W is obtained by gluing a disc to $E_{\pm 1}$) so that W is a copy of $\mathbf{P}^2(\mathbf{C})$.

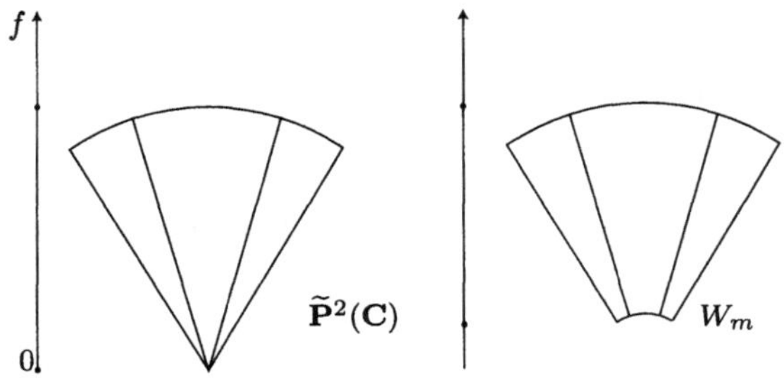

Figure 10

Theorem IV.5.5. *A compact connected symplectic 4-manifold endowed with an effective symplectic* SU(2)*-action with a somewhere submersive momentum mapping is equivariantly diffeomorphic with*

- *either* $\mathbf{P}^2(\mathbf{C})$ *(with the linear* SU(2)*-action) if there is a fixed point,*
- *or to the Hirzebruch surface* W_m *(diffeomorphic with* $\widetilde{\mathbf{P}}^2(\mathbf{C})$*) for some odd positive integer* m*, the order of the principal stabilizer* $\mathbf{Z}/m$ *if there is no fixed point.* □

This can be schematized as in Figure 10, a figure which should seem even more accurate after having solved Exercise IV.6.

Remark IV.5.6. The function $f = \frac{1}{2}\|\mu\|^2$, when

$$\mu : W \longrightarrow \mathfrak{g}^\star$$

is the momentum mapping of the action of a compact Lie group G on the symplectic manifold W, is not, in general, a Morse–Bott function (although, it is one in the examples considered here). It nevertheless has minimal degeneracies (in a precise sense), investigated by Frances Kirwan [**85**] who used it as the main tool in, for instance, her proof of the noncommutative convexity theorem [**86**].

Exercises

Exercise IV.1. The function defined on $\mathbf{P}^2(\mathbf{C})$ by the formula

$$f([x,y,z]) = \frac{1}{2}\frac{m|x|^2 + n|y|^2}{|x|^2+|y|^2+|z|^2}$$

belongs to the class studied in Example IV.1.2. Prove that, if m and $n \in \mathbf{Z}$, f is a periodic Hamiltonian.

Exercise IV.2. Determine the momentum mapping and its image for the $\mathbf{T}^2$-action on $\mathbf{P}^1(\mathbf{C}) \times \mathbf{P}^1(\mathbf{C})$ by

$$(u, v) \cdot ([a, b], [x, y]) = ([a, ub], [x, vy]).$$

Exercise IV.3. Determine the momentum mapping and its image[(10)] for the $\mathbf{T}^2$-action on $\mathbf{P}^2(\mathbf{C})$ by

$$(u, v) \cdot [x, y, z] = [x, uy, v^2 z].$$

Describe the fixed points and their stabilizers.

***Exercise IV.4* (Hirzebruch surfaces).** Let $k \in \mathbf{N}$ be an integer. Consider the subset of $\mathbf{P}^1(\mathbf{C}) \times \mathbf{P}^2(\mathbf{C})$

$$W_k = \left\{([a, b], [x, y, z]) \in \mathbf{P}^1(\mathbf{C}) \times \mathbf{P}^2(\mathbf{C}) \mid a^k y = b^k x\right\}.$$

Prove that W_k is a complex submanifold of $\mathbf{P}^1(\mathbf{C}) \times \mathbf{P}^2(\mathbf{C})$ and that the restriction of the projection $\mathbf{P}^1(\mathbf{C}) \times \mathbf{P}^2(\mathbf{C}) \to \mathbf{P}^1(\mathbf{C})$ to W_k is a bundle over $\mathbf{P}^1(\mathbf{C})$ with fiber $\mathbf{P}^1(\mathbf{C})$.

Endow $\mathbf{P}^1(\mathbf{C})$, $\mathbf{P}^2(\mathbf{C})$ and hence $\mathbf{P}^1(\mathbf{C}) \times \mathbf{P}^2(\mathbf{C})$ with the standard symplectic forms, so that W_k is a symplectic submanifold.

Let the torus $\mathbf{T}^2$ act on $\mathbf{P}^1(\mathbf{C}) \times \mathbf{P}^2(\mathbf{C})$ by

$$(u, v) \cdot ([a, b], [x, y, z]) = ([ua, b], [u^k x, y, vz]).$$

Prove that this is a Hamiltonian action with momentum mapping

$$\mu([a, b], [x, y, z]) = \left(\frac{1}{2}\left(\frac{|a|^2}{|a|^2 + |b|^2} + k\frac{|x|^2}{|x|^2 + |y|^2 + |z|^2}\right), \frac{1}{2}\frac{|z|^2}{|x|^2 + |y|^2 + |z|^2}\right).$$

Check that the action restricts to a $\mathbf{T}^2$-action on W_k and determine the image of the momentum mapping

$$\mu : W_k \longrightarrow \mathbf{R}^2.$$

Hint: determine the fixed points of the action. The result is shown in Figure 11.

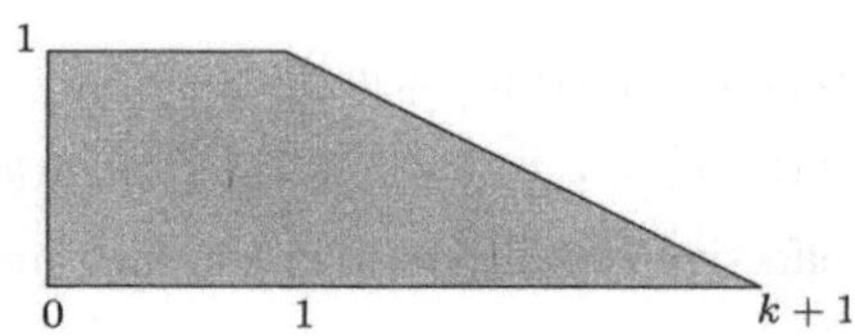

Figure 11

[(10)] See Figure 7.

***Exercise IV.5* (Blowing up the projective space).** The complex projective space $\mathbf{P}^n(\mathbf{C})$ is the space of all lines through 0 in $\mathbf{C}^{n+1}$. Consider

$$\widetilde{\mathbf{P}}^n(\mathbf{C}) = \left\{(d,\ell) \mid d \text{ is a line in } \mathbf{C}^n, \ell \subset \mathbf{C} \oplus d \subset \mathbf{C} \oplus \mathbf{C}^n = \mathbf{C}^{n+1}\right\},$$

a subset of $\mathbf{P}^{n-1}(\mathbf{C}) \times \mathbf{P}^n(\mathbf{C})$. Using homogeneous coordinates

$$([x_1,\ldots,x_n],[y_0,\ldots,y_n]) \in \mathbf{P}^{n-1}\mathbf{C} \times \mathbf{P}^n(\mathbf{C}),$$

check that $\widetilde{\mathbf{P}}^n(\mathbf{C})$ is the complex algebraic submanifold defined by the $n-1$ equations $x_1y_i - x_iy_1 = 0$ and thus that this is a symplectic submanifold (as usual, the two projective spaces are endowed with their standard symplectic forms). Consider the projection

$$\begin{aligned} \pi : \widetilde{\mathbf{P}}^n(\mathbf{C}) &\longrightarrow \mathbf{P}^n(\mathbf{C}) \\ (d,\ell) &\longmapsto \ell. \end{aligned}$$

Describe all its fibers. It is said that $\widetilde{\mathbf{P}}^n(\mathbf{C})$ is $\mathbf{P}^n(\mathbf{C})$ *blown up*(11) at the point $[1,0,\ldots,0]$. Notice that the Hirzebruch surface W_1 of exercise IV.4 is the case $n=2$, namely the projective plane blown up at a point.

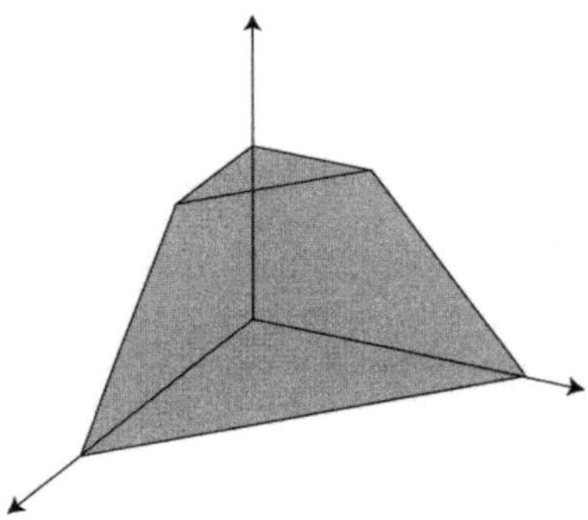

Figure 12

The torus $\mathbf{T}^n$ acts on $\widetilde{\mathbf{P}}^n(\mathbf{C})$ by

$$\begin{aligned}(u_1,\ldots,u_{n-1},v)\cdot([x_1,\ldots,x_n],[y_0,\ldots,y_n]) \\ = ([u_1x_1,\ldots,u_{n-1}x_{n-1},x_n],[vy_0,u_1y_1,\ldots,u_{n-1}y_{n-1},y_n]).\end{aligned}$$

Prove that this is an effective Hamiltonian action with momentum mapping

$$\mu([x_1,\ldots,x_n],[y_0,\ldots,y_n]) = \frac{1}{2}\left(\left(\frac{|x_j|^2}{\sum |x_i|^2} + \frac{|y_j|^2}{\sum |y_i|^2}\right)_{1\leqslant j\leqslant n-1}, \frac{|y_0|^2}{\sum |y_i|^2}\right).$$

Determine the image of this momentum mapping (Figure 12).

(11) See also Figure 2 in Chapter VI.

***Exercise IV.6* (Blowing up a fixed point of an SU(2)-action).** Let SU(2) act on $\mathbf{C}^2$ in the standard (linear) way. Prove that there exists an SU(2)-action on the blow up $\widetilde{\mathbf{C}}^2$ such that

- the map $\widetilde{\mathbf{C}}^2 \to \mathbf{C}^2$ is equivariant,
- the group SU(2) has no fixed point in $\widetilde{\mathbf{C}}^2$.

Let W be a 4-dimensional symplectic SU(2)-manifold. Prove that there exists a symplectic manifold $\widetilde{W}$ with an equivariant mapping $\widetilde{W} \to W$ such that SU(2) has no fixed point in $\widetilde{W}$. Compare the momentum mappings for the two manifolds.

***Exercise IV.7* (Symplectic cuts and images of momentum mappings).** We use the notation and results of Exercise III.19. Assume the manifold W we are cutting is endowed with the Hamiltonian action of a Lie group G which commutes with the $\mathbf{S}^1$-action used to cut. Prove that the symplectic cuts are still endowed with a Hamiltonian G-action.

Assume now that G is a torus $\mathbf{T}$. Let

$$\mu : W \longrightarrow \mathfrak{t}^\star$$

be the momentum mapping. Let $X \in \mathfrak{t}$ be an element generating a circle $\mathbf{S}_X \subset \mathbf{T}$. We use the $\mathbf{S}_X$-action to cut our manifold, at the level, t, of the periodic Hamiltonian $H = \langle \mu, X \rangle$. Call $W_{X,t}$ the symplectic cut. Prove that the image of the momentum mapping for the $\mathbf{T}$-action on $W_{X,t}$ is the *cut* polyhedron

$$\mu(W) \cap \{\xi \in \mathfrak{t}^\star \mid \langle \xi, X \rangle \geqslant t\}.$$

Symplectic versus complex blowing up. Assume now that the action is effective and completely integrable. Let p be a vertex of the polyhedron $P = \mu(W)$. Choose t close enough to $\langle \mu(p), X \rangle$ so that the polyhedron

$$P \cap \{\xi \in \mathfrak{t}^\star \mid \langle \xi, X \rangle \leqslant t\}$$

is an n-simplex(12) with a vertex at p. What is the corresponding symplectic cut?

The polyhedron obtained in Exercise IV.5 and shown in Figure 12 can be obtained by cutting the standard simplex by a hyperplane. Check that $\widetilde{\mathbf{P}}^n(\mathbf{C})$ is, indeed a symplectic cut of $\mathbf{P}^n(\mathbf{C})$... these were just two different points of view on the projective space blown up at a point.

Construct a 4-dimensional symplectic manifold endowed with an effective completely integrable torus action the image of the momentum mapping of which is the hexagon shown in Figure 13.

Recall from the proof of Theorem IV.4.25 that the symplectic volume of a manifold endowed with an effective completely integrable action is the volume of

(12) Recall that P is primitive (Proposition IV.4.17).

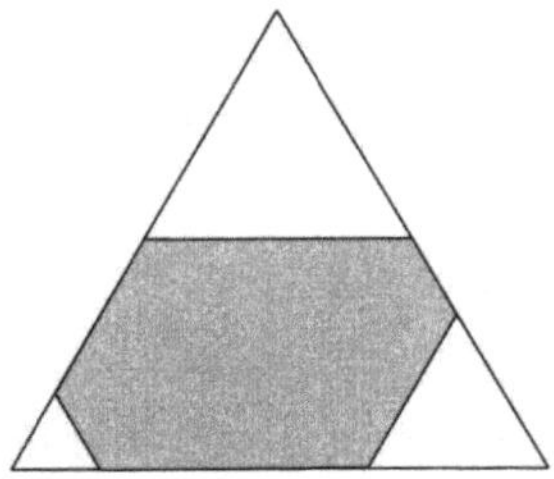

Figure 13

the image polyhedron. What can be said of the volume of a symplectic blow up[13] and more generally of a symplectic cut?

Exercise IV.8. The torus

$$\mathbf{T}^2 = \left\{ \begin{pmatrix} \cos\theta & -\sin\theta & & \\ \sin\theta & \cos\theta & & 0 \\ & 0 & \cos\varphi & -\sin\varphi \\ & & \sin\varphi & \cos\varphi \end{pmatrix} \mid \theta, \varphi \in \mathbf{R} \right\}$$

is a maximal torus in SO(4). Prove that the inclusion $\mathbf{T}^2 \subset \mathrm{SO}(4)$ induces the projection map

$$\begin{aligned} \mu : \mathfrak{so}(4)^\star &\longrightarrow (\mathfrak{t}^2)^\star = \mathbf{R}^2 \\ \textstyle\sum_{i<j} a_{i,j} e_i^\star \wedge e_j^\star &\longmapsto (a_{1,2}, a_{3,4}). \end{aligned}$$

This torus $\mathbf{T}^2$ acts on the (co-)adjoint orbits by conjugation. Let $\mathcal{O}$ be an orbit with invariants

$$\|A\|^2 = f_1 \text{ and } \mathrm{Pf}(A) = f_2$$

(see, if necessary, Exercise I.20) or

$$\|A_+\|^2 = \lambda_1^2 = f_1 + 2f_2 \text{ and } \|A_-\|^2 = \lambda_2^2 = f_1 - 2f_2.$$

Determine the image of the momentum mapping for the $\mathbf{T}^2$-action.

Exercise IV.9. More generally (this is in reference to Exercise IV.8 above), consider the quadric

$$Q_n : \sum_{j=0}^{n+1} z_j^2 = 0 \text{ in } \mathbf{P}^{n+1}(\mathbf{C}).$$

[13] in contradiction with this unavoidable terminology

Recall (from Exercises I.21 and III.15) that this is the Grassmannian $\widetilde{G}_2(\mathbf{R}^{n+2})$ and that this is also the coadjoint orbit of the matrix

$$A = \left(\begin{array}{cc|c} 0 & -1 & \\ 1 & 0 & 0 \\ \hline & & \\ 0 & & 0 \\ & & \end{array}\right) \in \mathfrak{so}(n+2)^\star.$$

Let $k = \left[\dfrac{n+2}{2}\right]$, so that $\mathbf{R}^{2k} \subset \mathbf{R}^{n+2}$ (as the $2k$ first factors if n is odd) and similarly

$$\mathrm{SO}(2)^{2k} \hookrightarrow \mathrm{SO}(n+2),$$

endowing our $2n$-dimensional symplectic manifold with a Hamiltonian $\mathbf{T}^k$-action. Prove that the momentum mapping of this action is given by

$$[z_0, \ldots, z_{n+1}] \longmapsto \frac{1}{\sum |z_j|^2}(x_0 y_1 - x_1 y_0, \ldots, x_{2k-2} y_{2k-1} - y_{2k-2} x_{2k-1})$$

(in which expression, $z_j = x_j + i y_j$) in the first description and by

$$A = (a_{i,j}) \longmapsto (a_{1,2}, a_{3,4}, \ldots, a_{2k-1,2k})$$

in the second.

Determine the fixed points of the action and prove that the image of the momentum mapping is the "octahedron", convex hull of the $2k$ points $\pm e_i \in \mathbf{R}^k$.

Let $0 < \lambda_1 < \cdots < \lambda_k$ be k real numbers. Prove that the function

$$f = \frac{2}{\sum |z_j|^2}\left(\lambda_1(x_0 y_1 - x_1 y_0) + \cdots + \lambda_k(x_{2k-2} y_{2k-1} - y_{2k-2} x_{2k-1})\right)$$

is a perfect Morse function and that it has

- if n is odd, exactly one critical point of index $2i$ for any i between 0 and n, while,
- if n is even, there are two points with the same index, which is half the dimension.

Exercise IV.10. Fix three real numbers $\lambda_1 < \lambda_2 < \lambda_3$. Let $\lambda = (\lambda_1, \lambda_2, \lambda_3)$, so that $\mathcal{H}_\lambda$ is the manifold of complete flags in $\mathbf{C}^3$. Check that the action of the torus

$$\mathbf{T}^2 = \left\{ \begin{pmatrix} t_1 & 0 & 0 \\ 0 & t_2 & 0 \\ 0 & 0 & t_3 \end{pmatrix} \mid |t_i| = 1 \text{ and } t_1 t_2 t_3 = 1 \right\}$$

by conjugation on $\mathcal{H}_\lambda$ is effective and that the image of the momentum mapping is indeed the hexagon shown in Figure 5. Look now at the (more precise) Figure 14, in which you should find the inspiration to answer the following questions.

- Determine the regular values of the momentum mapping.

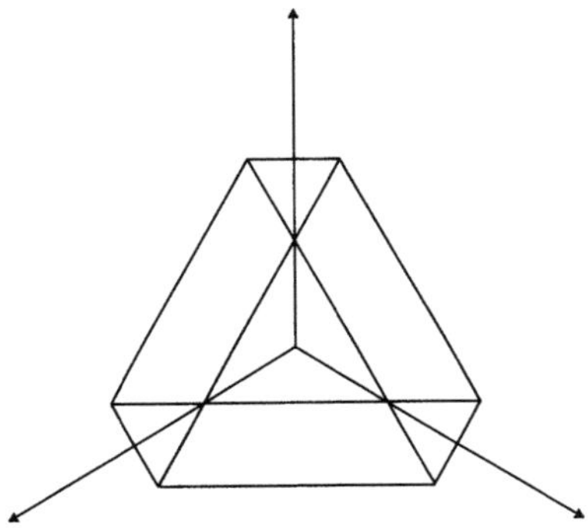

Figure 14

– What are the pre-images of all the segments (either on the boundary of the hexagon or not) drawn in Figure 14?

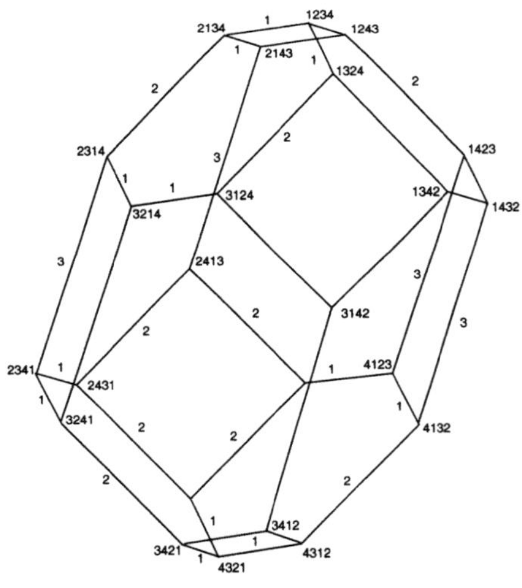

Figure 15. Permutahedron

***Exercise IV.11* (Permutahedra).** The permutahedron P_n is the convex hull of the points[(14)] of coordinates $(\sigma(1), \ldots, \sigma(n))$, $\sigma \in \mathfrak{S}_n$. Prove that this is a polyhedron of dimension $n-1$ with $n!$ vertices. Draw P_2 (this is a segment) and P_3 (this is a hexagon, actually one of the hexagons shown in Figure 16). Check that Figure 15, in which

---

(14) For more information on permutahedra, see for instance [**52, 99**].

– the label at a vertex is the coordinates of this vertex in $\mathbf{R}^4$,
– the label on an edge is the length of this segment (the unit being $\sqrt{2}$),
– the hexagonal faces are of the type shown in Figure 16

shows the actual permutahedron P_4 while the hexagon on the left in Figure 16 depicts P_3. Check that the three pictures here represent the diagonals of the matrices

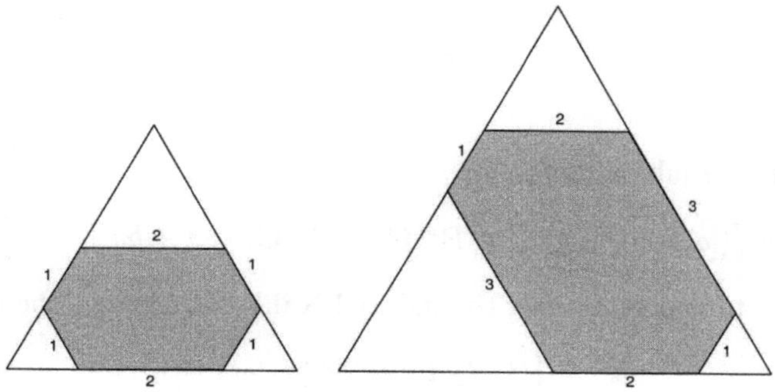

Figure 16. Faces of the permutahedron

in $\mathcal{H}_{1,2,3,4}$, $\mathcal{H}_{1,2,3}$, $\mathcal{H}_{1,2,4}$.

Exercise IV.12. Let $\mu : W \to \mathbf{R}^n$ be the momentum mapping for the effective completely integrable action of a torus on a compact connected symplectic manifold. Prove that μ can be considered as the quotient map

$$W \longrightarrow W/\mathbf{T} \cong P = \mu(W)$$

of the torus action.

Exercise IV.13. Determine all the primitive triangles and quadrilaterals in $\mathbf{R}^2$. Assume W is a 4-dimensional symplectic manifold endowed with an effective Hamiltonian $\mathbf{T}^2$-action with not more than four fixed points. Prove that W is (equivariantly) diffeomorphic with a complex projective plane or to a Hirzebruch surface (Exercise IV.4).

Exercise IV.14. Figure 17 shows a tetrahedron the vertices of which are some of the vertices of a unit cube. Prove that this tetrahedron is not a primitive polyhedron.

Exercise IV.15. Let $\mathcal{D}$ be the manifold of complete flags

$$0 \subset \ell \subset P \subset \mathbf{C}^3$$

in $\mathbf{C}^3$. Check that $\mathcal{D}$ can be identified with the submanifold

$$\mathcal{D} = \left\{([v],[\varphi]) \in \mathbf{P}(\mathbf{C}^3) \times \mathbf{P}(\mathbf{C}^3)^\vee \mid \varphi(v) = 0\right\}$$

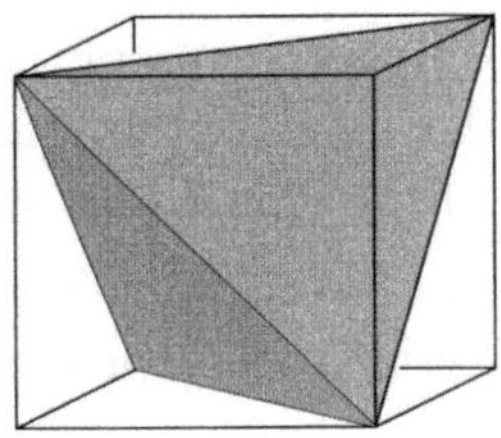

Figure 17

and thus with the submanifold

$$\mathcal{D} = \left\{([a,b,c],[x,y,z]) \in \mathbf{P}^2(\mathbf{C}) \times \mathbf{P}^2(\mathbf{C}) \mid ax+by+cz=0\right\}.$$

This allows us to endow $\mathcal{D}$ with the induced Kähler structure. The torus

$$\mathbf{T}^2 = \left\{(t_1,t_2,t_3) \in \mathbf{T}^3 \mid t_1t_2t_3 = 1\right\}$$

acts on $\mathcal{D}$ by

$$(t_1,t_2,t_3)\cdot([a,b,c],[x,y,z]) = ([t_1a,t_2b,t_3c],[t_1^{-1}x,t_2^{-1}y,t_3^{-1}z]).$$

Check that this is a Hamiltonian action with momentum mapping

$$\begin{aligned}\mu(&[a,b,c],[x,y,z])\\ &= \frac{1}{2}\left(\frac{|a|^2}{|a|^2+|b|^2+|c|^2} - \frac{|x|^2}{|x|^2+|y|^2+|z|^2}, \frac{|b|^2}{|a|^2+|b|^2+|c|^2} - \frac{|y|^2}{|x|^2+|y|^2+|z|^2},\right.\\ &\qquad\qquad \left.\frac{|c|^2}{|a|^2+|b|^2+|c|^2} - \frac{|z|^2}{|x|^2+|y|^2+|z|^2}\right)\end{aligned}$$

and determine the image of μ. Explain why you find the same image as in Figure 14.

Prove that the (complexified) action of the complex torus

$$(\mathbf{C}^\star)^2 = \left\{(z_1,z_2,z_3) \in \mathbf{C}^3 \mid z_1z_2z_3 = 1\right\}$$

is by

$$(z_1,z_2,z_3)\cdot([a,b,c],[x,y,z]) = ([z_1a,z_2b,z_3c],[z_1^{-1}x,z_2^{-1}y,z_3^{-1}z]).$$

What is the image of the closure of the orbit of the point $([1,1,0],[1,-1,1])$? Investigate the relation with the Hirzebruch surface W_1 of Exercise IV.4 and the images given by Figures 11 and 12.

Exercise IV.16. Using Exercise III.8, write an explicit formula for the momentum mapping for the action of $\mathrm{SU}(2)$ on W_m depicted in § IV.5.a.

***Exercise IV.17* (Hirzebruch surfaces—continuation).** Using the notation of § IV.5.a, prove that the mapping

$$\begin{aligned} E_{-m} &\longrightarrow \mathbf{P}^1(\mathbf{C}) \times \mathbf{P}^2(\mathbf{C}) \\ [(a,b),z] &\longmapsto ([a,b],[z^{-m}a, z^{-m}b, 1]) \end{aligned}$$

extends to W_m and identifies it with the Hirzebruch surface

$$W_m = \{([a,b],[x,y,z]) \mid a^m y - b^m x = 0\}.$$

Prove that W_k (Exercise IV.4) may be identified with $\mathbf{P}(\mathcal{O}(k) \oplus \mathbf{1})$. Deduce that $\mathbf{P}(\mathcal{O}(1) \oplus \mathbf{1})$ is $\mathbf{P}^2(\mathbf{C})$ blown up at one point. In the local chart $b \neq 0$ in $\mathbf{P}^1(\mathbf{C})$, put $u = a/b$ ($\in \mathbf{C}$) and

$$f(t) = \frac{1}{1+t^{-k}}.$$

Prove that

$$\begin{aligned} \varphi : \mathbf{C} \times \mathbf{P}^1(\mathbf{C}) &\longrightarrow W_k \\ (u,[v,w]) &\longmapsto ([u,1],[f(|u|)\overline{u}^k v, f(|u|)v, w]) \end{aligned}$$

defines a local trivialization of the bundle $W_k \to \mathbf{P}^1(\mathbf{C})$. Similarly, write a trivialization ψ in "the other" local chart ($a \neq 0$), in such a way that

$$\psi^{-1} \circ \varphi(u,[v,w]) = \left(u, \left[\frac{\overline{u}^k}{|u|^{-k}} v, w\right]\right).$$

Thus W_k is obtained by gluing two copies of $D^2 \times \mathbf{P}^1(\mathbf{C})$ by

$$\begin{aligned} g : \mathbf{S}^1 \times \mathbf{P}^1(\mathbf{C}) &\longrightarrow \mathbf{S}^1 \times \mathbf{P}^1(\mathbf{C}) \\ (z,[v,w]) &\longmapsto (z,[\overline{z}^k v, w]). \end{aligned}$$

Show that

$$z \longmapsto ([v,w] \longmapsto [\overline{z}^k v, w])$$

defines a loop in SO(3), homotopic to the constant loop when k is odd. Applying the result to $m-n$, show that W_m is diffeomorphic to W_n if $m \equiv n \bmod 2$. Deduce that the manifold $\mathbf{P}(\mathcal{O}(k) \oplus \mathbf{1})$ is diffeomorphic to $\mathbf{S}^2 \times \mathbf{S}^2$ when k is even and to $\mathbf{P}^2(\mathbf{C})$ blown up at one point when k is odd.

Exercise IV.18. Consider the Hamiltonian SO(3)-action on $\mathbf{P}^2(\mathbf{C})$ described in Exercise III.14 (with $n = 3$). Prove that $\mathbf{P}^2(\mathbf{R})$ is a Lagrangian submanifold (an illustration of Exercise II.31) and an orbit of the SO(3)-action. Describe the critical points of the function

$$f = \frac{1}{2}\|\mu\|^2 : \mathbf{P}^2(\mathbf{C}) \longrightarrow \mathbf{R}.$$

Imitating the proof of Theorem IV.5.5, prove that a compact connected symplectic manifold of dimension 4 endowed with an effective Hamiltonian action of SO(3) (the momentum mapping of which is somewhere submersive) is diffeomorphic either with $\mathbf{P}^2(\mathbf{C})$ or with $\mathbf{P}^1(\mathbf{C}) \times \mathbf{P}^1(\mathbf{C})$, and is equivariantly diffeomorphic with $\mathbf{P}^2(\mathbf{C})$ or with one of the Hirzebruch surfaces W_m for m even. See **[70]**.

Exercise IV.19. Let V be a compact symplectic manifold endowed with a periodic Hamiltonian H. Consider the Hamiltonian $\mathbf{S}^1$-action on $\mathbf{C}^2 \times V$,

$$u \cdot (z, x) = (uz, u^m \cdot x),$$

the Hamiltonian of which is the function

$$\widetilde{H}(z, x) = \frac{1}{2} |z|^2 + mH(x).$$

Check that the SU(2)-action on $\mathbf{C}^2 \times V$ defined by the linear action on $\mathbf{C}^2$:

$$A \cdot (z, x) = (A \cdot z, x)$$

preserves the levels of $\widetilde{H}$ and that it defines a Hamiltonian SU(2)-action on the reduced manifolds $W_\alpha = \widetilde{H}^{-1}(\alpha)/\mathbf{S}^1$ of the regular levels of $\widetilde{H}$. What is the momentum mapping μ of this action? Assume that $\alpha > m \sup_{x \in V} H(x)$. Prove that the critical values of $\|\mu\|^2$ are those of H.

Check that, in dimension 4, all the SU(2)-manifolds depicted in Theorem IV.5.5 can be obtained in this way.

Prove that, if $n \geqslant 3$ and $k \geqslant 2$, there exists a compact symplectic manifold of dimension $2n$ endowed with a Hamiltonian SU(2)-action such that $\|\mu\|^2$ has k critical values.

CHAPTER V

MODULI SPACES OF FLAT CONNECTIONS

In this chapter, we consider the moduli space of flat connections on a (trivial) bundle over a surface (defined in § V.1). There is a Hamiltonian group action that allows us to endow this space with a Poisson structure, as we show in § V.2 (this is due to Atiyah and Bott [**10**] in the case of a closed surface and to Fock and Roslyi [**50**] in the general case of a surface with boundary). We then look at a special case in § V.3, in which there is an integrable system (due to Goldman [**56**]) on the moduli space. Following Jeffrey and Weitsman [**77**], we exhibit a torus action and its momentum mapping, that is, action-angle variables for this system.

V.1. The moduli space of flat connections

The moduli space that I want to consider is that of flat connections on the trivial principal G-bundle on a surface Σ, namely on the bundle

$$P = \mathrm{G} \times \Sigma \longrightarrow \Sigma$$

endowed with the natural action of G on itself by left translations. The surface Σ is supposed to be compact and oriented, but may (and will often) have a boundary $\partial\Sigma$. The genus of Σ will be denoted by g and the number of boundary components by d.

The group G is a Lie group. I will use (and thus assume the existence of) a nondegenerate symmetric bilinear form $\mathcal{B}$ on the Lie algebra $\mathfrak{g}$ which is invariant under the adjoint G-action. Such a form certainly exists when G is a compact Lie group. The example I have in mind is mainly that of $\mathrm{SU}(n)$, so that I will usually assume in addition that G is simply connected—although I will also use $\mathbf{S}^1$.

I will consider only the *trivial* bundle $P = \mathrm{G} \times \Sigma$. Notice that this is not a very big restriction: when the group G is simply connected, all the principal G-bundles on Σ are trivializable. They are classified[(1)] by the homotopy classes

[(1)] See § VI.1 for the definition and construction of the classifying space BG.

of maps $\Sigma \to BG$. If $\pi_1(G)$ is trivial, then because π_2 of a group is always trivial (this is a theorem of Cartan [**33**]), G is 2-connected and BG is 3-connected. For a *surface* Σ, any continuous map $\Sigma \to BG$ is thus homotopic to a constant map and any principal G-bundle is trivializable. However, in the case of $\mathbf{S}^1$, we know perfectly well that there *are* nontrivial principal $\mathbf{S}^1$-bundles over surfaces. In any case, to simplify notation, we will assume that a trivialization is given and fixed.

Being trivialized, our bundle carries a canonical "trivial" connection[(2)], the horizontal subspace of which at $(g, x) \in G \times \Sigma$ is

$$\{0\} \times T_x\Sigma \subset T_gG \times T_x\Sigma = T_{(g,x)}(G \times \Sigma).$$

Notice that, although the space of connections has only the natural structure of an affine space, the space of connections on the trivial bundle can be considered as a vector space (see § V.4): the chosen trivialization is used as an origin.

The curvature mapping. Following Atiyah and Bott, we will consider the curvature as a mapping F from the vector space of connections on P to that of 2-forms with values in $\mathfrak{g}$,

$$F : \Omega^1(\Sigma, \mathfrak{g}) \longrightarrow \Omega^2(\Sigma, \mathfrak{g}).$$

We shall need to compute its tangent mapping.

Proposition V.1.1. *Let A be a connection and $\varphi \in \Omega^1(\Sigma, \mathfrak{g})$. Then*

$$F(A + \varphi) = F(A) + d_A\varphi + [\varphi, \varphi].$$

Corollary V.1.2. *For any connection A and any $\mathfrak{g}$-valued 1-form φ, one has*

$$T_A(\varphi) = d_A\varphi. \qquad \square$$

Proof of the proposition. Using the definition of F *via* the vertical projection V (see § V.4),

$$F(A + \varphi)(X, Y) = V_{A+\varphi}([h_{A+\varphi}(X), V_{A+\varphi}(Y)]).$$

Obviously (see Figure 11), for any vector field X on Σ,

$$h_{A+\varphi}(X) = h_A(X) + \varphi(X)$$

and for any invariant vector field Z on P,

$$V_{A+\varphi}(Z) = V_A(Z) - \varphi(\pi(Z)).$$

(2) The basic definitions and results used here are gathered in § V.4

A direct computation gives, for $F(A+\varphi)(X,Y)$,

$$\begin{aligned}
&V_{A+\varphi}\left([h_A(X),h_A(Y)]+[h_A(X),\varphi(Y)]+[\varphi(X),h_A(Y)]+[\varphi(X),\varphi(Y)]\right)\\
&= V_A[h_A(X),h_A(Y)]-\varphi[X,Y]\\
&\qquad + V_A[h_A(X),\varphi(Y)]-\varphi\circ\pi[h_A(X),\varphi(Y)]\\
&\qquad + V_A[\varphi(X),h_A(Y)]-\varphi\circ\pi[\varphi(X),h_A(Y)]\\
&\qquad + [\varphi(X),\varphi(Y)]\\
&= F(A)(X,Y)+V_A[h_A(X),\varphi(Y)]-V_A[h_A(Y),\varphi(X)]\\
&\qquad - \varphi[X,Y]+[\varphi(X),\varphi(Y)]\\
&= F(A)(X,Y)+\nabla^A_X\varphi(Y)-\nabla^A_Y(X)-\varphi[X,Y]+[\varphi(X),\varphi(Y)]\\
&= F(A)(X,Y)+d_A\varphi(X,Y)+[\varphi,\varphi](X,Y).
\end{aligned}$$

□

Flat (integrable) connections. A *flat* connection is a connection without curvature (that is, an A such that $F(A)=0$). Notice that this means that the bracket of any two horizontal vector fields is again a horizontal vector field, so that the horizontal distribution is integrable in the (Frobenius) sense that it is at each point, the tangent distribution to a submanifold through this point. This is why flat connections are sometimes called *integrable*.

The gauge group. The *gauge group* is the group $\mathcal{G}$ of automorphisms of the bundle. Still using the trivialization, we can identify it with the (functional) group of all the mappings $\Sigma \to \mathrm{G}$. Similarly, its Lie algebra $\mathbf{g}$ is that of all mappings $\Sigma \to \mathfrak{g}$. The gauge group acts on $\mathcal{A}$, (traditionally on the right) by

$$g\cdot A = g^{-1}Ag + g^{-1}dg$$

where "$g^{-1}dg$" is a symbolic notation for the 1-form on Σ with values in $\mathfrak{g}$ defined by

$$\mathsf{T}_x\Sigma \ni X \longmapsto \mathsf{T}_x g(X) \in \mathsf{T}_{g(x)}\mathrm{G} \xrightarrow{\text{“}g^{-1}\text{”}} \mathfrak{g}.$$

To get the fundamental vector field $\underline{\alpha}$ associated to both the element $\alpha \in \mathbf{g}$ and the $\mathcal{G}$-action, we only need to differentiate. We find that this is the vector field, defined on $\mathcal{A}$ by

$$\underline{\alpha}_A = d_A\alpha \in \Omega^1(\Sigma,\mathfrak{g}) = \mathsf{T}_A\mathcal{A}.$$

The holonomy. Assume we are given a connection A on our bundle P. Let p be any point in P and x be its image in Σ. For any differentiable loop γ based at x in Σ, the existence and uniqueness theorem for differential equations gives a unique lift $\widetilde{\gamma}$ starting at p and horizontal in the sense that the tangent vector $\dot{\widetilde{\gamma}}(t)$

is horizontal for any t. This lifted path ends at some point of the fiber of x, which can be written $g \cdot p$ for some element $g \in \mathrm{G}$, which is the *holonomy* of the loop γ.

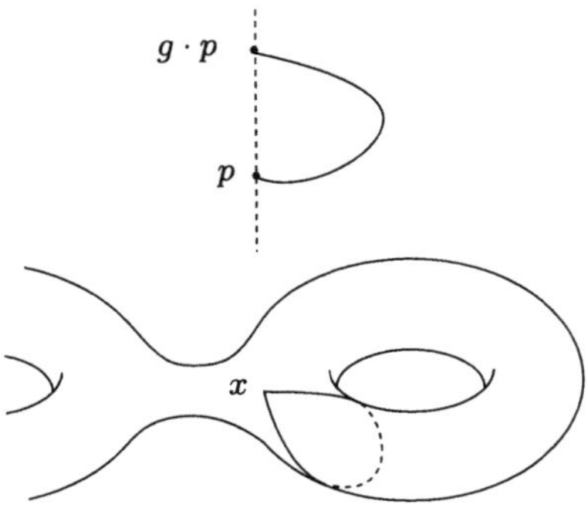

Figure 1

The subgroup of G consisting of all elements obtained this way is called the *holonomy group* of the connection A at the point p. Using the fact that our spaces are path-connected, it is easily seen that all the holonomy groups (when p varies in P) are conjugated subgroups of G. This is why it is usually spoken of as "the" holonomy group.

Assume now that the connection A is *flat*. As we have noticed above, the horizontal distribution is integrable, so that any differentiable map from a disc to the surface Σ can be lifted horizontally. Hence any loop γ which is homotopic to a constant loop in Σ lifts to a *loop* in P and thus has trivial holonomy.

The holonomy (or monodromy) representation is thus a map

$$\rho : \pi_1(\Sigma, x) \longrightarrow \mathrm{G}$$

well-defined up to conjugation.

Conversely, it is not hard to construct, from any conjugacy class of such map, a gauge equivalence class of flat connections over $\Sigma \times \mathrm{G}$.

The moduli space. The space $\mathcal{M}$ we want to consider is the quotient of the space $\mathcal{A}_{fl}$ of flat connections by the gauge group action:

$$\mathcal{M} = \mathcal{A}_{fl}/\mathcal{G}.$$

Example V.1.3. Assume $\mathrm{G} = \mathbf{S}^1$, so that $\mathfrak{g} = \mathbf{R}$, the Lie bracket is trivial and d_A is just the ordinary exterior derivative d, the curvature is dA. Remember that the bundle is trivial, so that the curvature of any connection must be an exact form on Σ (see § V.4 and more precisely Remark V.4.6). Flat connections are closed 1-forms, the gauge group acts by translations

$$g \cdot A = A + g^{-1}dg = A + g^\star \sigma$$

where σ is the form dz/z on $\mathbf{S}^1$. Thus $\mathcal{M}$ is the quotient of the space of closed 1-forms by the equivalence relation defined by the above translations, namely $\mathcal{M} = H^1(\Sigma; \mathbf{S}^1)$.

From the set-theoretical or topological viewpoints, the space is quite easy to describe (even if this does not mean that it is very simple). We assign to each flat connection its holonomy as explained above. This gives a one-to-one correspondence between the moduli space $\mathcal{M}$ and the quotient $\operatorname{Hom}(\pi_1 S, \mathrm{G})/\operatorname{Ad}\mathrm{G}$. This even describes $\mathcal{M}$ as a topological space[(3)], at least if the group G is compact... and I will only use the case of $\mathbf{S}^1$ and that of $\mathrm{SU}(n)$.

Singularities. The space $\operatorname{Hom}(\pi_1\Sigma, \mathrm{G})$ is itself often singular if the surface Σ is closed ($d = 0$). In this case, the fundamental group has a presentation with $2g$ generators and one relation

$$\pi_1\Sigma = \langle a_1, \ldots, a_g, b_1, \ldots, b_g; a_1 b_1 a_1^{-1} b_1^{-1} \cdots a_g b_g a_g^{-1} b_g^{-1} \rangle,$$

so that $\operatorname{Hom}(\pi_1\Sigma, \mathrm{G})$ is the hypersurface

$$\left\{(A_1, \ldots, A_g, B_1, \ldots, B_g) \in \mathrm{G}^{2g} \mid A_1 B_1 A_1^{-1} B_1^{-1} \cdots A_g B_g A_g^{-1} B_g^{-1} = \mathrm{Id}\right\} \subset \mathrm{G}^{2g},$$

a singular hypersurface in general.

Even when the boundary of Σ is not empty (that is, when $d \geqslant 1$) and $\operatorname{Hom}(\pi_1\Sigma, \mathrm{G})$ is smooth, the G-action is not locally free, so that $\mathcal{M}$ is almost always singular, although it is Hausdorff, according to Proposition I.1.11 as we assume that G is compact.

Regular points correspond to representations ρ, the centralizer in G of which has the same dimension as the center of G. In the case of $\mathbf{S}^1$ or $\mathrm{SU}(n)$, these correspond to irreducible representations. I will ignore singularities in what follows and refer the interested readers to Goldman's paper [**55**].

There are points at which the action is principal. The dimension of the moduli space is then

$$(2g + d - 1)\dim \mathrm{G} - \dim \mathrm{G} = (2g - 2 + d)\dim \mathrm{G}.$$

Remark V.1.4. If $\partial\Sigma$ is not empty, $\pi_1(\Sigma)$ is a *free* group on $2g + d - 1$ generators (recall that g is the genus of Σ and $d \geqslant 1$ is the number of components of the boundary) and $\operatorname{Hom}(\pi_1\Sigma, \mathrm{G})$ is the product of $2g + d - 1$ copies of G.

This shows, among other things, that different surfaces may give moduli spaces which are homeomorphic: it depends only on $2g + d - 1$. For instance the sphere with three holes ($g = 0$, $d = 3$) and the torus with one hole ($g = 1$, $d = 1$) shown in Figure 2 both give $\mathrm{G} \times \mathrm{G}/\operatorname{Ad}\mathrm{G}$ as moduli spaces. We shall see, however,

(3) Strictly speaking, to use $\pi_1(\Sigma)$, we need to have chosen a base point. However, the space $\operatorname{Hom}(\pi_1 S, \mathrm{G})/\operatorname{Ad}\mathrm{G}$ is independent of this choice, because we have factored out by conjugation.

that these two spaces are endowed with natural Poisson structures which are quite different.

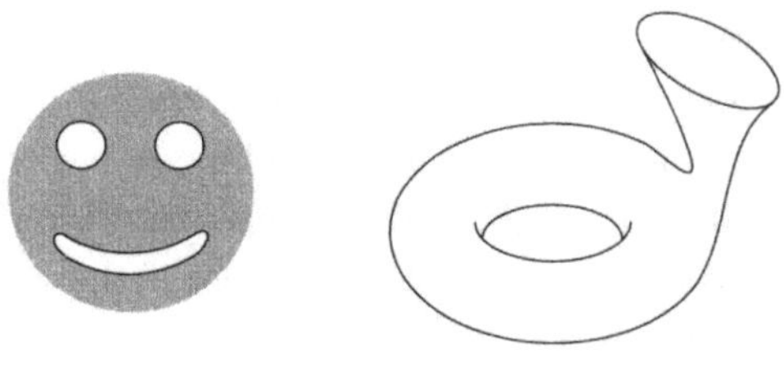

Figure 2

I will use these simple examples of surfaces, but also the simple example of the group $\mathbf{S}^1$ already mentioned to illustrate the constructions.

Example V.1.5. Let us now come back to a general surface of genus g with d holes and suppose that $\mathrm{G} = \mathbf{S}^1$. The commutativity of the group leads to two important simplifications. Firstly, the $\mathrm{Ad}\,\mathrm{G}$ action is trivial, so that the moduli space is simply $\mathrm{Hom}(\pi_1\Sigma, \mathbf{S}^1)$. Secondly, the group $\pi_1\Sigma$ has a presentation

$$\pi_1\Sigma = \left\langle \alpha_1, \ldots, \alpha_g, \beta_1, \ldots, \beta_g, \mu_1, \ldots, \mu_d; \prod_{j=1}^{d} \mu_j \prod_{i=1}^{g} (\alpha_i \beta_i \alpha_i^{-1} \beta_i^{-1}) \right\rangle$$

with a single relation, involving commutators, so that

$$\mathrm{Hom}(\pi_1\Sigma, \mathbf{S}^1) = \begin{cases} (\mathbf{S}^1)^{2g} & \text{if } d = 0 \\ (\mathbf{S}^1)^{2g} \times (\mathbf{S}^1)^{d-1} & \text{if } d \geqslant 1. \end{cases}$$

This is of course the same thing as above (Example V.1.3), where we had found that the moduli space $\mathcal{M}$ was $H^1(\Sigma; \mathbf{S}^1)$ in this case, except that here we have used a basis.

Remark V.1.6. This moduli space can be considered as a topological version of the Jacobian. See for instance [**119, 60**].

The case of $\mathrm{SU}(2)$**, the polyhedra of Jeffrey and Weitsman.** Let us concentrate on the case where $\mathrm{G} = \mathrm{SU}(2)$. We begin with a few specific features of the moduli space $\mathcal{M}$ in this case:

- Recall that all the elements of $\mathrm{SU}(2)$ are diagonalizable. More precisely, for any $A \in \mathrm{SU}(2)$, there is a unitary basis of $\mathbf{C}^2$ in which A takes the form

$$A \sim \begin{pmatrix} e^{i\theta} & 0 \\ 0 & e^{-i\theta} \end{pmatrix}$$

for some real number $\theta \in [0, \pi]$, and, provided $A \neq \pm \mathrm{Id}$ (which is equivalent to $\operatorname{tr} A \neq \pm 2$), the stabilizer of A in SU(2) is simply the $\mathbf{S}^1$ involved in the choice of the first eigenvector.

- As the group SU(2) is compact, the regular points of $\mathcal{M}$ correspond to the irreducible representations of $\pi_1\Sigma$.
- Moreover, there are rather few singular points: if all the matrices have a common eigenvector, they must have a second one, since the second eigenline is just the orthogonal complement of the first. In other words, they are diagonalizable in the same basis—and in particular they commute.

Let me make the last remark more precise. Assume that

$$\rho : \pi_1\Sigma \longrightarrow \mathrm{SU}(2)$$

is a reducible representation. Let $M_1, \ldots, M_d$ be the images in SU(2) of the loops around the holes of Σ. As the image of ρ is commutative, the relation in $\pi_1\Sigma$ gives only $M_1 \cdots M_d = \mathrm{Id}$. Write now all the M_j's in a common basis of eigenvectors

$$M_j = \begin{pmatrix} e^{i\varepsilon_j\theta_j} & 0 \\ 0 & e^{-i\varepsilon_j\theta_j} \end{pmatrix}$$

for $\varepsilon_j = \pm 1$ (recall that I assume $\theta_j \in [0, \pi]$), so that the relation becomes simply

$$\sum \varepsilon_j\theta_j \equiv 0 \mod 2\pi.$$

***Example V.1.7* (The three holed sphere).** Consider the case where Σ is a sphere with three holes ($g = 0$, $d = 3$). The moduli space is

$$\mathcal{M}_{0,3} = \{(M_1, M_2, M_3) \in \mathrm{SU}(2) \times \mathrm{SU}(2) \times \mathrm{SU}(2) \mid M_1M_2M_3 = \mathrm{Id}\} / \mathrm{SU}(2).$$

By conjugation, it may be assumed that

$$M_1 = \begin{pmatrix} e^{i\theta_1} & 0 \\ 0 & e^{-i\theta_1} \end{pmatrix} \qquad 0 \leqslant \theta_1 \leqslant \pi.$$

We can then conjugate M_2 by a diagonal element of SU(2). So we may assume that

$$M_2 = \begin{pmatrix} a & b \\ -\overline{b} & \overline{a} \end{pmatrix} \qquad \text{with } b \in \mathbf{R}^+$$

in other words, there exists $\beta \in [0, \pi]$ such that

$$\begin{cases} a = \cos\theta_2 - i\sin\theta_2\cos\beta \\ b = \sin\beta. \end{cases}$$

The condition that

$$M_1M_2 \text{ be conjugate to } \begin{pmatrix} e^{i\theta_3} & 0 \\ 0 & e^{-i\theta_3} \end{pmatrix}$$

is then $\left(e^{i\theta_1}a\right) = \cos\theta_3$, or

$$\cos\theta_1\cos\theta_2 + \sin\theta_1\sin\theta_2\cos\beta = \cos\theta_3.$$

This is reminiscent of spherical trigonometry: one may solve this for $\cos\beta$ if and only if θ_1, θ_2 and θ_3 are the lengths of the edges of a spherical triangle (see Figure 3). The representation is irreducible if and only if the triangle is an actual triangle, so that we can state:

Proposition V.1.8. *The moduli space $\mathcal{M}_{0,3}(\mathrm{SU}(2))$ can be identified with the set of isometry classes of spherical triangles. The map*

$$\begin{aligned}\mathcal{M}_{0,3} &\longrightarrow \mathbf{R}^3\\ (M_1, M_2, M_3) &\longmapsto \frac{1}{\pi}(\theta_1,\theta_2,\theta_3)\end{aligned}$$

defines a diffeomorphism onto the tetrahedron

$$0 \leqslant t_i \leqslant 1, \quad |t_1 - t_2| \leqslant t_3 \leqslant t_1 + t_2, \quad \sum t_i \leqslant 2.$$

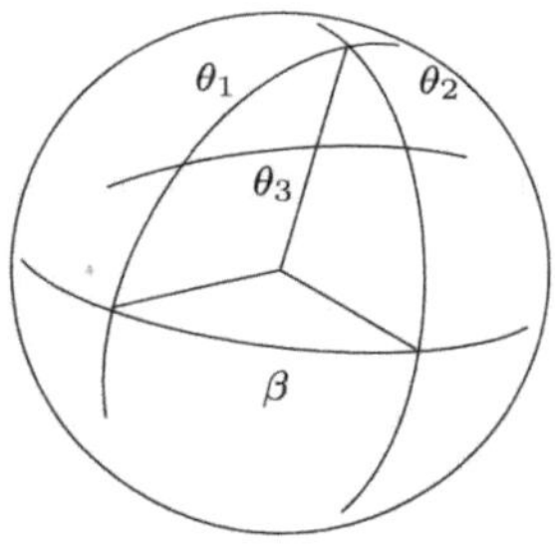

Figure 3

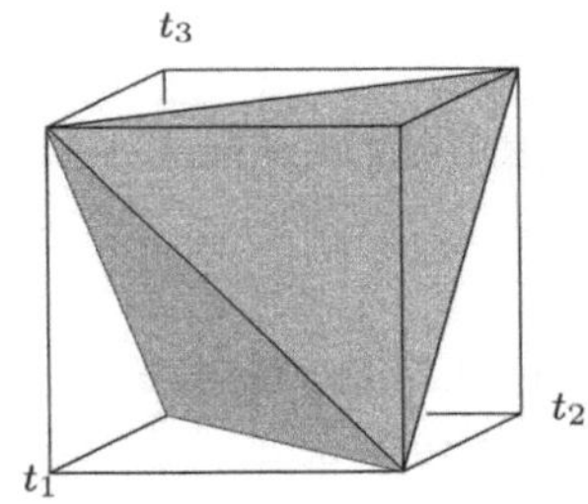

Figure 4. $\mathcal{M}_{0,3}(\mathrm{SU}(2))$

Lest us make a remark about this seemingly trivial example: the singularities of $\mathcal{M}_{0,3}$ are indeed those of the tetrahedron (namely the boundary points).

V.2. A Poisson structure on the moduli space of flat connections

We prove now that the space of flat connections, modulo gauge equivalence, is endowed with a Poisson structure.

A summary of notation. The Lie group is G, its Lie algebra is $\mathfrak{g}$, the latter is endowed with an invariant nondegenerate symmetric bilinear form, that is, a linear map

$$\mathcal{B} : \mathfrak{g} \otimes \mathfrak{g} \longrightarrow \mathbf{R}.$$

The gauge group is $\mathcal{G}$, its Lie algebra $\mathbf{g}$. If $\alpha \in \mathbf{g} = \Omega^1(\Sigma; \mathfrak{g})$, $\underline{\alpha}$ denotes the associated fundamental vector field on $\mathcal{A}$. The moduli space is $\mathcal{M}$ and also $\mathcal{A}_{fl}/\mathcal{G}$, but this $\mathcal{M}$ is only a shorthand for $\mathcal{M}_\Sigma(G)$, or $\mathcal{M}_{g,d}(G)$ (if g is the genus of Σ and d the number of components of the boundary $\partial\Sigma$), or $\mathcal{M}_{g,d}$ if the group is clear.

V.2.a. The symplectic form on $\mathcal{A}$. Following Atiyah and Bott, one defines a skew-symmetric bilinear form on the vector space $\mathcal{A}$, that can be written, as a differential form, as:

$$\omega_A(\varphi,\psi) = \int_\Sigma \mathcal{B}_\star(\varphi\wedge\psi).$$

Notice that ω is "constant" (this is just a bilinear form, ω_A does not depend on A) so that it is *closed:* this is a good reason to use the infinite dimensional description here; in the usual(4) finite dimensional descriptions, it is rather hard to prove that the forms obtained are closed (see [**56**] and [**82**]).

In the above formula, one considers the product of forms

$$\wedge : \Omega^\star(\Sigma,\mathfrak{g})\otimes\Omega^\star(\Sigma,\mathfrak{g}) \longrightarrow \Omega^\star(\Sigma,\mathfrak{g}\otimes\mathfrak{g})$$

and $\mathcal{B}_\star$ as a mapping

$$\mathcal{B}_\star : \Omega^\star(\Sigma,\mathfrak{g}\otimes\mathfrak{g}) \longrightarrow \Omega^\star(\Sigma,\mathbf{R}).$$

It is obvious that ω is nondegenerate (being defined by two nondegenerate objects: the form $\mathcal{B}$ and the pairing of 1-forms on the surface). Now, $\mathcal{A}$ is a symplectic (vector or affine) space.

The purpose of this section is to understand what structure the form ω defines on $\mathcal{M}$. We will prove:

Theorem V.2.1. *The symplectic form ω on $\mathcal{A}$ defines a Poisson structure on the moduli space $\mathcal{M}$, the symplectic leaves of which are obtained by fixing the conjugacy classes of the holonomies along the components of the boundary of Σ.*

V.2.b. Symplectic properties of the gauge group action. The gauge group action preserves the symplectic form (because it is defined by the invariant form $\mathcal{B}$):

$$\begin{aligned}(g^\star\omega)_A(\varphi,\psi) &= \omega_{g\cdot A}\left(\mathsf{T}_A g(\varphi),\mathsf{T}_A g(\psi)\right)\\ &= \omega_A\left(g^{-1}\varphi g, g^{-1}\psi g\right) \quad \text{as the } \mathcal{G}\text{-action on } \mathcal{A} \text{ is affine}\\ &= \int_\Sigma \mathcal{B}_\star((g^{-1}\varphi g)\wedge(g^{-1}\psi g))\\ &= \omega_A(\varphi,\psi).\end{aligned}$$

What we would like to do now is to write that the action is Hamiltonian and to describe its momentum mapping. We thus want to find, for any $\alpha\in\mathfrak{g}$,

(4) See also the Fock and Roslyi description in [**50, 17**].

a function H_α, the Hamiltonian vector field of which is the fundamental vector field $\underline{\alpha}$ and such that the mapping

$$\begin{aligned} \mathbf{g} &\longrightarrow \mathcal{C}^\infty(\mathcal{A}) \\ \alpha &\longmapsto H_\alpha \end{aligned}$$

is a morphism of Lie algebras (the Lie algebra structure on $\mathcal{C}^\infty(\mathcal{A})$ is that defined by the Poisson bracket associated with ω).

Atiyah and Bott have shown in **[10]** that, if $\partial\Sigma = \varnothing$, a solution to this problem is given by

$$H_\alpha(A) = \int_\Sigma \mathcal{B}_\star\left(\alpha \wedge F(A)\right).$$

In other words, when Σ has no boundary, the curvature is a momentum mapping for the gauge group action. This is a simple consequence of Corollary V.1.2. Here, some difficulties will arise from the boundary of Σ.

Proposition V.2.2. *For $\alpha \in \mathbf{g} = \Omega^0(\Sigma, \mathfrak{g})$ and $A \in \mathcal{A}$, let*

$$H_\alpha(A) = \int_\Sigma \mathcal{B}_\star\left(\alpha \wedge F(A)\right) + \int_{\partial\Sigma} \mathcal{B}_\star(\alpha \wedge A).$$

The Hamiltonian vector field of H_α is the fundamental vector field $\underline{\alpha}$, but

$$\left\{H_\alpha, H_\beta\right\}(A) = H_{[\alpha,\beta]}(A) + \int_{\partial\Sigma} \mathcal{B}_\star(\alpha \wedge d\beta).$$

Proof. Our candidate H_α should satisfy

$$\begin{aligned} \left(\mathsf{T}_A H_\alpha\right)(\varphi) &= \omega_A(\underline{\alpha}_A, \varphi) \\ &= \int_\Sigma \mathcal{B}_\star\left(\underline{\alpha}(A) \wedge \varphi\right) \\ &= \int_\Sigma \mathcal{B}_\star\left(d_A\alpha \wedge \varphi\right) \\ &= \int_\Sigma \mathcal{B}_\star d_A(\alpha \wedge \varphi) + \int_\Sigma \mathcal{B}_\star(\alpha \wedge d_A\varphi) \\ &= \int_{\partial\Sigma} \mathcal{B}_\star(\alpha \wedge \varphi) + \int_\Sigma \mathcal{B}_\star\left(\alpha \wedge d_A\varphi\right). \end{aligned}$$

But we know (from Corollary V.1.2) that

$$\left(T_A F\right)(\varphi) = d_A\varphi.$$

Now we have

$$H_{[\alpha,\beta]}(A) = \int_\Sigma \mathcal{B}_\star\left([\alpha,\beta] \wedge F(A)\right) + \int_{\partial\Sigma} \mathcal{B}_\star\left([\alpha,\beta] \wedge A\right)$$

so that the last assertion is a consequence of the invariance of $\mathcal{B}$. □

The central extension. The formula in the previous proposition shows that the mapping $\alpha \mapsto H_\alpha$ is not exactly a morphism of Lie algebras. If α, $\beta \in \mathbf{g}$, put

$$c(\alpha, \beta) = \int_{\partial\Sigma} \mathcal{B}_\star(\alpha \wedge d\beta).$$

This defines a skew symmetric bilinear form on $\mathbf{g}$, which has the additional property of being a cocycle, in the sense that

$$c([\alpha, \beta], \gamma) + c([\beta, \gamma], \alpha) + c([\gamma, \alpha], \beta) = 0$$

(direct verification). This gives the idea to consider the vector space $\widehat{\mathbf{g}} = \mathbf{g} \oplus \mathbf{R}$ and to endow it with the bracket

$$[(\alpha, t), (\beta, u)] = ([\alpha, \beta], c(\alpha, \beta)).$$

This gives it the structure of a Lie algebra. We have an exact sequence of Lie algebras

$$0 \longrightarrow \mathbf{R} \longrightarrow \widehat{\mathbf{g}} \longrightarrow \mathbf{g} \longrightarrow 0$$

so that $\widehat{\mathbf{g}}$ is an extension of $\mathbf{g}$. As the additional $\mathbf{R}$ obviously lies in the center of $\widehat{\mathbf{g}}$, this is called a *central extension* of $\mathbf{g}$. The structure on $\widehat{\mathbf{g}}$ was made so that, defining $H_{(\alpha,t)}(A) = H_\alpha(A) + t$, Proposition V.2.2 can be rephrased as:

Corollary V.2.3. *The mapping*

$$\begin{aligned} \widehat{\mathbf{g}} &\longrightarrow \mathcal{C}^\infty(\mathcal{A}) \\ (\alpha, t) &\longmapsto H_{(\alpha,t)} \end{aligned}$$

is a morphism of Lie algebras. Moreover, the Hamiltonian vector field of the function $H_{(\alpha,t)} : \mathcal{A} \to \mathbf{R}$ is the vector field $\underline{\alpha}$. □

The last assertion tells us that $\widehat{\mathbf{g}}$ acts infinitesimally on $\mathcal{A}$ *via* the infinitesimal $\mathbf{g}$-action we already know. This remark can be integrated to define a central extension $\widehat{\mathcal{G}}$ of the gauge group itself, acting on $\mathcal{A}$ *via* the gauge action of $\mathcal{G}$.

What I want to describe now is the momentum mapping for these actions. I must first say something about the dual of the Lie algebra $\widehat{\mathbf{g}}$. The proof of the next proposition is straightforward.

Proposition V.2.4. *The pairing of $\Omega^2(\Sigma, \mathfrak{g}) \oplus \Omega^1(\partial\Sigma, \mathfrak{g}) \oplus \mathbf{R}$ with $\widehat{\mathbf{g}}$ by*

$$(R, \varphi, z) \otimes (\alpha, t) \longmapsto \int_\Sigma \mathcal{B}_\star(\alpha \wedge R) + \int_{\partial\Sigma} \mathcal{B}_\star(\alpha \wedge \varphi) + zt$$

is nondegenerate. □

In this way, $\Omega^2(\Sigma, \mathfrak{g}) \oplus \Omega^1(\partial\Sigma, \mathfrak{g}) \oplus \mathbf{R}$ can be seen as a subspace of $\widehat{\mathbf{g}}^\star$ and the previous study can be summarized in:

Corollary V.2.5. *The mapping*

$$\begin{aligned} \mu : \mathcal{A} &\longrightarrow \widehat{\mathbf{g}}^\star \\ A &\longmapsto (F(A), A|_{\partial\Sigma}, 1) \end{aligned}$$

is a $\mathcal{G}$-equivariant momentum mapping for the gauge group action. □

Notice that the central part of $\widehat{\mathcal{G}}$ does not act on $\widehat{\mathbf{g}}^\star$, so the statement makes sense.

When the surface Σ is closed, the momentum mapping μ is just the curvature, so that the space of flat connections modulo the gauge group action is a reduced level set of a momentum mapping, and thus a symplectic manifold(5), as was explained in [**10**]. In the general case where the surface has boundary components, the curvature is the momentum mapping for the action of a smaller group. Let us consider the subgroup $\mathcal{G}_0$ of the gauge group $\mathcal{G}$ defined by

$$\mathcal{G}_0 = \{g \in \mathcal{G} \mid g|_{\partial\Sigma} = 1\}.$$

This is the kernel of the morphism of restriction to the boundary and, in particular, this is a normal subgroup.

Proposition V.2.6. *The normal subgroup $\mathcal{G}_0$ of $\mathcal{G}$ acts on $\mathcal{A}$ with momentum mapping $A \mapsto F(A)$. The space $\mathcal{M}_0$ of flat connections modulo the $\mathcal{G}_0$-action is symplectic.*

Proof. The Lie algebra of $\mathcal{G}_0$ is

$$\mathbf{g}_0 = \{\alpha \in \mathbf{g} \mid \alpha|_{\partial\Sigma} = 0\}$$

so that one can embed

$$\Omega^2(\Sigma, \mathfrak{g}) \oplus \mathbf{R} \hookrightarrow \widehat{\mathbf{g}_0}^\star$$

and conclude that the momentum mapping for the $\mathcal{G}_0$-action is the composed map:

$$p \circ \mu : A \longmapsto (F(A), A|_{\partial\Sigma}, 1) \longmapsto (F(A), 1) \in \widehat{\mathbf{g}}_0^\star.$$

Hence $\mathcal{M}_0 = (p \circ \mu)^{-1}(0,1)/\mathcal{G}_0$ is a symplectic reduced space. □

V.2.c. The Poisson structure and its symplectic foliation. As $\mathcal{G}_0$ is normal in $\mathcal{G}$, the gauge group $\mathcal{G}$ also acts on $\mathcal{M}_0$, the action preserves the symplectic structure, and the quotient $\mathcal{M}_0/\mathcal{G}$ is, of course, our moduli space $\mathcal{M}$. Being the quotient of a symplectic space by a symplectic group action, it inherits a natural (*i.e.*, coming from the form ω on $\mathcal{A}$) Poisson structure. The next aim is to describe its symplectic foliation. We apply Proposition III.2.22 to the action of $\mathcal{G}/\mathcal{G}_0$ on $\mathcal{M}_0$. There is an exact sequence

$$1 \longrightarrow \mathcal{G}_0 \longrightarrow \mathcal{G} \longrightarrow \Omega^0(\partial\Sigma, \mathrm{G})$$

(5) As long as this is a manifold, of course.

so that

$$\mathcal{G}/\mathcal{G}_0 \subset \oplus_{C_i \in \pi_0(\partial\Sigma)} \Omega^0(C_i, \mathrm{G}).$$

As usual, $\oplus_i \Omega^1(C_i, \mathfrak{g})$ can be identified with a subspace of the dual of the Lie algebra of $\mathcal{G}/\mathcal{G}_0$ by integration

$$\langle (\varphi_1, \ldots, \varphi_d), (\alpha_1, \ldots, \alpha_d) \rangle = \sum_{i=1}^{d} \int_{C_i} \mathcal{B}_\star(\varphi_i \wedge \alpha_i)$$

in such a way that the momentum mapping we are interested in is

$$\begin{aligned} \mathcal{M}_0 &\longrightarrow \bigoplus_{i=1}^{d} \Omega^1(C_i, \mathfrak{g}) \oplus \mathbf{R} \\ [A] &\longmapsto ((A|_{C_1}, \ldots, A|_{C_d}), 1). \end{aligned}$$

Let us now identify each component C_i with the circle $\mathbf{S}^1$, so that $\Omega^0(C_i; \mathfrak{g}) \cong \Omega^0(\mathbf{S}^1, \mathfrak{g})$ is the Lie algebra $L\mathfrak{g}$ of the loop group

$$L\mathrm{G} = \mathcal{C}^\infty(\mathbf{S}^1, \mathrm{G}) \cong \mathcal{C}^\infty(C_i, \mathrm{G}).$$

We thus have to understand the coadjoint action of the loop group $L\mathrm{G}$ on the dual $\widehat{L\mathfrak{g}}^\star$ of the central extension $\widehat{L\mathfrak{g}}$ defined by the same cocycle as $\widehat{\mathfrak{g}}$ above.

I shall postpone the description of these orbits, and thus the proof of the next proposition, to the end of this section. Here is the result:

Proposition V.2.7. *The space $\Omega^1(\mathbf{S}^1, \mathfrak{g}) \oplus \{1\}$ can be embedded as a subspace in the dual $\widehat{L\mathfrak{g}}^\star$. The map which associates to any form on $\mathbf{S}^1$ its holonomy along the circle gives a one-to-one correspondence from the set of coadjoint orbits of $L\mathrm{G}$ in $\Omega^1(\mathbf{S}^1, \mathfrak{g}) \oplus 1$ to the set of conjugacy classes in* G.

Notice that we will eventually have proven Theorem V.2.1. □

***Example V.2.8* (The case of $\mathbf{S}^1$).** Consider the case of the group $\mathrm{G} = \mathbf{S}^1$. Fixing a conjugacy class in $\mathbf{S}^1$ amounts to fixing an element of $\mathbf{S}^1$. The symplectic leaves in $\mathcal{M}_{g,d} = (\mathbf{S}^1)^{2g} \times (\mathbf{S}^1)^{d-1}$ are the $(\mathbf{S}^1)^{2g} \times \{c_1, \ldots, c_{d-1}\}$.

***Example V.2.9* (Different Poisson structures on the same space).** Let us turn now to our second favorite class of examples (Figure 2), the three holed sphere and the one holed torus, that is, the moduli spaces $\mathcal{M}_{0,3}$ and $\mathcal{M}_{1,1}$. As we have already noticed, both are copies of $\mathrm{G} \times \mathrm{G}/\operatorname{Ad}\mathrm{G}$.

An element of $\mathcal{M}_{0,3}$ consists of three elements M_1, M_2 and $M_3 \in \mathrm{G}$ such that $M_1 M_2 M_3 = 1$, up to simultaneous conjugation by group elements. To fix a symplectic leaf, we have to fix the conjugacy classes C_1, C_2 and C_3 of M_1, M_2 and M_3. It turns out (this is the subject of Exercise V.1) that the map

$$\begin{array}{rcl} C_1 \times C_2 \times C_3 & \longrightarrow & \mathrm{G} \\ (M_1, M_2, M_3) & \longmapsto & M_1 M_2 M_3 \end{array}$$

has maximal rank at a solution of the equation $M_1 M_2 M_3 = 1$ corresponding to an irreducible representation. The dimension of the symplectic leaf $\mathcal{M}_{0,3}^{C_1,C_2,C_3}$ is thus $\sum \dim C_i - 2\dim \mathrm{G}$.

As for an element of $\mathcal{M}_{1,1}$, it also consists of three elements A, B, $M \in \mathrm{G}$, but now the relation is $ABA^{-1}B^{-1}M = 1$ and, to fix a symplectic leaf, we need only fix the conjugacy class C of M. In the same way, it is shown that $\dim \mathcal{M}_{1,1}^{C} = \dim C$.

Generic conjugacy classes have the form $\mathrm{G}/\mathbf{T}$ for a maximal torus $\mathbf{T}$, and thus have dimension $\dim \mathrm{G} - \mathrm{rk}\, \mathrm{G}$, so that, for G a simple Lie group, the generic symplectic leaves have dimension $\dim \mathrm{G} - 3\,\mathrm{rk}\, \mathrm{G}$ in $\mathcal{M}_{0,3}$ and $\dim \mathrm{G} - \mathrm{rk}\, \mathrm{G}$ in $\mathcal{M}_{1,1}$, so that the Poisson spaces obtained are indeed very different.

***Example V.2.10* (The case of $\mathcal{M}_{0,3}(\mathrm{SU}(2))$).** According to Proposition V.1.8, this moduli space is simply the tetrahedron shown in Figure 4. the conjugacy classes have dimension 2, so that the symplectic leaves have dimension 0: the Poisson structure on the tetrahedron is trivial. Notice that in $\mathcal{M}_{1,1}(\mathrm{SU}(2))$, the leaves have dimension 2. We shall see in Example V.3.10 that these leaves are spheres.

Notice that these computations of dimension work only for a *simple* Lie group. Consider for instance $\mathcal{M}_{1,1}$ in the case of the commutative group $\mathbf{S}^1$. The only conjugacy class giving rise to a nonempty leaf is that of 1, so that $\mathcal{M}_{1,1}(\mathbf{S}^1)$ is actually symplectic.

Description of the central extension $\widehat{L\mathfrak{g}}$. Identify $\mathbf{S}^1$ with $\mathbf{R}/2\pi\mathbf{Z}$ and define, for α, $\beta : \mathbf{S}^1 \to \mathfrak{g}$,

$$c(\alpha, \beta) = \frac{1}{2\pi} \int_0^{2\pi} \mathcal{B}(\alpha(\theta), \beta'(\theta)) d\theta$$

(β' denotes the derivative of β with respect to θ). The Lie algebra $\widehat{L\mathfrak{g}}$ is defined by the cocycle c as above. The adjoint action of $\widehat{L\mathfrak{g}}$ on itself is actually an $L\mathfrak{g}$-action, by

$$\mathrm{ad}_\gamma(\alpha, t) = ([\gamma, \alpha], c(\gamma, \alpha))$$

and this is the infinitesimal version of the adjoint action of the group $L\mathrm{G}$. To write the latter, we need some notation: for $g : \mathbf{S}^1 \to \mathrm{G}$, we will consider $g^{-1}g'$ as a mapping $\mathbf{S}^1 \to \mathfrak{g}$ and hence as an element of $L\mathfrak{g}$; a bilinear form on $L\mathfrak{g}$, which will still be denoted by $\mathcal{B}$, is defined by

$$(\alpha, \beta) \longmapsto \frac{1}{2\pi} \int_0^{2\pi} \mathcal{B}(\alpha(\theta), \beta(\theta)) d\theta.$$

Now, the adjoint action is

$$\operatorname{Ad}_g(\alpha, t) = \left(\operatorname{Ad}_g \alpha, t + \mathcal{B}(g^{-1}g', \alpha)\right).$$

Let us describe now the coadjoint action, on the subspace $\Omega^1(\mathbf{S}^1, \mathfrak{g}) \oplus \mathbf{R}$ of $\widehat{L\mathfrak{g}}^\star$:

$$\begin{aligned}\langle \operatorname{Ad}_g^\star(\varphi, z), (\alpha, t)\rangle &= \langle(\varphi, z), \operatorname{Ad}_{g^{-1}}(\alpha, t)\rangle \\ &= \langle(\varphi, z), \left(\operatorname{Ad}_{g^{-1}} \alpha, t - \mathcal{B}\left(g'g^{-1}, \alpha\right)\right)\rangle \\ &= \int_{\mathbf{S}^1} \mathcal{B}\left(\varphi, \operatorname{Ad}_{g^{-1}} \alpha\right) + zt - z\int_{\mathbf{S}^1} \mathcal{B}\left(g'g^{-1}, \alpha\right)\end{aligned}$$

so that

$$\operatorname{Ad}_g^\star(\varphi, z) = \left(\operatorname{Ad}_g \varphi - zg'g^{-1}, z\right).$$

Remark V.2.11. This is an LG-action. However, notice that it depends, for each "level" z, on the actual value of z. The momentum mapping and the orbits we are interested in correspond to the value $z = 1$.

End of the proof of Proposition V.2.7. The last argument comes from [**117**, § 4.3]. Consider an element $\varphi \in \Omega^1(\mathbf{S}^1, \mathfrak{g})$ and write it as $\varphi = \alpha d\theta$ for some $\alpha \in L\mathfrak{g}$. Solve the differential equation

$$\begin{cases} f' \cdot f^{-1} = \alpha \\ f(0) = 1 \end{cases}$$

at least for a function $f : \mathbf{R} \to \mathrm{G}$. As α is a function on the circle $\mathbf{S}^1$, $f(\theta + 2\pi) = f(\theta)f(2\pi)$. Call $m(\varphi)$ (m is for *monodromy*, an alias for "holonomy") the element $f(2\pi) \in \mathrm{G}$. Let us make an element $g \in L\mathrm{G}$ act on $(\varphi, 1)$, obtaining $\widetilde{\varphi} = \operatorname{Ad}_g \varphi - (dg)g^{-1}$, that is,

$$\underline{\alpha} = \operatorname{Ad}_g \alpha - g'g^{-1}.$$

It is easy to check that $m(\operatorname{Ad}_g \varphi) = g(0)m(\varphi)g(0)^{-1}$ and this proves Proposition V.2.7. □

Remark V.2.12. The description of the symplectic leaves we have given was sketched by Atiyah in [**9**]. The symplectic form of the leaves is described in many papers (for instance in [**3**]). I have chosen a global description of the Poisson structure (which I learned from Volodya Fock (see [**49**])) because it shows *why* the symplectic leaves are obtained by fixing the monodromy along the boundary components, so that the symplectic foliation appears in a very natural way.

It is also possible to describe the Poisson structure in the infinite dimensional setting without using the loop group, as explained by Jeffrey in [**72**]: she considers connections A that are identically zero on some neighbourhood of the boundary, which makes the approach slightly simpler but less canonical than the one presented here, as she is forced to use a parametrization of the boundary.

V.3. Construction of commuting functions on $\mathcal{M}$

In this section, we define functions on $\mathcal{M}$, following Goldman [**56**], and prove that they are in involution.

V.3.a. Goldman's functions. They are usually described in the following way. Consider both an invariant function $f : \mathrm{G} \to \mathbf{R}$ and a simple closed curve C on Σ, and define a map

$$\begin{aligned} f_C : \mathcal{M} &\longrightarrow \mathbf{R} \\ [\rho] &\longmapsto f(\rho(C)) \end{aligned}$$

where ρ is a representative $\pi_1\Sigma \to \mathrm{G}$ of the class $[\rho]$ and C is considered as an element of $\pi_1\Sigma$ in any obvious way, the invariance of f implying that the result does not depend on the choices made.

For C a simple closed curve, let $\mathcal{G}_C$ be the subgroup of the gauge group $\mathcal{G}$ consisting of mappings $g : \Sigma \to \mathrm{G}$ that are identically equal to 1 on C. The subgroup $\mathcal{G}_C$ is, of course, normal. Call $\mathcal{M}_C = \mathcal{A}_{fl}/\mathcal{G}_C$. The map

$$\begin{aligned} \mu_C : \mathcal{M}_C &\longrightarrow \widehat{L\mathfrak{g}}^{\star} \\ A &\longmapsto (A|_C, 1) \end{aligned}$$

(with the notation of § V.2), defined using a diffeomorphism $C \to \mathbf{S}^1$, is the momentum mapping for the $\mathcal{G}_C$-action on $\mathcal{M}_C$ and thus defines a mapping

$$\Phi_C : \mathcal{M} \longrightarrow \widehat{L\mathfrak{g}}^{\star}/L\mathrm{G},$$

which associates, to the gauge equivalence class of the flat connection A, its holonomy along C. The Goldman functions are just the quantitative (numerical) versions of the mapping Φ_C: invariant functions on G more or less describe the conjugacy classes in G. The flow of a function f_C (as defined above) will preserve the levels of the map Φ_C. Such a level is a set of flat connections, the holonomy along C of which is in a given conjugacy class (notice that f_C depends only on the homotopy class of the free loop C). So we have the following proposition:

Proposition V.3.1. *Let C be a simple closed curve on Σ. The flow of f_C acts on the class $[A] \in \mathcal{M}$ of a flat connection by some element of the stabilizer $\mathrm{G}_{m(\Phi_C([A]))}$ of its holonomy along C.* □

After having defined these functions, Goldman has proved two beautiful theorems:

- He has computed the Poisson brackets of any two functions f_{C_1}, g_{C_2}, roughly speaking in terms of the intersections of the two curves C_1 and C_2.
- He has given an explicit formula for the flow of f_C, viewed as a Hamiltonian on $\mathcal{M}$.

We will not prove these theorems here, contenting ourselves to send the readers to the original paper [**56**].

***Theorem V.3.2* (Goldman [56]).** *Let C_1 and C_2 be two disjoint simple closed curves on the surface Σ. Then, for any two invariant functions f and g on the group* G,

$$\{f_{C_1}, g_{C_2}\} = 0.$$

Notice that Goldman considers a *closed* surface Σ, but this does not make any difference here.

Corollary V.3.3. *Let C be a simple closed curve on the surface Σ. Then for any two invariant functions f and g on the group* G, $\{f_C, g_C\} = 0$. □

As for the second theorem, I will use only a special case, which is a quantitative version of Proposition V.3.1, in § V.1.

V.3.b. Integrable systems?

Counting commuting Goldman functions. Recall that g is the genus and d the number of boundary components of Σ. Assume $d \geqslant 1$ if $g = 1$ and $d \geqslant 3$ if $g = 0$. Then the maximum number of disjoint (nontrivial) curves on Σ is $3g-3+d$.

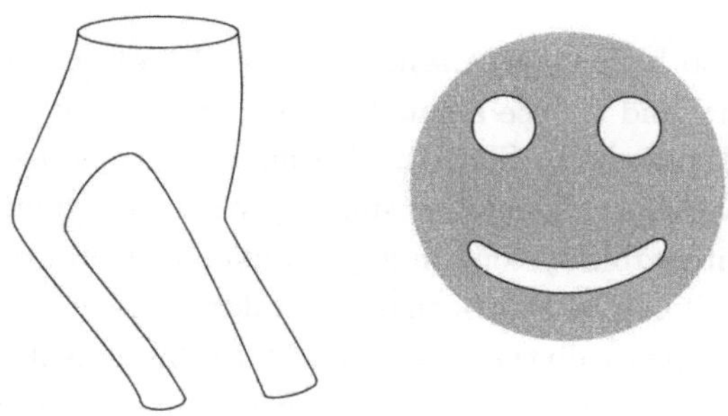

Figure 5. Trinions

This yields a *trinion*[(6)] decomposition, see Figure 6, so that Goldman gives us $(3g-3+d)\,\mathrm{rk}\,G$ (hopefully independent) commuting functions on $\mathcal{M}$. This will be an integrable system... provided it is defined on a symplectic manifold of dimension $2(3g-3+d)\,\mathrm{rk}\,G$. Notice that the Goldman functions defined by the boundary components are Casimir functions: this is more or less equivalent to saying that the symplectic leaves are obtained by fixing the conjugacy classes of

(6) The beautiful French terminology turns out to be very inconvenient when translated into English, as French people simply wear "pantalons" while the British need "*pairs* of pants". In any case, a trinion is just a sphere with three holes See Figure 5.

the holonomy along the boundary. This is the reason why I have not included them in the enumeration here.

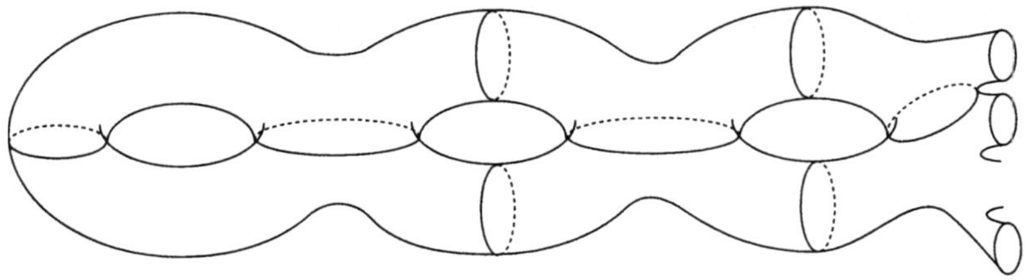

Figure 6. A trinion decomposition

In the case of a closed surface ($d = 0$) the moduli space itself is a symplectic manifold. If moreover the Lie group G is simple, we have seen that the dimension of $\mathcal{M}_{g,0}$ is $(2g-2)\dim \mathrm{G}$ and we have $(3g-3)\operatorname{rk}\mathrm{G}$ functions. Thus Goldman gives us an integrable system if and only if

$$\frac{\dim \mathrm{G}}{\operatorname{rk}\mathrm{G}} = 3.$$

This happens with $\mathrm{G} = \mathrm{SU}(2)$.

Another way to have enough functions would be to consider a surface with nonempty boundary and choose a small enough symplectic leaf. Consider for instance the moduli space $\mathcal{M}_{1,1}(\mathrm{SU}(n))$. Goldman allows us to use one curve and $n-1$ invariant functions... and this might give us (provided the functions are independent) an integrable system on any symplectic leaf of dimension $2(n-1)$.

In both cases, it can be shown that the Goldman functions are actually independent. I will concentrate on the case of SU(2). In this case, Jeffrey and Weitsman [**76, 77**] have exhibited a torus action that I describe now and which is nothing other than the action-angle coordinates for the Goldman system (see § III.3.b).

A point in $\mathcal{M}_{g,d}(\mathrm{SU}(2))$ is a conjugacy class of a representation $\rho : \pi_1(\Sigma) \to \mathrm{SU}(2)$. Let $M_1,\ldots,M_d$ be matrices corresponding to the d boundary components of Σ. Each M_j is an element of SU(2) so that its conjugacy class is well-defined by its eigenvalues $e^{i\theta_j}$, $e^{-i\theta_j}$. We have:

Proposition V.3.4. *Let $\theta_1,\ldots,\theta_d \in [0,\pi]$ be such that,*

$$\text{for all } \varepsilon_j \in \{\pm 1\}, \quad \sum \varepsilon_j\theta_j \not\equiv 0 \mod 2\pi.$$

Then the symplectic leaf defined by the conjugacy classes of the diagonal matrices

$$\begin{pmatrix} e^{i\theta_1} & 0 \\ 0 & e^{-i\theta_1} \end{pmatrix}, \ldots, \begin{pmatrix} e^{i\theta_d} & 0 \\ 0 & e^{-i\theta_d} \end{pmatrix}$$

consists of regular points of $\mathcal{M}_{g,d}(\mathrm{SU}(2))$ *and is a smooth compact symplectic manifold of dimension* $6g + 2d - 6$. □

Goldman's functions on SU(2). Let us turn to Goldman's functions in this case. The only invariant function is the trace, which defines a function f_C and a flow on $\mathcal{M}$ for any curve C. The first remark is a very special case of Proposition V.3.1:

Proposition V.3.5. *All the orbits of the flow of* f_C *are periodic.*

Proof. The stabilizer of a generic element of SU(2) is a circle. □

Remark V.3.6. The Hamiltonian system defined by the function f_C on $\mathcal{M}$ is *superintegrable*, in the sense that there are many ("count" the curves not intersecting C) functions commuting with f_C (of course they do not pairwise commute). The trajectories of the flow lie on the common level sets of all these functions. As there are enough functions, these common level sets are 1-dimensional. A good reason for the flow to be periodic.

Now we normalize so that this periodicity corresponds to an $\mathbf{S}^1$-action. Define $h_C : \mathcal{M} \to [0,1]$ by

$$h_C([\rho]) = \frac{1}{\pi} \arccos\left(\frac{1}{2} \operatorname{tr} \rho(C)\right)$$

(up to the factor $1/\pi$, this is just the θ in a matrix corresponding to C as above, namely, $e^{i\theta}$ is an eigenvalue). Let $\mathcal{U}_C \subset \mathcal{M}$ be the open subset consisting of all the classes of representations $\rho : \pi_1\Sigma \to \mathrm{SU}(2)$ such that $\rho(C) \neq \pm\,\mathrm{Id}$, so that h_C is smooth on $\mathcal{U}_C$ and $h_C(\mathcal{U}_C) \subset (0,1)$.

Proposition V.3.7. *The Hamiltonian* $h_C : \mathcal{U}_C \to \mathbf{R}$ *is periodic.*

Sketch of the proof. This is a quantitative version of Propositions V.3.1 and V.3.5 and a special case of Goldman's result described in § V.3. The function h_C is the Hamiltonian of the $\mathbf{S}^1$-action

$$(z \cdot \rho)(\gamma) = \begin{cases} \rho(\gamma) & \text{if } \gamma \cap C = \varnothing \\[2ex] \begin{pmatrix} z & 0 \\ 0 & \bar{z} \end{pmatrix} \rho(\gamma) & \text{if } \gamma \cdot C = 1 \end{cases}$$

in the case where the curve C does not disconnect the surface. When $\Sigma - C$ has two components Σ_1 and Σ_2 say, if $\gamma \cap C = \varnothing$, $(z \cdot \rho)(\gamma) = \rho(\gamma)$ if γ is homotopic to a curve in Σ_1 say, and

$$(z \cdot \rho)(\gamma) \text{ is the conjugate } \begin{pmatrix} z & 0 \\ 0 & \bar{z} \end{pmatrix} \rho(\gamma) \begin{pmatrix} \bar{z} & 0 \\ 0 & z \end{pmatrix}$$

if γ is homotopic to a curve in Σ_2. □

Trinion decompositions and torus actions. Let us now fix $3g+d-3$ curves on the surface Σ as in Figure 6. Together with the d boundary components of Σ, this gives us a set $\mathcal{C}$ of $3g+2d-3$ curves which divide the surface into $2g+d-2$ trinions (see Figure 6). To each curve $C \in \mathcal{C}$, associate the corresponding θ_C, h_C. Let

$$h : \mathcal{M} \longrightarrow \mathbf{R}^{\mathcal{C}} = \mathbf{R}^{3g+2d-3}$$

be the map the components of which are the functions h_C $(C \in \mathcal{C})$. To each trinion in the decomposition is associated a tetrahedron in the subspace $\mathbf{R}^3 \subset \mathbf{R}^{\mathcal{C}}$ corresponding to its three boundary curves. Let

$$\mathcal{U}_{\mathcal{C}} = \bigcap_{C \in \mathcal{C}} \mathcal{U}_C.$$

Proposition V.3.8. *The restriction to $\mathcal{U}_{\mathcal{C}}$ of the mapping*

$$h : \mathcal{M} \longrightarrow \mathbf{R}^{\mathcal{C}}$$

is the momentum mapping for a $\mathbf{T}^{3g+2d-3}$-action on the Poisson manifold $\mathcal{U}_{\mathcal{C}} \subset \mathcal{M}_{g,d}$. The closure of the image of h is the convex compact polyhedron $P_{\mathcal{C}}$ obtained by taking the product of the tetrahedra corresponding to the trinions defined by $\mathcal{C}$ and intersecting it with the hyperplanes corresponding to the gluing of two trinions.

Proof. That the image is included in the polyhedron $P_{\mathcal{C}}$ is clear. We have only to check that all the interior points are actually obtained. Now any point of $P_{\mathcal{C}}$ gives a point in all the tetrahedra corresponding to the trinions defined by $\mathcal{C}$, that is, according to V.1.8, a family of representations of the fundamental groups of the various trinions, or a gauge class of flat connections on each trinion (one even constructs local sections of the momentum mapping this way). One then checks that there exist either morphisms $\pi_1\Sigma \to \mathrm{SU}(2)$ or flat connections on Σ that extend the given data. $\square$

This proves that the h_C are the action coordinates corresponding to our angles (in the torus). This also proves that the Goldman functions under consideration are independent, as $P_{\mathcal{C}}$ is a polytope with nonempty interior.

Notice also that Proposition V.3.8 is not an application of the convexity theorem (Theorem IV.4.3) even when restricted to the intersection of a symplectic leaf with $\mathcal{U}_{\mathcal{C}}$, as this is not a compact smooth symplectic manifold.

Let me make this remark more precise. Choose

$$t = (t_1, \dots, t_d), t_i \in [0,1] \text{ such that } \sum \varepsilon_i t_i \not\equiv 0 \mod 2$$

and cut the polyhedron by the affine subspace defined by t (corresponding to the boundary curves). This gives a compact convex polyhedron $P^t_{\mathcal{C}}$ which is the image of the compact smooth symplectic leaf $\mathcal{M}^t$ under the mapping h. However, nobody claims that $P^t_{\mathcal{C}}$ is the image of the momentum mapping of a torus action on $\mathcal{M}^t$:

this is the image of h and also the closure of the image of the momentum mapping for a torus action on $\mathcal{M}^t \cap \mathcal{U}_{\mathcal{C}}$ (which is, indeed, the restriction of h). Obviously, if the genus of Σ is large enough, there can be some matrices equal to $\pm\,\mathrm{Id}$ in an irreducible representation of $\pi_1\Sigma$.

In the sequel of this §, I will discuss some examples, in order to explain the meaning of this result.

Example V.3.9 **(The three holed sphere).** Notice that Proposition V.1.8 is a special case of Proposition V.3.8, the case of $\mathcal{M}_{0,3}$. As the Poisson structure is trivial, a diffeomorphism can be a momentum mapping.

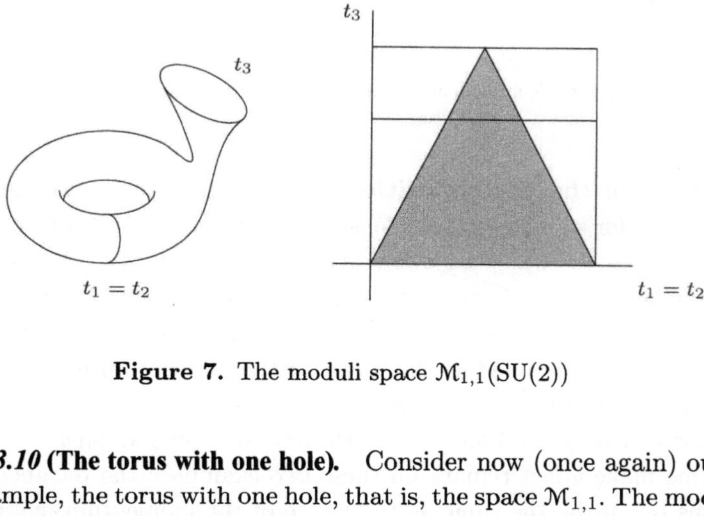

Figure 7. The moduli space $\mathcal{M}_{1,1}(\mathrm{SU}(2))$

Example V.3.10 **(The torus with one hole).** Consider now (once again) our second favorite example, the torus with one hole, that is, the space $\mathcal{M}_{1,1}$. The moduli space is 3-dimensional, as is $\mathcal{M}_{0,3}$. But now, the momentum mapping is for a $\mathbf{T}^2$-action and takes its values in $\mathbf{R}^2$. With the notation defined by Figure 7, the image of the momentum mapping is a triangle (intersection of the basic tetrahedron with the plane $t_1 = t_2$). The symplectic leaves are defined by t_3. Notice that all of them are smooth surfaces, provided $t_3 \neq 0, 1$. As this implies that $t_1 = t_2 \in (0,1)$, $\mathcal{U}_{\mathcal{C}} \cap \mathcal{M}^t = \mathcal{M}^t$. Thus the symplectic leaves have dimension 2 and are endowed with a Hamiltonian $\mathbf{S}^1$-action, and so they must be 2-spheres (see, if necessary, Exercise III.2), the image of the corresponding momentum mapping is the intersection of the triangle with the line $t_3 = \text{const}$ (Figure 7).

Example V.3.11 **(The torus with two holes).** The next example is a torus with two holes, that is, the moduli space $\mathcal{M}_{1,2}$. Using the trinion decomposition shown in Figure 8 and Proposition V.3.8, one gets a momentum mapping, the image of which is a 4-dimensional polytope. To obtain a symplectic leaf, fix the numbers t_3 and t_3' corresponding to the boundary components. The leaf is then mapped onto the rectangle shown in Figure 8 (and, again according to Theorem IV.4.20, this is an $\mathbf{S}^2 \times \mathbf{S}^2$).

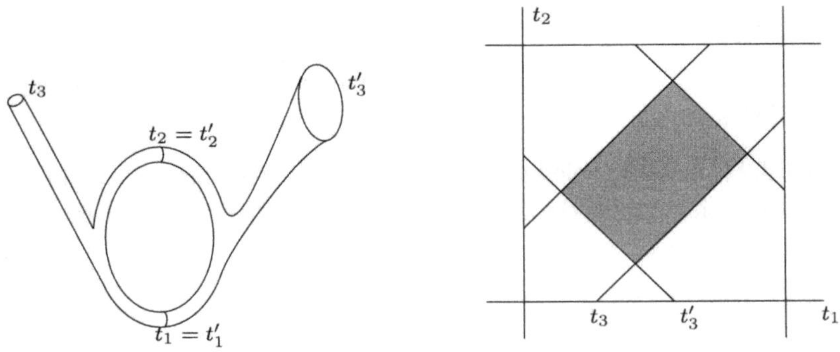

Figure 8. A symplectic leaf in the moduli space $\mathcal{M}_{1,2}(\mathrm{SU}(2))$

Dependence on the decomposition. There can exist many different trinion decompositions for a given surface. Consider the simple example of a closed surface of genus 2 ($g = 2$, $d = 0$). It has two trinion decompositions as shown in Figure 9, which also shows the closures of the images of the corresponding momentum mappings $h : \mathcal{M}_{2,0} \to \mathbf{R}^3$. The two polyhedra are quite different. For instance the pyramid cannot be obtained as the image of the momentum mapping of a torus action on a smooth compact symplectic manifold as it has a vertex with valency 4, which is forbidden by the linearization theorem for torus actions.

Let me make a last remark on these two examples: the tetrahedron in Figure 9 turns out to be the same as the image of the momentum mapping for the $\mathbf{T}^3$-action on $\mathbf{P}^3(\mathbf{C})$ (Example IV.4.6). To compare them, we must be rather careful with the integral lattices the tori define in $\mathbf{R}^3$. In the case of $\mathbf{P}^3(\mathbf{C})$, this is the standard integral lattice in $\mathbf{R}^3$. The $\mathbf{T}^3$-action on $\mathcal{M}_{2,0}$ I have written is not effective (see [**77**]), so that the lattice to consider in Figure 9 is... precisely the one generated by the vertices of the tetrahedron.

For the sake of completeness, let me recall that, following [**112**], $\mathcal{M}_{2,0}(\mathrm{SU}(2))$ is homeomorphic with the projective space $\mathbf{P}^3(\mathbf{C})$.

Applications. The system on $\mathcal{M}_{g,d}$ and the torus action on $\mathcal{U}_{\mathcal{C}}$ may help to understand the Poisson and/or symplectic geometry of the moduli space, as I have explained in the previous examples.

As a first application, Proposition V.3.8 and its symplectic specializations make it very easy to compute the volume of a symplectic moduli space, as this is simply the volume of the image polytope (with respect to the *ad hoc* lattice).

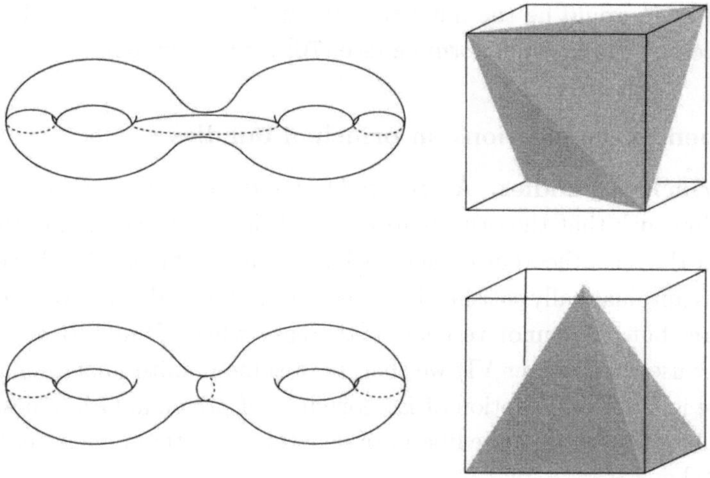

Figure 9. Torus actions on $\mathcal{M}_{2,0}$

For instance, an argument by induction on the genus g which is based on the decomposition of a surface with one hole shown in Figure 10 and on the volume of the rectangle of Figure 8 allows Jeffrey and Weitsman to give a simple proof of a theorem of Donaldson's, which states that the volume of the symplectic leaf defined by t in $\mathcal{M}_{g,1}$, as a function of t, is a Bernoulli polynomial (see [**78**]).

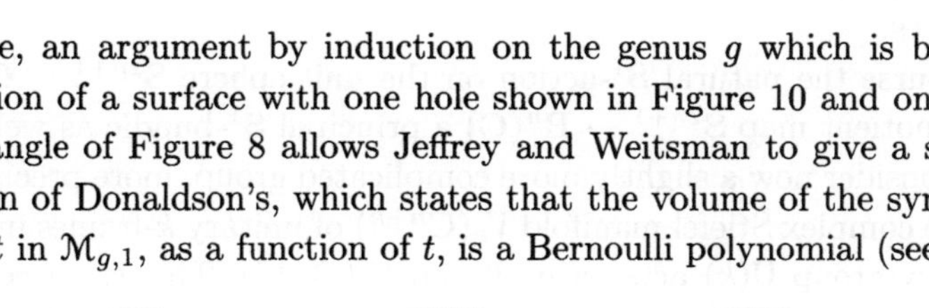

Figure 10

The torus actions on the moduli spaces are also used by Jeffrey and Weitsman to understand the cohomology of $\mathcal{M}$. Consider in particular the case of a closed surface, so that $\mathcal{M}$ is symplectic. They compute the number of integral points in $mP_{\mathbb{C}}$ (this is where you have to be very careful with the definition of the integral lattice—which I was not). It turns out that this number is given by the Verlinde formula[(7)]. Now this can be interpreted in the following way. The torus action defines a complex structure on $\mathcal{U}_{\mathbb{C}}$, and being integral, the symplectic form defines a complex line bundle $\mathcal{L}$ on $\mathcal{U}_{\mathbb{C}}$. Now, if we were in ordinary toric geometry (*i.e.*, with complete complex toric manifolds), the dimension of the space of holomorphic

[(7)] So that in particular, it does not depend on the actual trinion decomposition chosen!

sections of $\mathcal{L}^{\otimes m}$ would be the number of integral points in $mP_{\mathcal{C}}$ and thus would also be given by the Verlinde formula (see [**76**] for the details).

V.4. Appendix: connections on principal bundles

V.4.a. Principal bundles. A *principal* G-*bundle* is a space P endowed with a *free* G-action such that the orbit map $P \to B$ is a locally trivial fibration[(8)].

When the slice theorem (Theorem I.2.1) can be applied, the local triviality condition is automatically satisfied by a free action. We shall only consider compact groups here, but we cannot restrict to the case where P and B are manifolds, mainly because (in Chapter VI) we shall need some infinite dimensional spaces.

There is an obvious notion of *isomorphism* of principal G-bundles.

I shall, of course, take the liberty of denoting with the same letter the bundle and its total space.

Examples V.4.1

(1) In Chapter I, we have investigated principal $\mathbf{S}^1$-bundles over $\mathbf{S}^1$ and over surfaces.
(2) Of course the natural $\mathbf{S}^1$-action on the unit sphere $\mathbf{S}^{2n+1} \subset \mathbf{C}^{n+1}$ makes the quotient map $\mathbf{S}^{2n+1} \to \mathbf{P}^n(\mathbf{C})$ a principal $\mathbf{S}^1$-bundle as well.
(3) We consider now a slightly more complicated group, more precisely we look at the complex Stiefel manifold $V_k(\mathbf{C}^{n+k})$ of unitary k-frames in $\mathbf{C}^{n+k}$. The unitary group $\mathrm{U}(k)$ acts naturally on $V_k(\mathbf{C}^{n+k})$. This is a free action and the quotient is the complex Grassmannian[(9)] $G_k(\mathbf{C}^{n+k})$ of all k-dimensional vector subspaces in $\mathbf{C}^{n+k}$. The projection
$$V_k(\mathbf{C}^{n+k}) \longrightarrow G_k(\mathbf{C}^{n+k})$$
is a principal $\mathrm{U}(k)$-bundle (for $k=1$, this is $\mathbf{S}^{2n+1} \to \mathbf{P}^n(\mathbf{C})$).
(4) If the group G is discrete, a principal G-bundle is a Galois covering.

V.4.b. Connections. A *connection* on P is the choice of a horizontal subspace of T_pP for all points p of P, namely of a splitting of the exact sequence of vector bundles

$$0 \longrightarrow VP \underset{V_A}{\longrightarrow} \mathsf{T}P \xrightarrow{\mathsf{T}\pi} \pi^\star \mathsf{T}B \longrightarrow 0$$

which is also G-invariant. Here VP, the *vertical* bundle, is the kernel of $\mathsf{T}\pi$, intrinsically defined, contrary to the horizontal subbundles.

[(8)] Notice that the G-action on P is *free* and not only principal, a small incoherency in the terminology.

[(9)] See also Exercise III.16.

***Example V.4.2* (The trivial bundle).** Assume the principal bundle is (globally) trivial. Then there is an obvious connection on P, the *trivial* one, the horizontal subspace of which at $(g, x) \in \mathrm{G} \times B$ is

$$\{0\} \times \mathsf{T}_x B \subset \mathsf{T}_g \mathrm{G} \times \mathsf{T}_x B = \mathsf{T}_{(g,x)} (\mathrm{G} \times B).$$

For any other connection on the trivial bundle, the horizontal space $\mathcal{H}_{(g,x)}$ is the graph of a linear map $\mathsf{T}_x B \to \mathsf{T}_g \mathrm{G}$. The invariance property leads us to a linear map $\mathsf{T}_x B \to \mathfrak{g}$ (the Lie algebra of G), so that the space of all connections can be identified with the space $\Omega^1(B, \mathfrak{g})$ of 1-forms on B with values in $\mathfrak{g}$. This example is not only trivial, but also very important as any principal bundle is locally trivial.

More generally, a connection on P being fixed, the horizontal space of any other connection can be understood as the graph of a 1-form on B. The difference of two connections is an element of the vector space $\Omega^1(B, \mathfrak{g})$. The space $\mathcal{A}$ of all connections on P has thus the natural structure of an affine space modeled on this vector space. Notice that there exist, indeed, connections on principal G-bundles, as is asserted by Proposition V.4.3 below.

An important tool is the canonical 1-form on G. This can be considered as a complicated way to define a connection on the trivial bundle $\mathrm{G} \to \mathrm{pt}$ over a point. Concretely, this is the 1-form θ_0 with values in $\mathfrak{g}$ defined, at the point $g \in \mathrm{G}$ and on the vector $\mathsf{T}_1 g(X) \in \mathsf{T}_g \mathrm{G}$ (for $X \in \mathfrak{g}$) by

$$(\theta_0)_g(\mathsf{T}_1 g(X)) = \mathrm{Ad}_g(X) \in \mathfrak{g}.$$

This is an invariant form in the sense that

$$h^\star \theta_0 = \mathrm{Ad}_h \theta_0 \text{ for all } h \in \mathrm{G}.$$

Indeed, for $h, g \in \mathrm{G}$, $X \in \mathfrak{g}$, we have

$$\begin{aligned}
(h^\star \theta_0)_g(\mathsf{T}_g(X)) &= (\theta_0)_{h \cdot g}(\mathsf{T}_g h(\mathsf{T}_g(X))) \\
&= (\theta_0)_{h \cdot g}(\mathsf{T}_1(h \cdot g)(X)) \\
&= -\mathrm{Ad}_{h \cdot g}(X) \\
&= \mathrm{Ad}_h(\mathrm{Ad}_g(X)).
\end{aligned}$$

For instance, the horizontal distribution on the trivial bundle is the kernel of the form θ_0, viewed on $B \times \mathrm{G}$.

Proposition V.4.3. *On any principal* G*-bundle, there are connections.*

Proof. We construct a 1-form θ on P with values in $\mathfrak{g}$. The required horizontal distribution will be the kernel of θ. Let $(U_i)_{i \in I}$ be a covering of B such that

- the bundle is trivialized over U_i by diffeomorphisms $\varphi_i : p^{-1}(U_i) \to \mathrm{G} \times U_i$,
- there exists a partition of unity $(\psi_i)_{i \in I}$ with $\psi_i = 0$ out of U_i and $\sum_i \psi_i \equiv 1$.

We choose any connection on the trivial bundles $U_i \times \mathrm{G}$, for instance we choose the horizontal direction tangent to $\varphi_i(U_i \times \mathrm{G})$. Let now θ_i be the 1-form defined on $p^{-1}(U_i)$ by

$$\theta_i = \varphi_i^\star \theta_0$$

and θ be the 1 form defined on P by

$$(\theta)_x = \sum_i \psi(p(x))(\theta_i)_x.$$

Thanks to the invariance property of θ, the kernel of θ has the required invariance properties. □

Notice that we have constructed here a 1-form θ on P satisfying certain invariance conditions. This is actually equivalent to giving a connection (see Exercise V.4). A connection A being given, we have

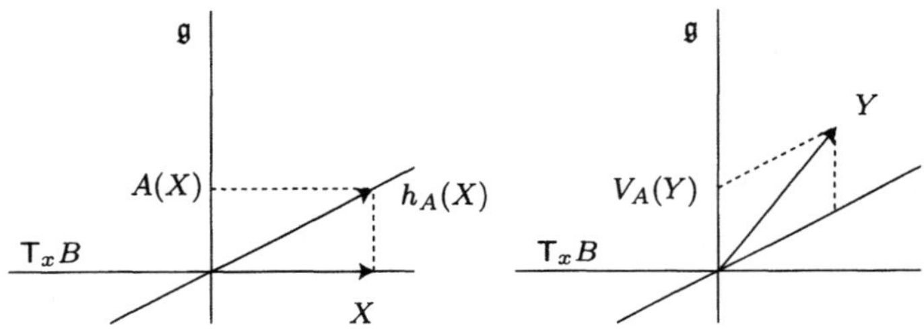

Figure 11

- a horizontal lift $X \mapsto h_A(X)$ for all vector fields on Σ
- a vertical projection $Y \mapsto V_A(Y)$ for all invariant vector fields on P. Notice that $V_A(Y)$ is nothing other than $\theta(Y)$ for the 1-form θ constructed in Proposition V.4.3.

They are shown in Figure 11.

Connections as derivations. The connections act on 0-forms with values in $\mathfrak{g}$ as:

$$\text{for } f \in \Omega^0(B, \mathfrak{g}), \qquad (d_A f)(X) = \nabla_X^A f = V_A[h_A(X), f].$$

Notice that

$$V_A[h_A(X), f] = V_A[h_A(X) \cdot f - f h_A(X)] = V_A(h_A(X) \cdot f).$$

For instance, for the trivial connection on the trivial bundle, this is simply the usual exterior derivative

$$(d_0 f)(X) = X \cdot f.$$

This extends to all $\mathfrak{g}$-valued forms, as the exterior derivative does. For instance, if $\varphi \in \Omega^1(P, \mathfrak{g})$, $d_A\varphi$ is the $\mathfrak{g}$-valued 2-form defined by

$$(d_A\varphi)(X,Y) = \nabla^A_X\varphi(Y) - \nabla^A_Y\varphi(X) - \varphi[X,Y].$$

The curvature. The *curvature* A of the connection A is defined as the 2-form on B with values in $\mathfrak{g}$ such that:

$$\text{for } X, Y \text{ vector fields on } B, \quad F(A)(X,Y) = V_A([h_A(X), h_A(Y)]).$$

V.4.c. The case of $\mathbf{S}^1$-bundles. Assume now that the group G is simply a circle $\mathbf{S}^1$. The Lie algebra is $\mathbf{R}$, the form θ_0 is the closed 1-form dz/z (better not call it $d\theta$ here). What Proposition V.4.3 gives us is an invariant 1-form, more precisely, a *connection form*, namely a 1-form α on P that satisfies, if X is the fundamental vector field of the $\mathbf{S}^1$-action,

- the invariance property $\mathcal{L}_X\alpha = 0$,
- the normalization condition $i_X\alpha \equiv 1$ (so that, for any vector field Z on P, $\alpha(Z)$ is simply $V_A(Z)$).

Example V.4.4. Let $\mathbf{S}^{2n-1} \subset \mathbf{C}^n$ denote, as usual, the unit sphere $\sum |z_i|^2 = 1$. Make $\mathbf{S}^1$ act diagonally by

$$u \cdot (z_1, \ldots, z_n) = (uz_1, \ldots, uz_n).$$

Recall that the fundamental vector field (associated with $\partial/\partial\theta$) of this action is

$$X_{(z_1,\ldots,z_n)} = (iz_1, \ldots, iz_n),$$

or, writing $z_j = q_j + ip_j$,

$$X = \sum_{j=1}^n \left(-p_j \frac{\partial}{\partial q_j} + q_j \frac{\partial}{\partial p_j}\right).$$

Let α be the 1-form (on $\mathbf{C}^n$, or on $\mathbf{S}^{2n-1}$) defined by

$$\alpha = \sum_{j=1}^n (-p_j dq_j + q_j dp_j).$$

By definition, we have $i_X\alpha \equiv 1$. On the other hand,

$$\begin{aligned} d\alpha &= \left(-\sum dp_j \wedge dq_j + \sum dq_j \wedge dp_j\right) \\ &= -2\omega \end{aligned}$$

where $\omega = \sum dp_j \wedge dq_j$ is the canonical symplectic form on $\mathbf{C}^n$. Since the vector field X preserves ω, we have

$$\begin{aligned}\mathcal{L}_X\omega &= di_X\alpha + i_X d\alpha \\ &= i_X(-2\omega) \\ &= 2dH = 0\end{aligned}$$

for $H = \frac{1}{2}\sum |z_i|^2$, constant on the sphere. In other words, α is a connection form on the principal $\mathbf{S}^1$-bundle $\pi : \mathbf{S}^{2n-1} \to \mathbf{P}^{n-1}(\mathbf{C})$.

Turning back to a general principal $\mathbf{S}^1$-bundle, let us look now at the form $d\alpha$. It satisfies $i_X d\alpha = 0$ and $\mathcal{L}_X d\alpha = di_X d\alpha = 0$. Thus there exists a 2-form η, the *curvature form*, such that $\pi^\star\eta = d\alpha$ and which is closed because $d\alpha$ is.

Example V.4.5. Consider once again the Hamiltonian $H = \frac{1}{2}\sum |z_i|^2$ and its reduction σ_a at the level a (see Example III.2.18). That is, consider the sphere of radius $\sqrt{2a}$ and its projection on $\mathbf{P}^{n-1}(\mathbf{C})$. As above, the form

$$\alpha_a = \frac{1}{2a}\sum_{j=1}^{n}(-p_j dq_j + q_j dp_j)$$

is a connection form, with

$$d\alpha_a = -\frac{\omega}{a} = -\pi^\star\frac{\sigma_a}{a}$$

with the notation of III.2.18, so that the curvature form, on $\mathbf{P}^{n-1}(\mathbf{C})$, is precisely the 2-form $-\dfrac{\sigma_a}{a}$.

The definition of the curvature used here is of course just another viewpoint on the same object as above, namely, for an $\mathbf{S}^1$-bundle, the curvature F_A and the curvature form σ agree (up to sign). Indeed, let Y, Z be two vector fields on B. Let us lift them to P as $h_A(Y)$, $h_A(Z)$ and compute

$$\begin{aligned}\sigma(Y,Z) &= (d\alpha)(h_A(Y), h_A(Z)) \text{ by definition of } \sigma \\ &= h_A(Y)\cdot\alpha(h_A(Z)) - h_A(Z)\cdot\alpha(h_A(Y)) - \alpha([h_A(Y), h_A(Z)]) \\ &= -\alpha([h_A(Y), h_A(Z)], \text{ horizontal vectors being in the kernel of } \alpha \\ &= -V_A([h_A(Y), h_A(Z)]) \text{ by definition of } \alpha.\end{aligned}$$

Remark V.4.6. If α, α' are two connection forms on the same principal $\mathbf{S}^1$-bundle $\pi : P \to B$, the 1-form $\alpha - \alpha'$ is invariant and vanishes on vertical vectors, since $i_X\alpha = i_X\alpha' = 1$ for the fundamental vector field. Hence there exists a 1-form β on the basis B, such that $\alpha - \alpha' = \pi^\star\beta\ldots$ so that we have, for the curvature forms,

$$\eta - \eta' = d\beta,$$

in particular, the de Rham cohomology class of the curvature form η does not depend on the choice of the connection form. In particular, if the bundle is trivial, choosing a trivialization gives a trivial connection, the connection form of which is simply the volume form on $\mathbf{S}^1$, a closed 1-form on $\mathbf{S}^1 \times B$, so that the curvature of *any* connection form on a trivial principal $\mathbf{S}^1$-bundle is an *exact* 2-form.

Exercises

Exercise V.1. Let M_1, M_2 and M_3 be three elements in $\mathrm{G} = \mathrm{SU}(n)$ such that

- $M_1 M_2 M_3 = 1$,
- there is no line in $\mathbf{C}^n$ which is invariant under M_1, M_2 and M_3.

Consider the two linear maps f and $g : \mathfrak{g} \to \mathfrak{g}$ defined by

$$f(X) = M_1 X M_1^{-1} - X \text{ and } g(X) = M_3^{-1} X M_3 - x$$

and prove that

$$\dim \operatorname{Ker} f + \dim \operatorname{Ker} g \leqslant \dim \mathfrak{g}.$$

Let C_1, C_2 and C_3 be the conjugacy classes in G of the matrices M_1, M_2 and M_3. Consider the differentiable map

$$\begin{aligned} h : C_1 \times C_2 \times C_3 &\longrightarrow \mathrm{G} \\ (A_1, A_2, A_3) &\longmapsto A_1 A_2 A_3. \end{aligned}$$

Check that its differential at (M_1, M_2, M_3) is given by

$$\begin{aligned} &\mathsf{T}_{(M_1,M_2,M_3)} h([X, M_1], [Y, M_2], [Z, M_3]) \\ &\qquad = M_1(Y - X)M_1^{-1} - (Y - X) + M_3^{-1}(Z - Y)M_3 - (Z - Y). \end{aligned}$$

Deduce that h has maximal rank at (M_1, M_2, M_3) and that

$$\dim \mathcal{M}_{0,3}^{C_1,C_2,C_3} = \sum \dim C_i - 2 \dim \mathrm{G}.$$

Exercise V.2. Fix conjugacy classes $\Gamma_1, \ldots, \Gamma_d$ in G and consider the symplectic leaf $\mathcal{M}^{\Gamma_1,\ldots,\Gamma_d} \subset \mathcal{M}_{g,d}$. Let C be a simple closed curve disjoint from the boundary. The map Φ_C of § *V*.3 can be considered as a map

$$\Phi_C : \mathcal{M}^{\Gamma_1,\ldots,\Gamma_d} \longrightarrow \widehat{L\mathfrak{g}}^{\star}/L\mathrm{G}.$$

Cut the surface along the curve C, getting a new surface with two more boundary components. Fix a conjugacy class Γ in G and consider the symplectic leaf

$$\mathcal{M}^{\Gamma_1,\ldots,\Gamma_d,\Gamma,\Gamma^{-1}} \subset \mathcal{M}_{g-1,d+2}.$$

Prove that the natural map

$$\Phi_C^{-1}(\Gamma) \longrightarrow \mathcal{M}^{\Gamma_1,\ldots,\Gamma_d,\Gamma,\Gamma^{-1}}$$

is the symplectic reduction of the level Γ in $\mathcal{M}^{\Gamma_1,\ldots,\Gamma_d}$.

Exercise V.3. For $A \in \mathcal{A}$, define

$$\begin{aligned} d_A : \Omega^0(\Sigma, \mathfrak{g}) &\longrightarrow \Omega^1(\Sigma, \mathfrak{g}) \\ \alpha &\longmapsto d\alpha + [A, \alpha] \end{aligned}$$

where $[A, \alpha]$ is the 1-form, defined on the tangent vectors to Σ by

$$X \longmapsto [A(X), \alpha],$$

and similarly

$$\begin{aligned} d_A : \Omega^1(\Sigma, \mathfrak{g}) &\longrightarrow \Omega^2(\Sigma, \mathfrak{g}) \\ \varphi &\longmapsto d\varphi + [A, \varphi] \end{aligned}$$

where $[A, \varphi]$ is the 2-form, defined on the pairs of tangent vectors to Σ by

$$[A, \varphi](X, Y) = [A(X), \varphi(Y)] - [A(Y), \varphi(X)].$$

Check that d_A is a derivation for the bracket of forms, namely that:

$$d_A[\varphi, \psi] = [d_A\varphi, \psi] \pm [\varphi, d_A\psi].$$

Prove that, for $\alpha \in \Omega^0(\Sigma, \mathfrak{g})$,

$$d_A \circ d_A(\alpha) = [d_A A, \alpha].$$

Deduce that

$$d_A \circ d_A : \Omega^0(\Sigma, \mathfrak{g}) \longrightarrow \Omega^2(\Sigma, \mathfrak{g})$$

is the action of the 2-form $F(A) = d_A A$ with values in $\mathfrak{g}$ (the curvature form).

Exercise V.4. Let $P \to B$ be a principal G-bundle. Prove that, associated to any connection on P, there is a a differentiable 1-form θ on P taking values in $\mathfrak{g}$, and which satisfies the two properties

- $\forall X \in \mathfrak{g}$, we have $i_{\underline{X}}\theta = X$,
- θ is invariant, in the sense that, $\forall g \in \mathrm{G}$, $g^\star\theta = \mathrm{Ad}_g \cdot\theta$.

Conversely, prove that such a form determines a connection on P.

Exercise V.5. Prove that, in the case of $\mathbf{S}^1$, the two properties of the form θ given in Exercise V.4 are equivalent to the two properties

$$\mathcal{L}_X\theta = 0 \text{ and } i_X\theta \equiv 1$$

(as in § V.4.c).

CHAPTER VI

EQUIVARIANT COHOMOLOGY AND THE DUISTERMAAT–HECKMAN THEOREM

Another famous and spectacular theorem in the theory of Hamiltonian torus actions is the Duistermaat–Heckman Theorem [**44**] (as we have learned since that time, similar results had been proved by Karasev [**81**] in 1981). This theorem has two versions, both asserting, once again, the importance of linear phenomena in the theory.

In the first theorem (here Theorem VI.2.3), the authors consider the reduced symplectic forms ω_ξ and ω_η on the quotients of two regular levels $\mu^{-1}(\xi)$ and $\mu^{-1}(\eta)$ which correspond to values ξ and η located in the same component of the set of regular values of the momentum mapping μ. One may then identify the quotients B_ξ and B_η and consider the difference of cohomology classes $[\omega_\xi] - [\omega_\eta] \in H^2(B_\xi)$. The theorem asserts that it is a *linear* function of $\xi - \eta$.

The second statement (here Theorem VI.3.18) is usually shortened to: a periodic Hamiltonian satisfies the "exact stationary phase formula".

It was Berline and Vergne [**22**] who first explained that this statement is a special case of a "localization theorem" in equivariant cohomology. This was a very productive idea: the language of equivariant cohomology actually fits very well with the investigation of Hamiltonian actions, as one might be readily convinced by looking at Proposition VI.2.1.

Later came two "survey" papers on the subject, one by Atiyah and Bott [**11**] in 1984 and one by Ginzburg [**53**] in 1987. From these come the methods and results I shall discuss in the present chapter.

As I have already said, the language of equivariant cohomology fits so well here that we shall see that the first of the Duistermaat–Heckman theorems becomes practically tautological (this will not detract from its beauty)... Of course we will have to pay for this, namely we will have to define equivariant cohomology, the Borel construction and classifying spaces... Thus we shall begin by recalling the classical results and constructions of Milnor and Dold, recommending the reader to consult these authors and [**69**] for details.

VI.1. Milnor joins, Borel construction and equivariant cohomology

VI.1.a. Principal and universal bundles. We shall consider *numerable* coverings, that is, locally finite coverings $(U_i)_{i\in I}$ such that there exists a partition of unity $(u_i)_{i\in I}$ with:

$$\overline{u_i^{-1}(]0,1])} \subset U_i, \quad \forall i \in I.$$

Notice, for instance, that any paracompact space has a numerable covering.

A principal bundle over B is said to be *numerable* if there exists a numerable covering of B which makes it locally trivial. In this chapter, all the principal bundles will be assumed to be numerable, even if I forget to mention it explicitly.

Let $p : E \to B$ be a principal G-bundle, and let $f : B' \to B$ be a continuous map, then the pull-back

$$f^\star E = \{(e, b') \mid p(e) = f(b')\}$$

or fibered product in the diagram

$$\begin{array}{ccc} f^\star E & \longrightarrow & E \\ \downarrow & & \downarrow \\ B' & \xrightarrow{\;f\;} & B \end{array}$$

is a principal G-bundle as well, and one shows[1] without too much difficulty:

Theorem VI.1.1. *Let $f_t : B' \to B$ be a homotopy. Then the two bundles $f_0^\star E$ and $f_1^\star E$ are isomorphic.* □

Universal bundles. We try now to construct a principal G-bundle which is universal. In other words, we want a bundle $\mathcal{E} \to \mathcal{B}$ such that *any* principal G-bundle is induced by a map $B \to \mathcal{B}$. More precisely, we will say that a numerable principal G-bundle $\mathcal{E} \to \mathcal{B}$ is *universal* if

- For any numerable principal G-bundle $E \to B$, there exists a map $f : B \to \mathcal{B}$ such that E is isomorphic to $f^\star\mathcal{E}$.
- Two maps $f, g : B \to \mathcal{B}$ induce isomorphic bundles if *and only if* they are homotopic.

Milnor [**107**] has given a very beautiful (and, furthermore, explicit) construction of universal bundles, called the *Milnor join*. Let

$$E\mathrm{G} = \mathrm{G} \star \mathrm{G} \star \cdots \star \mathrm{G} \star \cdots$$

be the "infinite join", or more explicitly:

$$E\mathrm{G} = \varinjlim E\mathrm{G}(n)$$

[1] See for example [**69**].

where $EG(n)$ is the quotient of the product

$$G^{n+1} \times \Delta^n = \left\{ (x_0, t_0; x_1, t_1; \ldots; x_n, t_n) \mid x_i \in G, t_i \in [0,1], \sum t_i = 1 \right\}$$

by the equivalence relation

$$(x_0, t_0; \ldots; x_n, t_n) \sim (x'_0, t'_0; \ldots; x'_n, t'_n) \text{ if and only if } \begin{cases} t_i = t'_i & \forall i \text{ and} \\ t_i = t'_i \neq 0 \Rightarrow x_i = x'_i. \end{cases}$$

We shall write $\langle x_0, t_0; \ldots; x_n, t_n \rangle$ for the equivalence class of the element under consideration.

Examples VI.1.2

(1) Assume first that $G = \mathbf{Z}/2 = \{\pm 1\}$. Then $EG(n)$ can be identified with $\mathbf{S}^n$ by the map

$$\begin{array}{ccc} EG(n) & \longrightarrow & \mathbf{S}^n \\ \langle x_0, t_0; \ldots; x_n, t_n \rangle & \longmapsto & (\sqrt{t_0}x_0, \ldots, \sqrt{t_n}x_n). \end{array}$$

(2) Let now G be the circle $\mathbf{S}^1$. The very same formulas identify $EG(n)$ with the sphere $\mathbf{S}^{2n+1}$.

The limit EG is defined by the inclusion maps

$$\begin{array}{ccc} EG(n) & \longrightarrow & EG(n+1) \\ \langle x_0, t_0; \ldots; x_n, t_n \rangle & \longmapsto & \langle x_0, t_0; \ldots; x_n, t_n; x_{n+1}, 0 \rangle. \end{array}$$

Any element in EG will thus be written

$$e = \langle x_0, t_0; \ldots; x_n, t_n; \ldots \rangle$$

or for brevity $e = \langle x, t \rangle$. Notice that, for any element of EG, all the t_i's, except a finite number, are zero. We endow this set with the least expensive *topology*(2) such that all the maps

$$\begin{array}{rcl} t_i : EG & \longrightarrow & [0,1] \\ \langle x_0, t_0; \ldots; x_n, t_n \rangle & \longmapsto & t_i \end{array}$$

and

$$\begin{array}{ccc} t_i^{-1}(]0,1]) & \longrightarrow & G \\ \langle x_0, t_0; \ldots; x_n, t_n \rangle & \longmapsto & x_i \end{array}$$

are continuous. Notice that the group G acts on $EG(n)$ and EG by

$$g \cdot \langle x, t \rangle = \langle gx, t \rangle,$$

(2) This the *direct limit topology*. See the comments on the possible names for this topology in [**109**, p. 63].

clearly free actions. The quotients will be denoted respectively $BG(n)$ and BG. Since G does not act on the coordinate t_i, there is also a continuous map

$$t_i : BG \longrightarrow [0,1].$$

It turns out that the open subsets $V_i = t_i^{-1}(]0,1])$ constitute a numerable covering of BG and that the principal bundle $EG \to BG$ is trivialized on these open sets (see [**40**] for detail).

Examples VI.1.3

(1) As we have seen, for $\mathrm{G} = \mathbf{Z}/2$, $E\mathrm{G}(n) = \mathbf{S}^n$. Thus

$$E\mathbf{Z}/2 = \mathbf{S}^\infty = \left\{(x_0, \ldots, x_n, \ldots) \mid \text{ almost all } x_i = 0 \text{ and } \sum x_i^2 = 1\right\}.$$

The group acts by the antipodal map, so that $BG(n) = \mathbf{P}^n(\mathbf{R})$ and $BG = \mathbf{P}^\infty(\mathbf{R})$.

(2) In the same way, for $\mathrm{G} = \mathbf{S}^1$, $BG(n) = \mathbf{P}^n(\mathbf{C})$, and $BG = \mathbf{P}^\infty(\mathbf{C})$.

We shall now explain (without too much proof) that these spaces were *made* to be universal. One begins by showing that all numerable bundles have a *numerable* partition of unity which makes them locally trivial. Granted this, we consider the t_i's on EG as a "universal partition" of unity ($\sum t_i = 1$), to prove:

Proposition VI.1.4. *For any numerable principal* G*-bundle, E over B, there exists a map $f : B \to BG$ such that $f^\star EG$ is isomorphic with E.*

Proof. We construct two maps f and g such that the diagram

$$\begin{array}{ccc} E & \xrightarrow{\ g\ } & EG \\ \downarrow & & \downarrow \\ B & \xrightarrow{\ f\ } & BG \end{array}$$

commutes. For that, we use a partition of unity $(u_n)_{n\geqslant 0}$ on B such that $E|_{u_n^{-1}(]0,1])}$ is trivial. Put $U_n = u_n^{-1}(]0,1])$. Let

$$\begin{array}{ccc} U_n \times \mathrm{G} & \xrightarrow{\quad h_n \quad} & E|_{U_n} \\ & \searrow \quad \swarrow & \\ & U_n & \end{array}$$

be a local trivialization, and call q_n the projection $U_n \times \mathrm{G} \to \mathrm{G}$. The map g we want to define is simply

$$g(z) = \langle q_0 h_0^{-1}(z), u_0(p(z)); \ldots ; q_n h_n^{-1}(z), u_n(p(z)); \ldots \rangle.$$

This is better defined than one might think: of course, $h_n^{-1}(z)$ is defined only for $z \in p^{-1}(U_n)$, but, if this is not the case, $u_n(p(z)) = 0$.

The map g is necessarily an isomorphism. We have more generally:

Lemma VI.1.5. *Let* $E_1 \to B_1$ *and* $E_2 \to B_2$ *be two principal* G*-bundles. Any morphism*

$$\begin{array}{ccc} E_2 & \xrightarrow{g} & E_1 \\ \downarrow & & \downarrow p \\ B_2 & \xrightarrow{f} & B_1 \end{array}$$

induces an isomorphism $g' : E_2 \to f^\star E_1$.

Of course, this lemma ends the proof of the proposition. □

Proof of the lemma. Recall that $f^\star E_1$ can be defined as the fibered product

$$f^\star E_1 = \{(e_1, b_2) \mid f(b_2) = p(e_1)\}$$

so that the map $g' : E_2 \to f^\star E_1$, defined by

$$g'(e_2) = (g(e_2), p(e_2))$$

is a morphism of principal G-bundles over B_2. It is easy to check that g' is *injective.* If $g'(e) = g'(e')$, then $p(e) = p(e')$ and $e' = x \cdot e$ for some $x \in$ G; applying g' again, one sees that $x = 1$. Moreover g' is *surjective* as well: let $e \in f^\star E_1$, and let e' be a point in the fiber of the image of e in B_2. One can find an $x \in$ G such that $g'(xe') = e$. □

It is slightly more technical, but not much more difficult, to prove that the Milnor join is indeed universal. From this universality, some kind of uniqueness follows: if $E_1 \to B_1$ and $E_2 \to B_2$ are two universal principal G-bundles, writing explicitly this property for the former gives a morphism

$$\begin{array}{ccc} E_2 & \longrightarrow & E_1 \\ \downarrow & & \downarrow \\ B_2 & \xrightarrow{f} & B_1 \end{array}$$

and similarly, writing that $E_2 \to B_2$ is universal gives a morphism $g : B_1 \to B_2$. As $f \circ g : B_1 \to B_1$ induces an isomorphism, it is homotopic to the identity, and the same is true for $g \circ f$: thus the homotopy type of the universal space BG is well defined. As we shall use these spaces only in algebraic topology calculations, the uniqueness of their homotopy type will be sufficient for our purposes.

In the same mood, one proves (see [**40**]) the following useful result.

Proposition VI.1.6. *Let* $E \to B$ *be any numerable principal* G*-bundle such that the total space* E *is* contractible, *then this is a universal principal* G*-bundle.* □

In [**107**], Milnor has proved that his spaces are *weakly* contractible (this means that their homotopy groups are trivial), but in the mentioned paper, Dold has shown that Milnor joins are actually contractible. Consider for example the infinite sphere $\mathbf{S}^\infty$, namely our example of space EG for G $= \mathbf{Z}/2$ or G $= \mathbf{S}^1$. Using the fact that $\mathbf{S}^n$ is $(n-1)$-connected immediately gives the fact that $\mathbf{S}^\infty$ is weakly contractible. The reason why the infinite sphere is actually contractible is that any sphere $\mathbf{S}^n$ can be contracted to a point in $\mathbf{S}^{n+1}$. To derive a complete proof is a little bit more elaborate (see Exercise VI.1). We shall admit the general fact that the spaces EG we have described are indeed contractible. This has readily a few applications.

Examples VI.1.7

(1) If H is a subgroup of G, it acts freely on EG, so that EG/H is a model for BH.
(2) If G and H are two groups, the product G × H acts on EG × EH in the way you imagine and the quotient is BG × BH which is a model for B(G × H).
(3) If f : G → H is a group morphism, it induces a continuous map BG → BH: this is easy to construct on the "Milnor join" models.

From these "theoretical" examples, one gets concrete examples of universal bundles.

Examples VI.1.8

(1) From the inclusion of the cyclic group $\mathbf{Z}/m \subset \mathbf{S}^1$, we deduce that the quotient of $\mathbf{S}^{2n+1}$ by the equivalence relation

$$(z_0, \ldots, z_n) \sim (\zeta z_0, \ldots, \zeta z_n) \text{ if } \zeta^m = 1,$$

namely the lens space L_m, is a finite approximation to $B(\mathbf{Z}/m)$, the latter then being the "infinite lens space" $\mathbf{S}^\infty/(\mathbf{Z}/m)$.
(2) In the same way, the torus $\mathbf{T}^m = \mathbf{S}^1 \times \cdots \times \mathbf{S}^1$ acts diagonally on the product $\mathbf{S}^{2n+1} \times \cdots \times \mathbf{S}^{2n+1}$, which is thus found to be an approximation to $E\mathbf{T}^m$ and $B\mathbf{S}^1 \times \cdots \times B\mathbf{S}^1$ becomes therefore a model for $B\mathbf{T}^m$.
(3) Consider the join $E\,\mathrm{U}(k)(n) = \{\langle A_0, t_0; \ldots; A_n, t_n\rangle \mid A_i \in \mathrm{U}(k)\}$ and the map into the Stiefel manifold $V_k(\mathbf{C}^{k(n+1)})$ which, with any $\langle A, t\rangle$, associates the k column vectors in the $k(n+1)$ rows matrix

$$\begin{pmatrix} \sqrt{t_0}A_0 \\ \vdots \\ \sqrt{t_n}A_n \end{pmatrix}.$$

Taking limits and passing to quotients, this induces a morphism

$$\begin{array}{ccc} E\,\mathrm{U}(k) & \longrightarrow & V_k(\mathbf{C}^\infty) \\ \downarrow & & \downarrow \\ B\,\mathrm{U}(k) & \longrightarrow & G_k(\mathbf{C}^\infty). \end{array}$$

As for the infinite sphere (which is the case $k = 1$), it is easy to prove that the infinite Stiefel manifold $V_k(\mathbf{C}^\infty)$ is weakly contractible: the homotopy exact sequence of the fibration[(3)]

$$V_{k-1}(\mathbf{C}^N) \longrightarrow V_k(\mathbf{C}^N) \longrightarrow \mathbf{S}^{2N-1}$$

gives that the map

$$\pi_n(V_{k-1}(\mathbf{C}^N)) \longrightarrow \pi_n(V_k(\mathbf{C}^N))$$

is an isomorphism for N large enough, from what the expected result is deduced by induction on k. We conclude that the infinite Grassmann manifold $G_k(\mathbf{C}^\infty)$ is a model for $B\,\mathrm{U}(k)$.

VI.1.b. The Borel construction. If W is a space acted on by the group G, we know perfectly well that the orbit space W/G may be rather complicated. The idea in the Borel construction is to try to get a reasonable substitute for this quotient.

Make G act on $E\mathrm{G} \times W$ by

$$g \cdot (e, x) = (g \cdot e, g \cdot x).$$

This action is free, simply because it is free on the first factor. The quotient space $W_\mathrm{G} = E\mathrm{G} \times_\mathrm{G} W$ is called the *Borel construction* on W.

Remark VI.1.9. In this way we have only defined the homotopy type of this space.

Consider the two projection maps from the product $E\mathrm{G} \times W$:

– the projection to $E\mathrm{G}$

$$\begin{array}{ccc} E\mathrm{G} \times W & \longrightarrow & E\mathrm{G} \\ \downarrow & & \downarrow \\ W_\mathrm{G} & \longrightarrow & B\mathrm{G} \end{array}$$

induces a fibration $W_\mathrm{G} \to B\mathrm{G}$ with fiber W,

– the projection $E\mathrm{G} \to W$ induces a map

$$\sigma : W_\mathrm{G} \longrightarrow W/\mathrm{G}.$$

The latter is not a fibration in general, but we have nevertheless:

(3) See Exercise VI.2.

Proposition VI.1.10. *Let* G *be a compact Lie group acting* freely *on a space W. Then the map*

$$\sigma : W_{\mathrm{G}} \longrightarrow W/\mathrm{G}$$

is a fibration with contractible fibers, and in particular a homotopy equivalence.

Proof. Denote by [ ] the classes, either in $E\mathrm{G} \times_{\mathrm{G}} W$ or in W/G. Let us investigate first the fibers of σ in the general case. Let $x \in W$. The fiber of $[x]$ is

$$\sigma^{-1}([x]) = \{[e, y] \in E\mathrm{G} \times_{\mathrm{G}} W \mid \mathrm{proj}[e, y] = [x]\}.$$

Let us look for the classes $[e, x]$ for $e \in E\mathrm{G}$. Of course, $[e', x] \sim [e, x]$ if and only if there exists an element $g \in \mathrm{G}$ such that $e' = g \cdot e$ and $x = g \cdot x$, that is, such that $g \in \mathrm{G}_x$. Hence $\sigma^{-1}([x]) = E\mathrm{G}/\mathrm{G}_x$, which is a model for $B\mathrm{G}_x$.

When the G-action is free, all the fibers are thus spaces $E\mathrm{G}$, in particular they are contractible. The local triviality is a consequence of that of $W \to W/\mathrm{G}$, that is, of the slice theorem. □

It is in this sense that one may say that W_{G} is the *homotopy theoretical quotient* of W:

- this is a "good" quotient,
- when the genuine quotient is "good", it has the same homotopy type.

On the contrary, if the G-action on W is trivial, then $W_{\mathrm{G}} = B\mathrm{G} \times W$; this is the case, for instance, when W is a point, in which case $W_{\mathrm{G}} = B\mathrm{G}$.

VI.1.c. Equivariant cohomology. Let W be a topological space acted on by a group G. The *equivariant cohomology* of W is the cohomology of the Borel construction W_{G}. It is written $H^\star_{\mathrm{G}}(W)$.

There are as many possible theories as there are cohomology theories. Here we shall mainly use de Rham cohomology. That is to say that we shall assume W to be a smooth manifold, with a G-action which is smooth as well. Unfortunately, neither $E\mathrm{G}$ nor *a fortiori* W_{G} are manifolds, which causes some trouble: what could be a differentiable form on W_{G}? All the groups we shall be interested in are subgroups[(4)] of some $\mathrm{U}(n)$ This allows us to use the infinite Stiefel manifold $V_n(\mathbf{C}^\infty) = \cup_k V_n(\mathbf{C}^{n+k})$ of unitary k-frames in $\mathbf{C}^\infty$ (or any other union of manifolds that would fit) as a model for $E\mathrm{G}$. A differentiable form on such a space is simply a family of differentiable forms, that agree *via* the inclusions[(5)].

Remark VI.1.11. The reader may ask why we are doing such complicated things. A simpler way to define an "equivariant cohomology" would seem to be to consider the cohomology of G-invariant forms on W. Actually, when G is compact and

(4) Every compact Lie group can be embedded in some unitary group $\mathrm{U}(n)$ for n large enough.
(5) In general, we shall content ourselves with finite approximations of $E\mathrm{G}$ or W_{G}.

connected, this is nothing else than the usual de Rham cohomology of W, as Exercise VI.4 shows. This fake "equivariant cohomology" has nothing more to say than the cohomology of W. On the contrary, we have very good reasons to think that the cohomology of W_G says more: for example, if W is a point, then $W_G = BG$ and $H^\star_G(\mathrm{pt}) = H^\star(BG)$, which is never zero.

On the other hand, $H^\star_G(W)$, being the cohomology of a space W_G, is, in particular, a ring. It has even more structure: using the projection map

$$W_G \xrightarrow{\ \pi\ } BG,$$

it receives the structure of an $H^\star_G(\mathrm{pt})$-module. We shall mainly concentrate on this structure, which has a lot to say about the action. For instance if the action is free, $H^\star_G(W) \cong H^\star(W/G)$ is a *torsion* $H^\star_G(\mathrm{pt})$-module.

The first thing to understand is what kind of a ring $H^\star_G(\mathrm{pt})$ is. We limit ourselves to the case where G is a torus, and we begin of course with the case of the group $G = \mathbf{S}^1$.

Theorem VI.1.12. *The cohomology ring $H^\star(B\mathbf{S}^1)$ is a polynomial ring, on a generator u of degree 2.*

Remark VI.1.13. This is true over $\mathbf{Z}$ and therefore over any ring of coefficients, as will be the case for all statements in this chapter which are not given with more precision.

We have said that the infinite complex projective space is a model for $B\mathbf{S}^1$. We only need to calculate its cohomology. This is deduced from the cohomology of the (finite) projective spaces. Recall the classical proposition:

Proposition VI.1.14. *The cohomology ring $H^\star(\mathbf{P}^n(\mathbf{C}); \mathbf{Z})$ is a truncated polynomial ring $\mathbf{Z}[u]/u^{n+1}$ on a generator u of degree 2. The inclusion $j : \mathbf{P}^n(\mathbf{C}) \subset \mathbf{P}^{n+1}(\mathbf{C})$ induces the obvious (truncation) map*

$$j^\star : \mathbf{Z}[u]/u^{n+2} \longrightarrow \mathbf{Z}[u]/u^{n+1}.$$

The theorem follows easily from the proposition (the cohomology of a direct limit of spaces is the inverse limit of the cohomologies). As a benefice to innocent readers, there is a proof of the proposition in § VI.4. Merely recall here that u is the *Euler class* of the complex line bundle $\mathcal{O}(1)$, the *canonical* bundle over $\mathbf{P}^\infty(\mathbf{C})$.

Consider now a torus, written as a product of circles

$$\mathbf{T}^n = \mathbf{S}^1 \times \cdots \times \mathbf{S}^1$$

to derive:

Corollary VI.1.15. *Let $\mathbf{T}$ be a torus of dimension m. Then $H^\star(B\mathbf{T})$ is a polynomial ring on m variables of degree 2.* □

Generators for de Rham cohomology. Using the previous results, we exhibit now "the" generator of $H^2(B\mathbf{S}^1;\mathbf{R})$.

Consider the connection form α given in Example V.4.4 for the $\mathbf{S}^1$-principal bundle $\mathbf{S}^{2n-1}\to\mathbf{P}^{n-1}(\mathbf{C})$. Its curvature form η is a closed 2-form on $\mathbf{P}^{n-1}(\mathbf{C})$. This form is not exact, as it is, up to a scalar factor, the reduced symplectic form, precisely

$$\eta=-\frac{\sigma_a}{a}$$

for σ_a the reduced form obtained from the sphere of radius $\sqrt{2a}$ (see Example V.4.5). It thus represents a *nonzero* element in $H^2(\mathbf{P}^{n-1}(\mathbf{C});\mathbf{R})$.

The cohomology classes defined are compatible *via* the inclusion maps $\mathbf{C}^n\subset\mathbf{C}^{n+1}$, so that what we have actually defined is a nonzero element $[\eta]$ in $H^2(B\mathbf{S}^1;\mathbf{R})$. We still have to compare it with u, which is characterized by the fact that $\langle u,[\mathbf{P}^1(\mathbf{C})]\rangle=1$. Recall from Example V.4.5 and Exercise III.11 that

$$\int_{\mathbf{P}^1(\mathbf{C})}\sigma_a=2\pi a,\ \text{ so that }\int_{\mathbf{P}^1(\mathbf{C})}\eta=-2\pi.$$

Thus the cohomology class of $-\eta/2\pi$ coincides with u.

Let us now consider the cohomology of $B\mathbf{T}$. Once a basis $(X_1,\dots,X_m)$ of the kernel of the exponential in $\mathfrak{t}\cong\mathbf{R}^m$ has been chosen, we get an isomorphism

$$\varphi:\mathbf{S}^1\times\cdots\times\mathbf{S}^1\longrightarrow\mathbf{T}$$

a homotopy equivalence

$$\Phi:B\mathbf{S}^1\times\cdots\times B\mathbf{S}^1\longrightarrow B\mathbf{T}$$

and a ring isomorphism

$$\Phi^\star:H^\star B\mathbf{T}\longrightarrow H^\star B\mathbf{S}^1\otimes\cdots\otimes H^\star B\mathbf{S}^1\cong\mathbf{R}[u_1,\dots,u_m]$$

(using de Rham cohomology). The torus $\mathbf{T}$ acts on $(\mathbf{C}^n)^m$ by

$$(t_1,\dots,t_m)\cdot(Z_1,\dots,Z_m)=(t_1Z_1,\dots,t_mZ_m)$$

where $Z_i\in\mathbf{C}^n$. We can thus use the product of m copies of the unit sphere $\mathbf{S}^{2n-1}\times\cdots\times\mathbf{S}^{2n-1}$ as a finite approximation to $E\mathbf{T}$. On this product, we have m forms $\alpha_1,\dots,\alpha_m$ with $i_{X_j}\alpha_i=\delta_{i,j}$ and the isomorphism

$$H^2(B\mathbf{T})\longrightarrow\mathfrak{t}^\star$$

which sends $(u_1,\dots,u_m)$ to the basis dual to $(X_1,\dots,X_m)$. It is easily checked that this isomorphism does not depend on the choice of the basis $(X_1,\dots,X_m)$. Hence we can identify $H^2(B\mathbf{T};\mathbf{R})$ with $\mathfrak{t}^\star$ and the polynomial ring $\mathbf{R}[u_1,\dots,u_m]$ becomes the algebra $S(\mathfrak{t}^\star)$ of multilinear symmetric forms on $\mathfrak{t}$.

Remark VI.1.16. The reader may find that this is not very elegant to

– choose a basis,

– then use it to define an isomorphism
– and then prove the latter does not depend on the basis.

If this is the case, she or he should re-write this paragraph in a more intrinsic way.

Euler classes for fixed point free T-actions. Using equivariant cohomology, let us define the *Euler class* for an $\mathbf{S}^1$-bundle or a $\mathbf{T}$-bundle, thus generalizing the definition given in Chapter I.

To begin with, we consider a *free* $\mathbf{S}^1$-action on a space W, and the quotient B of this action. We have seen (in Proposition VI.1.10) that the natural map

$$\sigma^\star : H^\star(B) \longrightarrow H^\star_{\mathbf{S}^1}(W)$$

is an isomorphism. There exists thus a unique element in $H^2(B;\mathbf{Z})$ that is mapped to $-u$. This is called the *Euler class* of the principal $\mathbf{S}^1$-bundle $W \to B$. Notice that this Euler class is *natural*, in the sense that, in a pull-back diagram

$$\begin{array}{ccc} W' & \longrightarrow & W \\ \big\downarrow & & \big\downarrow \\ B' & \xrightarrow[f]{} & B \end{array}$$

of principal bundles, $f^\star e = e'$. Indeed, it is sufficient to use the diagram

$$\begin{array}{ccc} H^\star_{\mathbf{S}^1}(W) & \longrightarrow & H^\star_{\mathbf{S}^1}(W') \\ \big\uparrow & & \big\uparrow \sigma'^\star \\ H^\star(B) & \longrightarrow & H^\star(B'). \end{array}$$

Example VI.1.17. The Euler class of the principal $\mathbf{S}^1$-bundle $\mathbf{S}^{2n+1} \to \mathbf{P}^n(\mathbf{C})$ is $-u$ (one can either use this example as an exercise or look for a proof in § VI.5). Notice that $-u$ is the cohomology class of $\eta/2\pi$, where η is the curvature form for a connection on the principal bundle.

The last property is fairly general, as the next proposition asserts.

Proposition VI.1.18. *Let $\pi : P \to B$ be a principal $\mathbf{S}^1$-bundle. Let α be a connection form on P, with curvature η. The cohomology class of the form $\eta/2\pi$ in $H^2(B;\mathbf{R})$ is the Euler class of the bundle.*

Proof. Recall that, according to Remark V.4.6, the cohomology class of η does not depend on the connection form η comes from. We may thus choose any convenient connection form and prove the result for it. We know that there exists a "classifying map" $B \to \mathbf{P}^n(\mathbf{C})$ for n large enough and a commutative diagram

$$\begin{array}{ccc} P & \xrightarrow{F} & \mathbf{S}^{2n+1} \\ \downarrow & & \downarrow \\ B & \xrightarrow[f]{} & \mathbf{P}^n(\mathbf{C}). \end{array}$$

We pull back our old α and η on the model bundle, so that $F^\star\alpha$ is a connection form on P, the curvature of which is $f^\star\eta$. This gives the result, by naturality of the Euler classes. □

There is a more precise statement in Exercise VI.5.

The definition of the Euler class can be extended to all *fixed point free* $\mathbf{S}^1$-actions by the very same method as in § I.3.d. Let W be a manifold acted on by $\mathbf{S}^1$ and assume the action has no fixed points. Choose a common multiple n of all the orders of the exceptional orbit stabilizers and consider the quotient W' of W by the $\mathbf{Z}/n$-action. It is endowed with an $\mathbf{S}^1/(\mathbf{Z}/n) \cong \mathbf{S}^1$-action. Call p the projection map $p : W \to W'$.

Although W' is rather an orbifold than a manifold, we may apply the previous definition to W' (as long as we are dealing with de Rham cohomology, there is no difficulty). Look at the diagram

$$\begin{array}{ccc} H^2(B) & \xrightarrow{(\sigma')^\star} & H^2_{\mathbf{S}^1/(\mathbf{Z}/n)}(W') \\ & & \downarrow{\scriptstyle p^\star} \\ & & H^2_{\mathbf{S}^1}(W) \end{array}$$

and at the Euler class e' of the principal $\mathbf{S}^1$-bundle $W' \to B$. This is the unique element in $H^2(B)$ which satisfies $\sigma'^\star e' = -u'$. Let us prove that, in fact, $p^\star u' = nu$. This is an assertion that may be checked at the classifying space level, where we must show that the map induced by the n-fold covering map

$$p : \mathbf{S}^1 \longrightarrow \mathbf{S}^1$$

is the multiplication by n in $H^2(B\mathbf{S}^1)$. But the induced map from $B\mathbf{S}^1$ onto itself is precisely the one which classifies the bundle $\mathcal{O}(n)$ as will be shown in § VI.5.

We thus have a unique class $e \in H^2(B)$ (namely e'/n) such that $\sigma^\star e = -u$. Notice that the result does not depend on n. We call it the *Euler class* of the $\mathbf{S}^1$-action on W. We still have to prove (having used the same name for two apparently different notions) that if $\dim W = 3$, it is the same Euler class as in Chapter I. This will be done in § VI.5, which will be specially devoted to the study of these various Euler classes.

In the same way, dealing with a principal (*resp.* without singular orbits) torus $\mathbf{T}$-action ($\dim \mathbf{T} = m$) on the manifold W with quotient B, one can define an Euler class $e \in H^2(B)^m$ (*resp.* with coefficients in a ring containing the inverses of the orders of the stabilizers). If we do not insist on being intrinsic, this can be defined as the unique element $e = (e_1, \ldots, e_m) \in (H^2(B))^m$ such that

$$\sigma^\star e = (-u_1, \ldots, -u_m) \in H^2_{\mathbf{T}}(W).$$

Using de Rham cohomology, $e \in H^2(B; \mathbf{R}) \otimes \mathfrak{t}$ is the unique class such that, if we identify $H^2(B\mathbf{T})$ with $\mathfrak{t}^\star$ in the diagram

$$\begin{array}{ccc} H^2(B) & \xrightarrow{\sigma^\star} & H^2_{\mathbf{T}}(W) \\ & & \uparrow \\ & & H^2(B\mathbf{T}) \end{array}$$

we have $\sigma^\star \langle e, \xi \rangle = \xi$ (where $\langle e, \xi \rangle$ is the element of $H^2(B)$ we get using duality in $\mathfrak{t}$ in an obvious way).

For example, any regular level of the momentum mapping $\mu : W \to \mathfrak{t}^\star$ of a Hamiltonian $\mathbf{T}$-action on the manifold W (assumed to be symplectic!) defines an Euler class e in the cohomology of the symplectic reduced orbifold (see § III.2.f), at least with rational coefficients.

VI.2. Hamiltonian actions and the Duistermaat–Heckman theorem

VI.2.a. Relationship between equivariant cohomology and Hamiltonian actions.

As we have already had the opportunity to mention, equivariant cohomology is a very convenient language in which to speak of Hamiltonian group actions. This is clear in the statement of the next easily proven proposition, in which θ denotes a connection form on EG as constructed in Proposition V.4.3.

Proposition VI.2.1. *Let (W, ω) be a symplectic manifold endowed with a symplectic action of the Lie group* G. *Let $\mu : W \to \mathfrak{g}^\star$ be any differentiable map. The formula*

$$\omega^\sharp = \omega + d\langle \theta, \mu \rangle$$

defines a closed *2-form on the Borel construction W_{G} if and only if the* G*-action is Hamiltonian with momentum mapping μ.*

Notice that $\theta \otimes \mu$ is a 1-form taking values in $\mathfrak{g} \otimes \mathfrak{g}^\star$, and may be contracted to give a 1-form taking values in $\mathbf{R}$, which we denote $\langle \theta, \mu \rangle$.

Proof. Assume the G-action on W is Hamiltonian and that μ is its momentum mapping. Consider the 2-form $\widetilde{\omega} = \omega + d\langle \theta, \mu \rangle$ on $EG \times W$. It is G-invariant by definition. Moreover, we have, for all X in $\mathfrak{g}$,

$$\begin{aligned} i_{\underline{X}}\widetilde{\omega} &= i_{\underline{X}}\omega + i_{\underline{X}}d\langle\theta,\mu\rangle \\ &= i_{\underline{X}}\omega - d i_{\underline{X}}\langle\theta,\mu\rangle \\ &= i_{\underline{X}}\omega - d\langle i_{\underline{X}}\theta,\mu\rangle \\ &= i_{\underline{X}}\omega - d\langle X,\mu\rangle \\ &= i_{\underline{X}}\omega - d\widetilde{\mu}_X = 0. \end{aligned}$$

Hence $\widetilde{\omega}$ satisfies

$$\mathcal{L}_{\underline{X}}\widetilde{\omega} = 0 \text{ and } i_{\underline{X}}\widetilde{\omega} = 0.$$

Moreover, this form is obviously closed. It thus comes from a closed 2-form $\omega^\sharp$ on the Borel construction W_{G}.

Conversely, consider again the pull back $\widetilde{\omega}$ on $E\mathrm{G}\times W$. The same computation gives, for $X \in \mathfrak{g}$,

$$i_{\underline{X}}\widetilde{\omega} = i_{\underline{X}}\omega + i_{\underline{X}}\langle\theta,\mu\rangle.$$

Using Cartan's formula, we get

$$i_{\underline{X}}\widetilde{\omega} = i_{\underline{X}}\omega + \langle \operatorname{ad}_{\underline{X}}\theta,\mu\rangle + \langle\theta,\underline{X}\cdot\mu\rangle - d\langle X,\mu\rangle$$

Now, $\underline{X}$ is a fundamental vector field, so that

$$\langle\theta,\underline{X}\cdot\mu\rangle = \langle\theta,\operatorname{ad}^\star_X\mu\rangle \text{ and } \langle\operatorname{ad}_X\theta,\mu\rangle = -\langle\theta,\operatorname{ad}^\star_X\mu\rangle. \qquad \square$$

VI.2.b. Variation of the reduced symplectic forms. Consider now a torus $\mathbf{T}$ acting on a symplectic manifold (W,ω) with momentum mapping $\mu : W \to \mathfrak{t}^\star$. Let ξ be a regular value of μ. We consider, as usual, the two maps

$$\begin{array}{ccc} V_\xi & \xrightarrow{\;j_\xi\;} & W \\ {\scriptstyle\sigma_\xi}\big\downarrow & & \\ B_\xi & & \end{array}$$

namely the inclusion

$$j_\xi : V_\xi \subset W$$

of the corresponding level $V_\xi = \mu^{-1}(\xi)$ and the quotient map

$$\sigma_\xi : V_\xi \longrightarrow B_\xi.$$

As the map j_ξ is $\mathbf{T}$-equivariant, it induces a map $1 \times_{\mathbf{T}} j_\xi$

$$\begin{array}{ccc} E\mathbf{T}\times_{\mathbf{T}} V_\xi & \xrightarrow{\;1\times_{\mathbf{T}} j_\xi\;} & E\mathbf{T}\times_{\mathbf{T}} W \\ {\scriptstyle\sigma_\xi}\big\downarrow & & \\ B_\xi & & \end{array}$$

(the vertical map is, since $\mathbf{T}$ acts freely on V_ξ, the homotopy equivalence induced by the fibration σ_ξ; we slightly abuse notation and call it σ_ξ). The form

$$\omega_\xi^\sharp = (1 \times_{\mathbf{T}} j_\xi)^\star \omega^\sharp$$

is the unique closed form on $E\mathbf{T} \times_T V_\xi$ which can be lifted to $E\mathbf{T} \times V_\xi$ as $j_\xi^\star \omega + \langle d\theta, \xi \rangle$, since we have

$$\begin{aligned}\omega^\sharp &= \omega + d\langle \theta, \mu \rangle \\ &= \omega + \langle d\theta, \mu \rangle - \langle \theta, d\mu \rangle,\end{aligned}$$

and, by the definition of V_ξ,

$$(1 \times j_\xi)^\star \omega^\sharp = j_\xi^\star \omega + \langle d\theta, \xi \rangle.$$

From this, we deduce immediately the inverse image of the class $[\omega_\xi^\sharp]$ by the morphism

$$\sigma_\xi^\star : H^2(B_\xi) \longrightarrow H^2_{\mathbf{T}}(V_\xi).$$

As $j_\xi^\star \omega$ gives the reduced symplectic form and $d\theta$ the Euler class of the $\mathbf{T}$-bundle $V_\xi \to B_\xi$, we have

$$(\sigma_\xi^\star)^{-1}[\omega_\xi^\sharp] = [\omega_\xi] + \langle e, \xi \rangle.$$

Let now $\mathcal{U}$ be a convex open subset of $\mathfrak{t}^\star$, all the points of which are regular values of μ, so that the $\mathbf{T}$-action on the open subset $\mu^{-1}(\mathcal{U})$ of W has no singular orbit. Consider then, for $\xi \in \mathcal{U}$, the diagram of quotients and inclusions

$$\begin{array}{ccc} V_\xi & \xrightarrow{\ j_\xi\ } & \mu^{-1}(\mathcal{U}) \\ {\scriptstyle \sigma_\xi}\downarrow & & \downarrow{\scriptstyle \sigma} \\ B_\xi & \xrightarrow[\ i_\xi\ ]{} & \mu^{-1}(\mathcal{U})/\mathbf{T} \end{array}$$

and its (equivariant) cohomological analogue

$$\begin{array}{ccc} H^2_{\mathbf{T}}(V_\xi) & \xleftarrow{\ j_\xi^\star\ } & H^2_{\mathbf{T}}(\mu^{-1}(\mathcal{U})) \\ {\scriptstyle \sigma_\xi^\star}\uparrow & & \uparrow{\scriptstyle \sigma^\star} \\ H^2(B_\xi) & \xleftarrow[\ i_\xi^\star\ ]{} & H^2(\mu^{-1}(\mathcal{U})/\mathbf{T}) \end{array}$$

where all the maps are isomorphisms (j_ξ is a homotopy equivalence, because this is the inclusion of a fiber over the contractible space $\mathcal{U}$).

We consider $[\omega^\sharp]$ as an element of $H^2_{\mathbf{T}}(\mu^{-1}(\mathcal{U}))$ by restriction. There is thus a unique (that is, independent of $\xi \in \mathcal{U}$) class

$$x = (\sigma^\star)^{-1}[\omega^\sharp] \in H^2(\mu^{-1}(\mathcal{U})/\mathbf{T})$$

the image of which by $\sigma_\xi^\star$ is $[\omega_\xi^\sharp]$. Hence, from the relation above, we deduce:

Proposition VI.2.2. *Let $\mathcal{U}$ be a convex open subset of $\mathfrak{t}^\star$ consisting of regular values of the momentum mapping $\mu : W \to \mathfrak{t}^\star$. There exists a class x in $H^2(\mu^{-1}(\mathcal{U})/T)$ such that*

$$\forall \xi \in \mathcal{U}, \ i_\xi^\star x = [\omega_\xi] + \langle e, \xi \rangle \in H^2(B_\xi)$$

where ω_ξ is the reduced symplectic form at the level ξ and e is the Euler class of any of the $\mathbf{T}$-bundles $V_\xi \to B_\xi$. □

Fixing an element ξ_0 in $\mathcal{U}$, we get an equivariant isomorphism $\mu^{-1}(\mathcal{U}) \cong \mathcal{U} \times V_{\xi_0}$ and, passing to quotients, a way to identify each $H^2(B_\xi)$ with $H^2(B_{\xi_0})$, and thus a way to compare the various classes $[\omega_\xi]$. As a corollary, this gives the famous Duistermaat–Heckman theorem.

***Theorem VI.2.3* (Duistermaat and Heckman [44]).** *Let $\mathcal{U}$ be a convex open subset of $\mathfrak{t}^\star$ consisting of regular values of the momentum mapping $\mu : W \to \mathfrak{t}^\star$. For ξ, $\xi_0 \in \mathcal{U}$,*

$$[\omega_\xi] - [\omega_{\xi_0}] = \langle -e, \xi - \xi_0 \rangle.$$ □

Remark VI.2.4. The sign in this statement comes from the orientation conventions in Chapter I where we managed to get -1 as Euler class for the Hopf bundle.

Example VI.2.5. Consider for example the case of a circle action on a 4-manifold. All the reduced regular levels are smooth surfaces—the same smooth surface—on which the symplectic form is characterized by its integral, the volume of the reduced space. The theorem asserts that the function

$$t \longmapsto \operatorname{Vol} B_t$$

is piecewise linear, more precisely that it is linear (affine) over each component of the set of regular values, the slope being the opposite of the Euler class. Consider for instance the $\mathbf{S}^1$-action on $\mathbf{P}^2(\mathbf{C})$ by

$$u \cdot [x, y, z] = [ux, y, z],$$

associated with the Hamiltonian

$$H([x, y, z]) = \frac{1}{2} \frac{|x|^2}{|x|^2 + |y|^2 + |z|^2}.$$

The fixed points are the projective line $x = 0$, the minimum of H, and the point $[1, 0, 0]$, its maximum. The regular values are all the points of $]0, \frac{1}{2}[$. The reduced regular levels are all 2-spheres.

Figure 1 shows the graph of this volume function. The slope is -1, according to the fact that the Euler class of the bundle $V_t \to B_t$ is $+1$. This is not by pure chance that this looks very much like one of the triangles shown in Figure 2 of

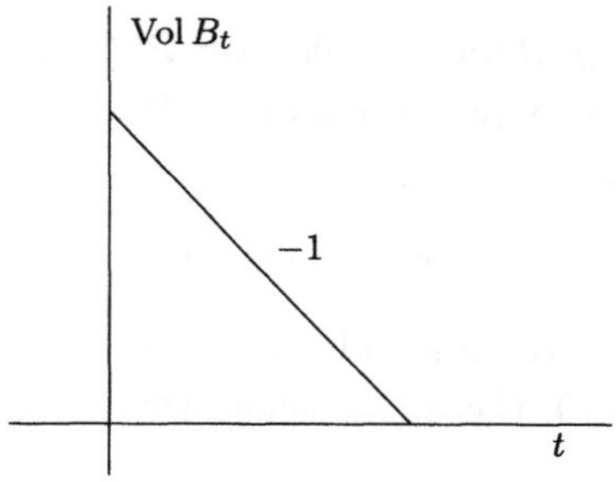

Figure 1

Chapter IV, which triangles represented the image of the momentum mapping for the standard $\mathbf{T}^2$-action.

VI.2.c. Duistermaat–Heckman with singularities. In this section, we describe a construction, independently due to Brion and Procesi [**29**] on the one hand and to Guillemin and Sternberg [**64**] on the other, and which generalizes the previous remarks.

For the sake of simplicity, we shall assume throughout that all actions are semi-free (see nevertheless Remarks VI.2.7). We recall first a useful operation, already met in Exercise III.19.

Blowing up 0 in $\mathbf{C}^n$. Consider the space one gets by replacing, in $\mathbf{C}^n$, the point 0 by the set of all straight lines through it. In other words consider

$$\widetilde{\mathbf{C}}^n = \{(v, \ell) \mid v \in \ell\} \subset \mathbf{C}^n \times \mathbf{P}^{n-1}(\mathbf{C}).$$

From the viewpoint of the projection on $\mathbf{P}^{n-1}(\mathbf{C})$, this can be considered as the

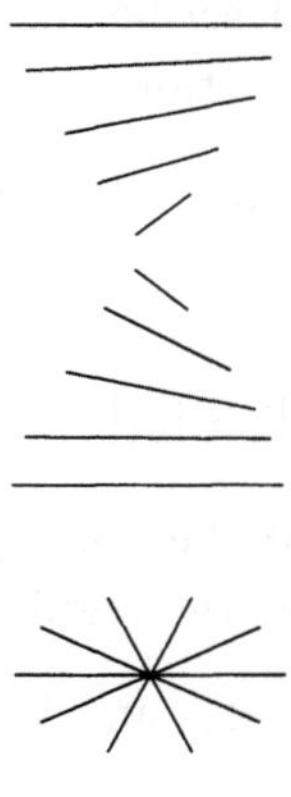

Figure 2. Blowing up a point

total space of the *tautological* line bundle over $\mathbf{P}^{n-1}(\mathbf{C})$, but we will rather look at it from the viewpoint of its projection onto $\mathbf{C}^n$:

$$\begin{aligned}\pi : \widetilde{\mathbf{C}}^n &\longrightarrow \mathbf{C}^n\\ (v,\ell) &\longmapsto v.\end{aligned}$$

If $v \neq 0$, the fiber $\pi^{-1}(v)$ contains only one point: the nonzero vector v defines a unique line. But, if $v = 0$, there is a whole $\mathbf{P}^{n-1}(\mathbf{C})$ above it. It is called the *exceptional divisor*(6).

The space $\widetilde{\mathbf{C}}^n$ is a complex manifold: it may be described by the algebraic equation

$$\widetilde{\mathbf{C}}^n = \{((x_1,\dots,x_n),[a_1,\dots,a_n]) \mid a_i x_j - a_j x_i = 0 \text{ for all } i,j\}.$$

It is also easy to find local coordinates for it:

– If $a_n \neq 0$, put $y_i = \dfrac{a_i}{a_n}$, then

$$\begin{aligned}\mathbf{C}^n &\longrightarrow \widetilde{\mathbf{C}}^n \subset \mathbf{C}^n \times \mathbf{P}^{n-1}(\mathbf{C})\\ (y_1,\dots,y_{n-1},x)) &\longmapsto ((xy_1,\dots,xy_{n-1},x),[y_1,\dots,y_{n-1},1])\end{aligned}$$

is a diffeomorphism onto its image.

– One would write just as easily local coordinates on the other affine charts $a_i \neq 0$.

As a complex submanifold, the blown-up $\widetilde{\mathbf{C}}^n$ has symplectic forms, induced by those of $\mathbf{C}^n \times \mathbf{P}^{n-1}(\mathbf{C})$. As we have seen in Exercise III.19 (see also Exercise VI.7), it is even possible to construct symplectic forms inducing the standard form of $\mathbf{C}^n$ outside any given disc centered at 0. This allows us, *via* Darboux, to blow up any point in any symplectic manifold and to obtain new symplectic manifolds (see more generally [**102**]).

Here is now another way to describe the blow up (see also Exercise III.19). Consider, on $W = \mathbf{C} \times \mathbf{C}^n$, the $\mathbf{S}^1$-action

$$u \cdot (x, y_1, \dots, y_n) = (ux, \overline{u}y_1, \dots, \overline{u}y_n).$$

Choose the Hamiltonian

$$H = \frac{1}{2}\left(|x|^2 - |y_1|^2 - \cdots - |y_n|^2\right)$$

so that 0 is a critical point and 0 the corresponding critical value. Let ε be a positive real number. Let us look at the regular levels:

$$V_- = H^{-1}(-\varepsilon), \text{ and } V_+ = H^{-1}(\varepsilon).$$

(6) *Divisor* simply means complex codimension 1 here.

They are respectively diffeomorphic to $\mathbf{S}^{2n-1} \times \mathbf{C}$ and to $\mathbf{S}^1 \times \mathbf{C}^n$. The quotients by the $\mathbf{S}^1$-action are diffeomorphic, respectively, to

$$B_- \cong \widetilde{\mathbf{C}}^n \text{ and } B_+ \cong \mathbf{C}^n.$$

Remark VI.2.6. This endows $\widetilde{\mathbf{C}}^n$ with various reduced symplectic forms (as noticed in Exercise III.19). I will come back to this in § VI.4.b.

Crossing a critical value, simple situation. Let H be a periodic Hamiltonian on a symplectic manifold W. We consider a critical submanifold Z of H with signature $(2, 2p)$. We study the situation in a neighborhood of the critical submanifold, so that, using an invariant metric, we can assume that, transversally to Z, $W = \mathbf{C} \times \mathbf{C}^p$ with

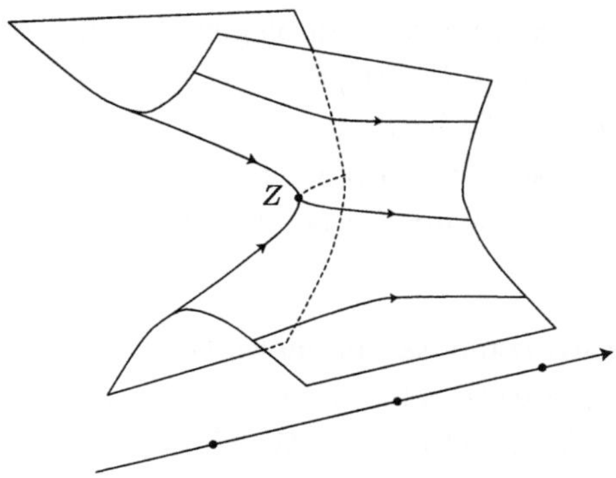

Figure 3

$$u \cdot (x, y_1, \ldots, y_p) = (ux, \overline{u}y_1, \ldots, \overline{u}y_p)$$

and

$$H = \frac{1}{2}\left(|x|^2 - |y_1|^2 - \cdots - |y_p|^2\right).$$

The gradient flow is given by

$$\varphi_t(x, y_1, \ldots, y_p) = (e^t x, e^{-t} y_1, \ldots, e^{-t} y_p).$$

The situation, transversally to Z, is described in Figure 3, where the gradient flow is indicated as well. Consider the map

$$\begin{aligned} \pi : \mathbf{C} \times \mathbf{C}^p &\longrightarrow \mathbf{C}^p \\ (x, y_1, \ldots, y_p) &\longmapsto (xy_1, \ldots, xy_p). \end{aligned}$$

Restricted to a level of H, this descends to the quotient

$$\begin{array}{ccc} H^{-1}(t) & \xrightarrow{\quad\pi\quad} & \mathbf{C}^p \\ & \searrow \quad \nearrow & \\ & H^{-1}(t)/\mathbf{S}^1 & \end{array}$$

Notice that, for $t > 0$, on the level t, we have

$$|x| = \sqrt{2t + \sum |y_i|^2}$$

so that the map $H^{-1}(t)/\mathbf{S}^1 \to \mathbf{C}^p$ can be considered as the quotient map (and the quotient is $\mathbf{C}^p$). On the other hand, for $t < 0$, the level $H^{-1}(t)$ is diffeomorphic with $\mathbf{C} \times \mathbf{S}^{2p-1}$ and the quotient space $H^{-1}(t)/\mathbf{S}^1$ is the blow up $\widetilde{\mathbf{C}}^p$ of $\mathbf{C}^p$ at 0, in other words the total space of the tautological bundle $\mathcal{O}(-1)$ over the quotient $\mathbf{P}^{p-1}(\mathbf{C})$ of $\mathbf{S}^{2p-1}$. The map $H^{-1}(t)/\mathbf{S}^1 \to \mathbf{C}^p$ contracts the projective space $[y_1, \ldots, y_p]$ and is precisely the blow up.

In other words, the surgery upstairs (from negative to positive levels) corresponds downstairs to the blow up of a point (transversally to Z).

Remarks VI.2.7

(1) A straightforward computation shows that at the $-\varepsilon$ level, the reduced symplectic form integrates to ε on the generator of the second cohomology group of the exceptional divisor. We thus see, using the Hamiltonian H, a (real) 1-parameter family of symplectic manifolds which, for negative values of the parameter ε are $\widetilde{\mathbf{C}}^p$'s such that the volume of their exceptional divisor decreases to 0 as $\varepsilon \to 0$. For nonnegative values of the parameter the exceptional divisor has disappeared and the symplectic manifolds are $\mathbf{C}^p$'s.

(2) Once we accept working with orbifolds as quotients, there is no essential difference with non semi-free actions... in particular with actions of the form

$$u \cdot (x, y_1, \ldots, y_p) = (ux, \overline{u}^m y_1, \ldots, \overline{u}^m y_p),$$

which will appear in the following, even after having started with semi-free actions.

(3) We will come back to this situation in the simple case where $p = 1$, on a symplectic manifold of dimension 4, a case where blowing up a point does not change the topology, in § VIII.1.

The general case. We consider now a critical submanifold Z of signature $(2p, 2q)$. Transversally to Z we can assume as before that $W = \mathbf{C}^{p+q}$, $Z = \{0\}$, and

$$u \cdot (y_1, \ldots, y_p, z_1, \ldots, z_q) = (uy_1, \ldots, uy_p, \overline{u}z_1, \ldots, \overline{u}z_q)$$

and the trick is to blow up W along Z in order to mimic the previous simple situation. We thus look at

$$\widetilde{\mathbf{C}}^{p+q} = \{([\eta,\zeta],y,z) \mid (y,z) \in [\eta,\zeta]\}$$

(here $[\eta,\zeta]$ is a line in $\mathbf{C}^{p+q}$ while (y,z) is a vector in the same $\mathbf{C}^{p+q}$). This is endowed with the $\mathbf{S}^1$-action extended linearly as usual:

$$u \cdot ([\eta,\zeta],y,z) = ([u\eta,\overline{u}\zeta],uy,\overline{u}z)\,.$$

The fixed "point" $Z = 0$ becomes two fixed submanifolds included in the exceptional divisor $y = z = 0$:

- a projective space $\mathbf{P}^{q-1}(\mathbf{C})$, that of equation $\eta = 0$,
- and a $\mathbf{P}^{p-1}(\mathbf{C})$, of equation $\zeta = 0$.

Note that all points of the exceptional divisor now have a $\mathbf{Z}/2$ as stabilizer.

Let $\widetilde{H}$ be a Hamiltonian for the $\mathbf{S}^1$-action (with respect to some invariant symplectic form on the blown up manifold). Look now at the normal bundles and indices of critical submanifolds. The projective space $\mathbf{P}^{q-1}(\mathbf{C})$ embeds in $\mathbf{P}^{p+q-1}(\mathbf{C})$ with normal bundle the sum of p copies of the line bundle $\mathcal{O}(1)$. The latter is in turn a submanifold of $\widetilde{\mathbf{C}}^{p+q}$, with normal bundle $\mathcal{O}(-1)$. The situation for

$$\mathbf{P}^{p-1}(\mathbf{C}) \subset \mathbf{P}^{p+q-1}(\mathbf{C}) \subset \widetilde{\mathbf{C}}^{p+q}$$

is quite similar.

What we want to understand is what happens when the critical value, 0, say, is crossed. Let us call V_-, V_+ respectively, regular levels just before 0, just after 0, the 0-level itself being denoted by V_0. Let us denote B_-, B_+, B_0, the corresponding quotients. In the blown-up manifold, let V' be a regular level between the two (new) critical submanifolds.

Figure 4 shows, schematically (it is not that easy to draw a blow up in two dimensions), the situation: we have split our critical value in two. The new critical submanifolds are joined by an "exceptional divisor" E. A better picture could be that in Figure 5.

Following the gradient or minus the gradient we get a diagram

$$\begin{array}{ccc} B' & \longrightarrow & B_+ \\ \downarrow & & \downarrow \\ B_- & \longrightarrow & B_0 \end{array}$$

as in **[29]** where B' was defined as a fiber product, a definition which could seem rather artificial from the topological viewpoint (if not from the algebraic one). Recall that V_0 was a critical level, thus B_0 is not smooth in general.

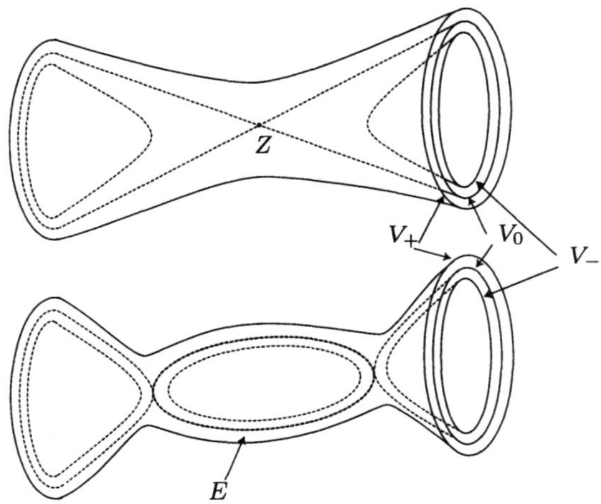

Figure 4

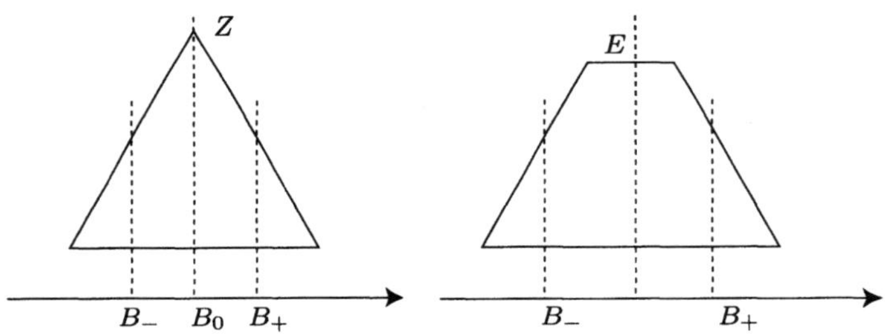

Figure 5

Let us now describe more precisely the topology of V' and B' and the map $B' \to B_0$. The regular level $V' = \widetilde{H}^{-1}(0)$ meets the exceptional divisor $\mathbf{P}^{p+q-1}(\mathbf{C})$ along the submanifold of equation

$$|\eta|^2 = |\zeta|^2 .$$

This submanifold of $\mathbf{P}^{p+q-1}(\mathbf{C})$ is the quotient of $\mathbf{S}^{2p-1} \times \mathbf{S}^{2q-1} \subset \mathbf{S}^{2p+2q-1}$ by the diagonal $\mathbf{S}^1$-action. Being the common boundary of tubular neighborhoods of $\mathbf{P}^{p-1}(\mathbf{C})$ and $\mathbf{P}^{q-1}(\mathbf{C})$ in $\mathbf{P}^{p+q-1}(\mathbf{C})$, this is a sphere bundle over $\mathbf{P}^{q-1}(\mathbf{C})$, precisely

$$S(p\mathcal{O}(-1)) \longrightarrow \mathbf{P}^{q-1}(\mathbf{C}),$$

as well as over $\mathbf{P}^{p-1}(\mathbf{C})$, this time

$$S(q\mathcal{O}(-1)) \longrightarrow \mathbf{P}^{p-1}(\mathbf{C}).$$

Look now at quotients: to construct B', we have removed the singular point and

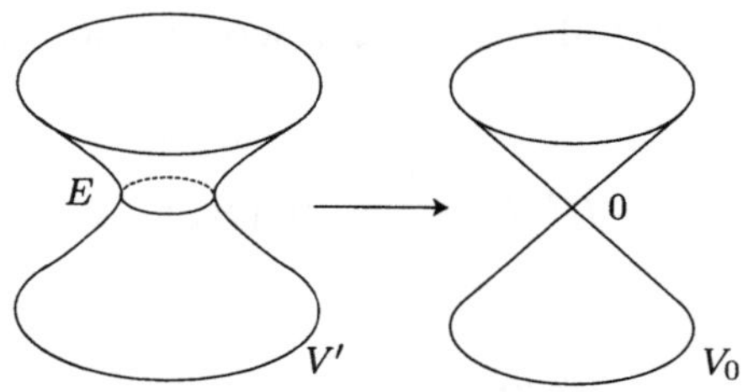

Figure 6

replaced it with $V' \cap E/\mathbf{S}^1$ (E is the exceptional divisor). This is the same as the quotient of the sphere bundle

$$S(p\mathcal{O}(1)) \longrightarrow \mathbf{P}^{q-1}(\mathbf{C})$$

by the equivalence relation

$$[\eta, \zeta] \sim [t\eta, \bar{t}\zeta] \sim [t^2\eta, \zeta].$$

Thus, this is nothing other than the projectivized bundle

$$\mathbf{P}(p\mathcal{O}(1)) \longrightarrow \mathbf{P}^{q-1}(\mathbf{C}).$$

Of course the projectivized bundle is trivial and the quotient is a $\mathbf{P}^{p-1}(\mathbf{C}) \times \mathbf{P}^{q-1}(\mathbf{C})$.

Remark VI.2.8. If q (or p) equals 1, we get the blowing up of a point (or a smooth submanifold) of B_0 which in this case is smooth as we have already remarked. In general the map $B' \to B_0$ is a blow up in the algebraic sense.

Application: the Duistermaat–Heckman problem at critical values. We are thus looking for a diagram like

$$\begin{array}{ccc} B' & \xrightarrow{\sigma_-} & B_+ \supset \mathbf{P}^{p-1}(\mathbf{C}) \\ {\scriptstyle \sigma_+}\downarrow & \searrow^{\sigma} & \downarrow \\ B_- & \longrightarrow & B_0 \\ \cup & & \\ \mathbf{P}^{q-1}(\mathbf{C}) & & \end{array}$$

In the Duistermaat–Heckman Theorem VI.2.3, all the regular quotients of $H^{-1}(\pm\varepsilon)$ were identified with the same manifold $B_\pm$ and this allowed us to compare the classes

$$[\sigma_\varepsilon] \in H^2(B_+), \quad [\sigma_{-\varepsilon}] \in H^2(B_-)$$

and to prove that in each case, staying in the same component of the set of regular values, $[\sigma_\varepsilon]$ was an affine function of ε. Our diagram will be used to compare $[\sigma_\varepsilon]$

and $[\sigma_{-\varepsilon}]$ even if 0 is a critical value. Let us look at the diagram we get by taking cohomology:

$$\begin{array}{ccc} H^2(B') & \xleftarrow{\sigma_-^\star} & H^2(B_+) \\ \sigma_+^\star \uparrow & & \\ H^2(B_-) & & \end{array}$$

which enables us to compare the classes of the reduced symplectic forms in $H^2(B')$. Recall (see Proposition VI.4.5 below) that the two maps σ_+ and σ_- induce inclusions at the H^2-level. Using the reduced symplectic forms σ_t at levels $t > 0$, $t < 0$ and these inclusions, we consider the curve

$$t \longmapsto [\sigma_t] - [\sigma_t']$$

defined, for all $t > 0$, as follows:

- $[\sigma_t]$ is the class of the reduced symplectic form at level t considered as an element in $H^2(B')$ *via* the inclusion.
- $[\sigma_t]$ for $t < 0$, being an affine function of t, is easy to extend as an affine function for all t, $[\sigma_t']$ denotes its value for the positive values of t (once again the inclusion is omitted).

***Theorem VI.2.9* (Brion and Procesi [29]).** *The curve $t \mapsto [\sigma_t] - [\sigma_t']$ is a half-line, directed by the class of the exceptional divisor of the map $B' \to B_0$.*

Proof. The exceptional divisor of the map $B' \to B_0$ is the sum of the two exceptional divisors of $B' \to B_\pm$. Since everything is affine it is thus sufficient to prove the theorem in the "simple" case where it is an obvious consequence of the remark on reduced symplectic forms in § VI.2.c. □

Example VI.2.10. Assume the symplectic manifold has dimension 4, so that the reduced levels B_- and B_+ are surfaces. To blow up a point in a surface does not change its topology. It is thus possible to compare σ_ε and $\sigma_{-\varepsilon}$ just by looking at the function

$$t \longmapsto \int_B \sigma_t.$$

Over each component of the regular set, the graph is a straight line (with an integral slope in the semi-free case we are considering). When crossing the critical level V_0, the slope will decrease by 1. In this situation, the change in the topology is trivial, but it is not at the level of reduced symplectic forms, the blow up reflects in the change of the slope.

Remark VI.2.11. As noticed by Guillemin and Sternberg in [**64**], everything we did, we could have done in an equivariant way for any group action which commutes with the $\mathbf{S}^1$-action. Consider for instance a Hamiltonian torus action with momentum mapping $\mu : W \to \mathfrak{t}^\star$ and try to compare the reduced symplectic forms when crossing a codimension-1 wall between two polyhedra of regular values of μ. It is sufficient to be able to cross it orthogonally and thus to consider a $\mathbf{S}^1 \times \mathbf{T}^{n-1}$ situation which can be studied as before. This is why a result like Theorem VI.2.9 is called a "wall crossing formula".

VI.3. Localization at fixed points and the Duistermaat–Heckman formula

VI.3.a. The support of an $H^\star B\mathbf{T}$-module.

As usual, we use de Rham cohomology. Having chosen a basis of the Lie algebra $\mathfrak{t}$ of a torus $\mathbf{T}$, we can identify

$$H^\star B\mathbf{T} \cong S(\mathfrak{t}^\star) \cong \mathbf{R}[u_1, \ldots, u_n]$$

(the last isomorphism depending on the choice of the basis). In particular, this will allow us to consider elements of $H^\star B\mathbf{T}$ as functions on $\mathfrak{t}$ or as polynomials in $(u_1, \ldots, u_n)$. For any $f \in H^\star B\mathbf{T} \cong S(\mathfrak{t}^\star)$, let V_f be the subset of zeros of the polynomial f

$$V_f = \{X \in \mathfrak{t} \mid f(X) = 0\}.$$

Let M be an $H^\star B\mathbf{T}$-module. Its *support* is defined as the intersection

$$\operatorname{Supp} M = \bigcap_{\{f \mid f \cdot M = 0\}} V_f \subset \mathfrak{t}.$$

Examples VI.3.1

(1) The module $\{0\}$ has empty support.
(2) Let M be a free module. If f annihilates M, then f is zero and V_f is $\mathfrak{t}$, so that $\operatorname{Supp} M = \mathfrak{t}$. Hence the modules having a proper support do have torsion.
(3) If $\mathbf{T} = \mathbf{S}^1$, the ring $H^\star B\mathbf{T}$ is a ring of polynomials in one variable, hence a principal ideal domain. If f is a generator of the ideal annihilating M, then $\operatorname{Supp} M$ consists of the zeros of f. In this example, one can imagine why it is often more convenient to use complex coefficients. It suffices for that to replace $H^\star B\mathbf{T}$ by $H^\star B\mathbf{T} \otimes \mathbf{C} \cong H^\star(B\mathbf{T}; \mathbf{C})$ and $\mathfrak{t}$ by $\mathfrak{t} \otimes \mathbf{C}$. Thus, when M does have torsion, there is a (proper) support. We shall see that in the Hamiltonian case there is no problem and we can "stay real": the supports we shall have to consider will be unions of vector subspaces.

Lemma VI.3.2. *If* $M' \xrightarrow{a} M \xrightarrow{b} M''$ *is an exact sequence of* $H^\star B\mathbf{T}$-*modules, then* $\operatorname{Supp} M \subset \operatorname{Supp} M' \cup \operatorname{Supp} M''$.

Proof. Put $S = \operatorname{Supp} M$, $S' = \operatorname{Supp} M'$ and $S'' = \operatorname{Supp} M''$. Let $x \notin S' \cup S''$. Then, as $x \notin S'$, there exists a polynomial f such that $f(x) \neq 0$ and $f \cdot M' = 0$. For the same reasons, there exists a polynomial g annihilating M'' but not x. But

$$b(g \cdot M) = g \cdot b(M) \subset g \cdot M'' = 0.$$

Thus

$$g \cdot M \subset a(M') \text{ and } f \cdot (g \cdot M) \subset a(f \cdot M') = 0,$$

hence

$$(fg) \cdot M = 0, \quad f(x)g(x) \neq 0, \text{ and so } x \notin S.$$ □

The next two results are proved similarly.

Lemma VI.3.3. *Let M and M' be $H^\star B\mathbf{T}$-algebras with units. Let $a : M' \to M$ be a morphism of algebras. Then* $\operatorname{Supp} M \subset \operatorname{Supp} M'$.

Proof. If $x \notin \operatorname{Supp} M'$, there is a polynomial f annihilating M' but not x. The unit $1_{M'}$ is killed by f,

$$a(f \cdot 1_{M'}) = f \cdot 1_M = 0$$

so that f kills M and $x \notin \operatorname{Supp} M$. □

Proposition VI.3.4. *If $0 \to M' \to M \to M'' \to 0$ is an exact sequence of $H^\star B\mathbf{T}$-modules, then* $\operatorname{Supp} M = \operatorname{Supp} M' \cup \operatorname{Supp} M''$. □

Supports of $H^\star_{\mathbf{T}}(\mathcal{U})$, examples. We begin to investigate now the structure of $H^\star B\mathbf{T}$-module of the equivariant cohomology of $\mathbf{T}$-manifolds.

Lemma VI.3.5. *Let* H *be a closed subgroup of the torus* $\mathbf{T}$. *Then*

$$\operatorname{Supp} H^\star_{\mathbf{T}}(\mathbf{T}/\mathrm{H}) = \mathfrak{h} \subset \mathfrak{t}.$$

Proof. We first notice that

$$E\mathbf{T} \times_{\mathbf{T}} \mathbf{T}/\mathrm{H} \sim E\mathbf{T}/\mathrm{H} \sim B\mathrm{H}.$$

We deduce that $H^\star_{\mathbf{T}}(\mathbf{T}/\mathrm{H})$ is simply $H^\star B\mathrm{H}$ with the $H^\star B\mathbf{T}$-module structure given by the restriction

$$H^\star B\mathbf{T} \longrightarrow H^\star B\mathrm{H}.$$

On the other hand, H is a closed subgroup of the torus $\mathbf{T}$, and so, up to the fact that it might not be connected, this is a torus. More precisely, this is the product of a torus H_0 by some finite group that de Rham cohomology will not even notice and hence we can suppose that H is a torus.

The structure of $H^\star B\mathrm{H}$ as an $H^\star B\mathbf{T}$-module is rather easy to understand. The inclusion of the Lie algebra

$$i : \mathfrak{h} \longrightarrow \mathfrak{t}$$

induces the restriction of polynomials

$$S({}^t i) : S(\mathfrak{t}^\star) \longrightarrow S(\mathfrak{h}^\star).$$

The torsion elements of $H^\star B\mathrm{H}$ are thus all the polynomials on $\mathfrak{t}$ the restriction to $\mathfrak{h}$ of which vanishes. Thus

$$\operatorname{Supp} H^\star_{\mathbf{T}}(\mathbf{T}/\mathrm{H}) = \bigcap_{f|_{\mathfrak{h}}=0} V_f = \mathfrak{h}. \qquad \square$$

Corollary VI.3.6. *Let V be a $\mathbf{T}$-manifold. Assume there exists an equivariant map $V \to \mathbf{T}/\mathrm{H}$. Then* $\operatorname{Supp} H^\star_{\mathbf{T}}(V) \subset \mathfrak{h}$.

Proof. The induced map

$$f^\star : H^\star_{\mathbf{T}}(\mathbf{T}/\mathrm{H}) \longrightarrow H^\star_{\mathbf{T}}(V)$$

is a ring morphism. Applying Lemma VI.3.3, we find that

$$\operatorname{Supp} H^\star_{\mathbf{T}}(\mathbf{T}/\mathrm{H}) \supset \operatorname{Supp} H^\star_{\mathbf{T}}(V). \qquad \square$$

This corollary may be applied in particular when V is an equivariant tubular neighborhood of any orbit of the smooth $\mathbf{T}$-action on a manifold W (thanks to the slice theorem) and gives:

Corollary VI.3.7. *Let V be an equivariant tubular neighborhood of an orbit of type* (H) *in the $\mathbf{T}$-manifold W. Then* $\operatorname{Supp} H^\star_{\mathbf{T}}(V) \subset \mathfrak{h}$. □

In the same way, we have also:

Corollary VI.3.8. *Assume $\mathbf{T}$ acts freely on W. Then* $\operatorname{Supp} H^\star_{\mathbf{T}}(W) = 0$. □

Considering now the union of all orbits in W the stabilizer of which is a proper subgroup of $\mathbf{T}$, that is, the complement of the fixed point set F, we are going to prove:

Proposition VI.3.9. *Let F be the fixed point set of a $\mathbf{T}$-action on a compact manifold W. Then*

$$\operatorname{Supp} H^\star_{\mathbf{T}}(W - F) \subset \bigcup_{\mathrm{H}} \mathfrak{h}$$

where H *describes the (finite) set of all the (proper) stabilizers of points in W.*

Remark VI.3.10. In particular, this support is a proper subspace of $\mathfrak{t}$, contained in a finite union of vector subspaces, thus $H^\star_{\mathbf{T}}(W - F)$ is a torsion $H^\star B\mathbf{T}$-module.

Proof of the proposition. The space $W - F$ has the same (equivariant) homotopy type as the complement of a small equivariant neighborhood of F. The complement may be covered by open subsets $\mathcal{U}_1, \dots, \mathcal{U}_h$ which are equivariant tubular neighborhoods of orbits, with proper subgroups of $\mathbf{T}$ as stabilizers.

We may apply Corollary VI.3.7 to each $\mathcal{U}_i$. This allows us to prove, by induction on h, that:

$$\operatorname{Supp} H^\star_{\mathbf{T}}(\mathcal{U}_1 \cup \cdots \cup \mathcal{U}_h) \subset \mathfrak{h}_1 \cup \cdots \cup \mathfrak{h}_h.$$

We start the induction by applying Corollary VI.3.7 to $\mathcal{U}_1$. Suppose then the result to hold up to order h, put

$$\mathcal{V}_h = \mathcal{U}_1 \cup \cdots \cup \mathcal{U}_h \text{ and thus } \mathcal{V}_{h+1} = \mathcal{V}_h \cup \mathcal{U}_{h+1}.$$

We have

$$\operatorname{Supp} H^\star_{\mathbf{T}}(\mathcal{V}_h) \subset \bigcup_{i=1}^{h} \mathfrak{h}_i$$

(this is the induction hypothesis). Hence we can use the Mayer–Vietoris exact sequence:

$$\begin{array}{ccccc} H^\star_{\mathbf{T}}(\mathcal{V}_h \cap \mathcal{U}_{h+1}) & \longrightarrow & H^\star_{\mathbf{T}}(\mathcal{V}_{h+1}) & \longrightarrow & H^\star_{\mathbf{T}}(\mathcal{V}_h) \oplus H^\star_{\mathbf{T}}(\mathcal{U}_{h+1}) \\ M' & \longrightarrow & M & \longrightarrow & M'' \end{array}$$

in which

- the intersection $\mathcal{V}_h \cap \mathcal{U}_{h+1} \subset \mathcal{U}_{h+1}$ is endowed with an equivariant map onto the central orbit of $\mathcal{U}_{h+1}$ in such a way that $\operatorname{Supp} M' \subset \mathfrak{h}_{h+1}$,
- we have the inclusion

$$\operatorname{Supp} M'' = \operatorname{Supp} H^\star_{\mathbf{T}}(\mathcal{V}_h) \cup \operatorname{Supp} H^\star_{\mathbf{T}}(\mathcal{U}_{h+1}) = \bigcup_{i=1}^{h} \mathfrak{h}_i \cup \mathfrak{h}_{h+1}$$

by Proposition VI.3.4.

This gives the result for $\operatorname{Supp} \mathcal{V}_{h+1}$. For $\operatorname{Supp} M$ it follows from Lemma VI.3.2. □

VI.3.b. The localization theorem. Our goal in this investigation is to prove rather precise versions of the fact that, forgetting torsion, the $H^\star B\mathbf{T}$-module $H^\star_{\mathbf{T}}(W)$ looks very much like the *free* $H^\star B\mathbf{T}$-module $H^\star_{\mathbf{T}}(F)$.

Theorem VI.3.11. *Let $i : F \hookrightarrow W$ be the inclusion of fixed points of the action of a torus $\mathbf{T}$ on a manifold W. Then the supports of both the kernel and cokernel of*

$$i^\star : H^\star_{\mathbf{T}}(W) \longrightarrow H^\star_{\mathbf{T}}(F)$$

are included in $\bigcup_{\mathrm{H} \text{ stabilizer } \neq T} \mathfrak{h}$.

Proof. Let $\mathcal{U}$ be an equivariant tubular neighborhood of the fixed point set F. We know that

$$\operatorname{Supp} H^\star_{\mathbf{T}}(W - \mathcal{U}) \subset \bigcup_{\mathrm{H} \neq T} \mathfrak{h}$$

and that the same is true of $\operatorname{Supp} H^\star_{\mathbf{T}}(\partial(W-\mathcal{U}))$. Using the long (equivariant) cohomology exact sequence of the pair $(W-\mathcal{U}, \partial(W-\mathcal{U}))$, the same is true for the support of $H^\star_{\mathbf{T}}(W-\mathcal{U}, \partial(W-\mathcal{U}))$.

Let $\mathcal{V}$ be another equivariant tubular neighborhood, a little larger than $\mathcal{U}$, such that $\mathcal{V}-\mathcal{U} \sim \partial(W-\mathcal{U}) = \partial\mathcal{U}$. We have isomorphisms:

$$H^\star_{\mathbf{T}}(W,F) \cong H^\star_{\mathbf{T}}(W,\mathcal{V}) \xrightarrow{\text{excision}} \cong H^\star_{\mathbf{T}}(W-\mathcal{U}, \mathcal{V}-\mathcal{U}) \cong H^\star_{\mathbf{T}}(W-\mathcal{U}, \partial(W-\mathcal{U})).$$

In particular,

$$\operatorname{Supp} H^\star_{\mathbf{T}}(W,F) \subset \cup \mathfrak{h}.$$

The exact cohomology sequence of the pair (W,F)

$$H^\star_{\mathbf{T}}(W,F) \longrightarrow H^\star_{\mathbf{T}}(W) \xrightarrow{i^\star} H^\star_{\mathbf{T}}(F) \longrightarrow H^\star_{\mathbf{T}}(W,F)$$

allows us to conclude. □

The rank of the $H^\star B\mathbf{T}$-module $H^\star_{\mathbf{T}}(W)$ is thus the same as the rank of the free module $H^\star_{\mathbf{T}}(F)$. It is possible to give an even more precise statement, using the notion of localization. If $f \in H^\star B\mathbf{T}$ is a nonzero element, consider the *localized* subring $(H^\star B\mathbf{T})_f$ of the ring of fractions $H^\star B\mathbf{T}$, consisting of all the fractions which have a power of f as denominator. In the same way, for any module M, consider the *localized* $(H^\star B\mathbf{T})_f$-module

$$M_f = M \otimes_{H^\star B\mathbf{T}} (H^\star B\mathbf{T})_f$$

obtained by extension of scalars.

Let f be a polynomial which vanishes on all the $\mathfrak{h}$'s. Consider the induced morphism

$$i^\star : (H^\star_{\mathbf{T}}(W))_f \longrightarrow (H^\star_{\mathbf{T}}(F))_f.$$

This is an isomorphism because of the next lemma.

Lemma VI.3.12. *Assume*

$$M' \xrightarrow{a} M \xrightarrow{b} M''$$

is an exact sequence of modules. Then the sequence

$$M'_f \xrightarrow{a'} M_f \xrightarrow{b'} M''_f$$

is exact.

In other words, the $H^\star(B\mathbf{T})$-module $(H^\star B\mathbf{T})_f$ is *flat*.

Proof. We have $b' \circ a' = 0$ since $b \circ a = 0$. Hence

$$\operatorname{Im} a' \subset \operatorname{Ker} b'.$$

Reciprocally, let $x \in \operatorname{Ker} b'$. By the definition of M_f, we can find an integer m such that $f^m x \in M$ and as we have $b'(x) = 0$, $b(f^m x) = f^m b'(x) = 0$ so that $f^m x \in \operatorname{Ker} b = \operatorname{Im} a$. We can thus write $f^m x = a(y)$ and

$$x = \frac{a(y)}{f^m} = a'\left(\frac{y}{f^m}\right) \in \operatorname{Im} a'. \qquad \square$$

According to the lemma, the kernel and the cokernel of the (localized) morphism $i^\star$ are the localizations of the kernel N and the cokernel C. On the other hand, we know that

$$\operatorname{Supp} N \subset V_f \text{ and } \operatorname{Supp} C \subset V_f$$

by the definition of f. Thus N_f and C_f are zero and $i^\star$ is actually an isomorphism.

Because of the adjunction formula (Proposition VI.4.6 below) relating $i^\star$ and the Gysin morphism $i_\star$,

$$i^\star i_\star 1 = e_T(\nu),$$

we suspect that $i_\star$ is more or less the inverse isomorphism, up to the inversion of some Euler classes. This is what we shall make precise now. Let us begin with two properties of $i_\star$ analogous to Theorem VI.3.11 for $i^\star$:

Proposition VI.3.13. *The kernel and cokernel of*

$$i_\star : H^\star_{\mathbf{T}}(F) \longrightarrow H^\star_{\mathbf{T}}(W)$$

both have support in the union $\cup\mathfrak{h}$ of the Lie algebras of the stabilizers $\neq \mathbf{T}$ of points of W.

Proof. We just have to consider the cohomology exact sequence of the pair $(W, W - F)$,

$$\begin{array}{ccccccc} H^\star_{\mathbf{T}}(W-F) & \longrightarrow & H^\star_{\mathbf{T}}(W, W-F) & \longrightarrow & H^\star_{\mathbf{T}}(W) & \longrightarrow & H^\star_{\mathbf{T}}(W-F) \\ & & \downarrow & \nearrow_{i_\star} & & & \\ & & H^\star_{\mathbf{T}}(F) & & & & \end{array}$$

and to deduce the result from the previous lemmas. $\square$

Let us now invert Euler classes. Any polynomial f vanishing on the support is suitable to define a localization. We are going to construct a minimal such f.

Let Z be a component of the fixed point set F of the $\mathbf{T}$-action on W. Consider its equivariant Euler class $e_T(\nu_Z) \in H^\star_{\mathbf{T}}(Z)$ (see § VI.4.b). As the $\mathbf{T}$-action on Z is trivial, $H^\star_{\mathbf{T}}(Z) \cong H^\star(Z) \otimes H^\star B\mathbf{T}$ and we may decompose

$$e_T(\nu_Z) = \sum_{i,j} a_i^j \otimes f_{2m-i}^j \in H^\star(Z) \otimes H^\star B\mathbf{T}$$

where the lower index denotes the degree of the cohomology class under consideration, and $2m$ is the codimension of the submanifold Z in W.

Of course, all the elements of positive degree in $H^\star(Z)$ are nilpotent, so that we have:

Lemma VI.3.14. *The Euler class, $e_T(\nu_Z)$ is invertible in $H^\star_{\mathbf{T}}(Z)$ if and only if its component $\sum_j a_0^j \otimes f_{2m}^j$ in $H^0(Z) \otimes H^{2m}(B\mathbf{T})$ is invertible.* □

Let $z \in Z$ be a fixed point. The sum $\sum a_0^j \otimes f_{2m}^j$ may be interpreted as the restriction of $e_T(\nu_Z)$ in $H^0(z) \otimes H^{2m} B\mathbf{T} \cong H^\star_{\mathbf{T}}(z)$. Let us compute this restriction. Write

$$\mathsf{T}_z W = \mathsf{T}_z Z \oplus L_1 \oplus \cdots \oplus L_m$$

where $\mathbf{T}$ acts on L_i by the morphism $\alpha_i : \mathbf{T} \to \mathbf{S}^1$. Also denote by $\alpha_i : \mathfrak{t} \to \mathbf{R}$ the linear form which is the derivative of α_i at 1. Then the restriction of $e_T(\nu_Z)$ is (up to sign) the product of the linear forms α_i (considered as a polynomial on $\mathfrak{t}$)(7).

Remark VI.3.15. It is clear that the kernel of α_i is one of the $\mathfrak{h}$'s in the support of $H^\star_{\mathbf{T}}(W - F)$. Indeed, the subgroup $\mathrm{H} = \operatorname{Ker} \alpha_i \subset \mathbf{T}$ fixes vectors in the complex subspace L_i, but the $\mathbf{T}$-action on L_i is nothing other than the derivative at z of that on W.

Let us put $f_Z = \prod \alpha_i$ (this is $e_T(\nu_z)$ for some z in Z) and $f = \prod_{Z \subset F} f_Z$. Then all the classes $e_T(\nu_Z)$ are invertible in $(H^\star_{\mathbf{T}}(F))_f$ and the map

$$Q = \sum_{Z \subset F} \frac{i_Z^\star}{e_T(\nu_Z)}$$

(where each i_Z is the inclusion of the component Z) defines a homomorphism

$$Q : (H^\star_{\mathbf{T}}(W))_f \longrightarrow (H^\star_{\mathbf{T}}(F))_f$$

inverse to $i_\star$. In other words, we have proved:

***Theorem VI.3.16* (Localization at fixed points).** *If $x \in H^\star_{\mathbf{T}}(W)$, in a suitable localization,*

$$x = \sum_{Z \subset F} \frac{i_\star^Z i_Z^\star x}{e_T(\nu_Z)}. \qquad \square$$

Remark VI.3.17. Actually, in the Hamiltonian case, using a well chosen projection of the momentum mapping and the ideas of [**51**], one shows that the $H^\star B\mathbf{T}$-module $H^\star_{\mathbf{T}} W$ is the free module generated by $H^\star W$. The restriction to the fiber

$$H^\star_{\mathbf{T}} W \longrightarrow H^\star W$$

(7) This is the definition of equivariant Euler classes, see § VI.4.b.

is surjective, while the restriction to the fixed points

$$i^\star : H^\star_{\mathbf{T}} W \longrightarrow H^\star_{\mathbf{T}} F$$

is injective.

VI.3.c. The Duistermaat–Heckman formula.

***Theorem VI.3.18* (Duistermaat and Heckman [44]).** *Let (W, ω) be a compact symplectic manifold of dimension $2n$ and let H be a periodic Hamiltonian on W with only isolated fixed points. Then*

$$\int_W e^{-Hu} \frac{\omega^{\wedge n}}{n!} = \sum_{Z \in F} \frac{e^{-uH(Z)}}{e_{\mathbf{S}^1}(\nu_Z)}.$$

Remarks VI.3.19

- It is not absolutely necessary to assume that the fixed points are isolated. There is a more general statement in [**44**].
- Similarly, it is not necessary to consider only circle actions, analogous results hold with higher dimensional tori.
- The formula must be read as an equality of formal series in the variables u and u^{-1}. We shall come back to this remark later on.

Proof of the theorem. As H is a periodic Hamiltonian, we know that the symplectic form ω may be "extended" to $W_{\mathbf{S}^1}$ (according to Proposition VI.2.1) as a closed 2-form $\omega^\sharp$. Consider the latter and all its powers. In order to deal with all of them at the same time, we use formally $\alpha = \exp[\omega^\sharp]$ as if this was an element of $H^\star_{\mathbf{S}^1}(W)$. Apply first the localization Theorem VI.3.16 to get:

$$\alpha = \sum_{Z \subset F} \frac{i^Z_\star i^\star_Z \alpha}{e_{\mathbf{S}^1}(\nu_Z)}.$$

Then "integrate in the fibers" (see § VI.4.c) of the fibration $\pi : W_{\mathbf{S}^1} \to B\mathbf{S}^1$, in other words apply $\pi_\star$. The right-hand side becomes

$$\sum_{Z \in F} \frac{i^\star_Z \alpha}{e_{\mathbf{S}^1}(\nu_Z)},$$

as $\pi_\star i_\star = 1$, according to § VI.4.c applied to the section

$$B\mathbf{S}^1 = E\mathbf{S}^1 \times_{\mathbf{S}^1} Z \hookrightarrow E\mathbf{S}^1 \times_{\mathbf{S}^1} W = W_{\mathbf{S}^1}.$$

We therefore calculate $i^\star_Z \alpha = \exp i^\star_Z \omega^\sharp$. The "form" $\omega^\sharp$ is the projection of

$$\omega + d(\theta H),$$

so that $i^\star_Z \omega^\sharp$ is the projection of

$$i^\star_Z(\omega + d\theta H - \theta \wedge dH).$$

As Z is a point, this is simply the projection of $d\theta H(Z)$, that is $-uH(Z)$. Our right-hand side is thus

$$\sum_{Z\in F} \frac{e^{-uH(Z)}}{e_{\mathbf{S}^1}(\nu_Z)}.$$

Let us look now at the left-hand side:

$$\pi_\star \alpha = \pi_\star \exp \omega^\sharp.$$

We thus have to consider

$$\exp(\omega + d\theta H - \theta \wedge dH) \text{ on } E\mathbf{S}^1 \times W,$$

to integrate over W and to project. During the integration process over W, all the terms in the exponential of degree different from $2n$ vanish. There remains only $\omega^{\wedge n}/n!$ and other terms which are all exact forms, as for instance $(dH)^{2n}/(2n)!$. Being exact, they all integrate to zero. The form $d\theta$ still projects onto $-u$. The left-hand side of our relation is thus

$$\int_W e^{-Hu} \frac{\omega^{\wedge n}}{n!}$$

as we have announced it. □

Image of the symplectic measure. Replacing the formal variable u by it and taking Fourier transforms gives another formulation of Theorem VI.3.18:

Corollary VI.3.20. *The density with respect to the Lebesgue measure of the direct image measure, under the momentum mapping $\mu : W \to \mathfrak{t}^\star$, of the symplectic measure, is a piecewise polynomial function.* □

Notice that we have only proved it for a 1-dimensional torus, but this makes no essential difference.

The stationary phase method. This is a method to compute, approximately, integrals of the form

$$I(u,t) = \left(\frac{t}{2\pi}\right)^n \int e^{it\varphi(x,u)} a(x,u,t)dx$$

which arise, for instance, in geometrical optics. The first term in an asymptotic expansion depends only on the behaviour of the function φ (the *phase*) near its critical (*stationary*) points. See [**41**].

In the case we are considering, namely that of

$$\left(\frac{t}{2\pi}\right)^n \int_W e^{-itH} \frac{\omega^{\wedge n}}{n!}$$

for a periodic Hamiltonian, the phase H is stationary at the fixed points of the circle action and the result consists in only one term

$$\left(\frac{t}{2\pi}\right)^n \sum_{Z\in F} \frac{e^{-itH(Z)}}{i^n e_{\mathbf{S}^1}(\nu_Z)}$$

(notice that t appears with the exponent n at the denominator as well). This is why the Duistermaat–Heckman formula is often qualified as an *exact* stationary phase formula.

VI.3.d. Examples of applications. The aim of this paragraph is to make Theorem VI.3.18 talk. The left-hand side of the formula is a formal series in u, but not *a priori* its right-hand side: if $\mathbf{S}^1$ acts near the fixed point Z by

$$u\cdot(z_1,\ldots,z_n) = (u^{a_1}z_1,\ldots,u^{a_n}z_n),$$

then

$$\pm a_1\cdots a_n u^n = e_{\mathbf{S}^1}(\nu_Z)$$

appears in the denominator. The Duistermaat–Heckman formula implies therefore that certain elements must vanish.

Here are some examples of these cancellations. Assume for simplicity that H is the Hamiltonian of a semi-free action, that is, that $a_i = \pm 1$ for all i. Let $\lambda(Z)$ be the number of minus signs (so that the index of Z as a critical point of H is $2\lambda(Z)$). The formula is then:

$$\int_W e^{-Hu}\frac{\omega^{\wedge n}}{n!} = \frac{1}{u^n}\sum_{Z\in F}(-1)^{\lambda(Z)}e^{-uH(Z)}.$$

Put "$u=0$" in this formula.

(1) The case where $n=1$. This gives

$$\int_W \omega = \left[\frac{1}{u}\left(\sum_{Z\in\min} e^{-uH(Z)} - \sum_{Z\in\max} e^{-uH(Z)}\right)\right]_{u=0}.$$

In other words, we have

$$\operatorname{Vol} W = H(\max) - H(\min)$$

as we know (from Theorem IV.3.1) that there are only one local minimum and one local maximum (see Figure 7). Our symplectic manifold is a sphere, its symplectic volume is the length of the segment image of the Hamiltonian (or momentum mapping). This is no surprise.

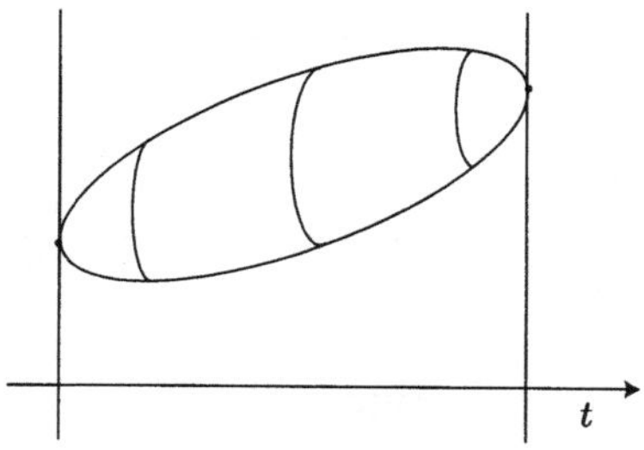

Figure 7

(2) The case where $n = 2$. Our symplectic manifold has now dimension 4. Let ρ_i be the number of critical points of index $2i$ and let F_i be the set of such points. Our formula gives

$$\int_W \frac{\omega^{\wedge 2}}{2} = \left\{ \frac{1}{u^2} \left[\rho_0 - \rho_1 + \rho_2 - u \left(\sum_{Z\in F_0} H(Z) - \sum_{Z\in F_1} H(Z) + \sum_{Z\in F_2} H(Z) \right) \right.\right.$$
$$\left.\left. + \frac{u^2}{2} \left(\sum_{Z\in F_0} H(Z)^2 - \sum_{Z\in F_1} H(Z)^2 + \sum_{Z\in F_2} H(Z)^2 \right) \right] \right\}_{u=0} .$$

The terms in the brackets give

- In degree 0, $\rho_0 - \rho_1 + \rho_2 = 0$ but we know that $\rho_0 = \rho_2 = 1$ and thus there are exactly two index-2 points. Call them a and b.
- In degree 1, $H(\min) - (H(a) + H(b)) + H(\max) = 0$, in other words the two intervals $[\min, \max]$ and $[a, b]$ have the same middle. We have already discussed these properties when we proved Theorem VI.2.3. We will come back to them in § VIII.1.c. In the meantime, look at Figure 8. Of course, with slopes 1 and -1, the middles of the segments must be the same.
- The degree-2 term computes the volume:

  $$\operatorname{Vol} W = \frac{1}{2} \left[H(\min)^2 - (H(a)^2 + H(b)^2) + H(\max)^2 \right].$$

  This is the surface of the trapezium in Figure 8! In other words this is the integral of the volume function $t \mapsto \operatorname{Vol} B_t$. Actually in this case, using the methods of Chapter VIII, we shall see that $W = \mathbf{S}^2 \times \mathbf{S}^2$, and that the $\mathbf{S}^1$-action extends to a $\mathbf{T}^2$-action... of course the trapezium is related with the image of the momentum mapping.

(3) The case where $n = 3$. With the same notation, and the same hypotheses, the same method gives for example

$$\rho_0 - \rho_1 + \rho_2 - \rho_3 = 0.$$

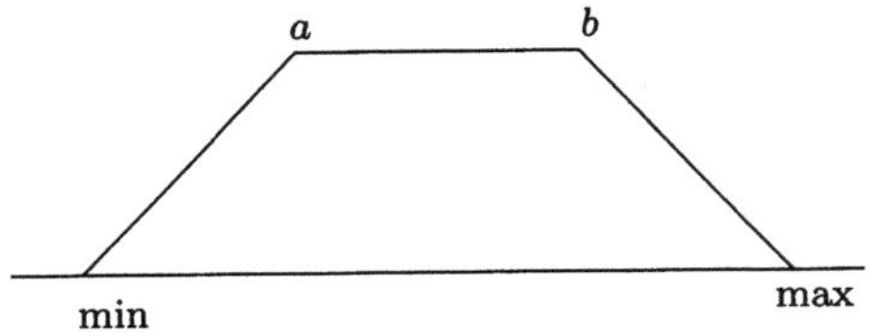

Figure 8

This, added to the fact that $\rho_0 = \rho_3 = 1$, gives $\rho_1 = \rho_2$. More generally, notice we always have $\rho_i = \rho_{n-i}$ in dimension $2n$ when the fixed points are isolated: this is Poincaré duality, as H is a "perfect" Morse function and so $\rho_k = \dim H^{2k}(W)$.

Remark VI.3.21. In dimension 6 ($n = 3$) and with the hypotheses above, it can be shown that $\rho_1 = \rho_2 = 3\ldots$ using essentially the same method, a localization formula near fixed points analogous to Theorem VI.3.16, but in *equivariant K-theory.* In some sense, this is even easier: the group $K_{\mathbf{T}}(W)$ is defined (starting with $\mathbf{T}$-vector bundles over W) much more simply than $H^{\star}_{\mathbf{T}}(W)$ for which we had to use classifying spaces and the Borel construction.

The functor "equivariant K-theory" $K_{\mathbf{T}}$ has properties analogous to that of the "equivariant cohomology" $H^{\star}_{\mathbf{T}}$, which allows us to prove a localization theorem. From this theorem, Hattori remarked in [**67**] that it was possible to deduce, among other things, with our hypotheses (periodic Hamiltonian, isolated critical points, semi-free action) that $\rho_i = \binom{n}{i}$. For $n = 2$, we have already explained this several times. For $n = 3$, remark that the method, even if analogous, is nevertheless more powerful, as we are not able to deduce $\rho_1 = \rho_2 = 3$ from Theorem VI.3.16 or Theorem VI.3.18. It is also with the help of the localization formula in equivariant K-theory, for Hamiltonian torus actions on toric varieties (see Chapter VII) that Brion gave very elegant proofs of some results about convex polyhedra (see [**28**]). The localization formula has many applications. There are also "non-Abelian" localization formulas (for other groups than the torus), due to Lisa Jeffrey and Frances Kirwan, see [**73, 74, 75**] with applications to the cohomology of symplectic quotients and in particular to that of moduli spaces.

VI.4. Appendix: some algebraic topology

We give here some hints of proofs and references for the main notions of algebraic topology used in this chapter. There are a lot of good books. We have already mentioned [**69**] when we have used it at the beginning of the chapter. I like very much and used a lot the one by Milnor [**109**] especially for this appendix.

VI.4.a. The Thom class of an oriented vector bundle. Let $E \to B$ be an oriented (for example, complex) vector bundle of (real) rank n. Denote E_0 the complement of the zero section in E, F a fiber and F_0 the intersection $F \cap E_0$. The fiber F is the vector space $\mathbf{R}^n$ and we have

$$H^n(F, F_0; \mathbf{Z}) = H^n(\mathbf{R}^n, \mathbf{R}^n - \{0\}; \mathbf{Z}) \cong H^{n-1}(\mathbf{R}^n - \{0\}; \mathbf{Z}) \cong H^{n-1}(\mathbf{S}^{n-1}; \mathbf{Z}) \cong \mathbf{Z}$$

so that the chosen orientation (which specifies the last isomorphism) gives, for each fiber F an element U_F which generates the infinite cyclic group $H^n(F, F_0; \mathbf{Z})$. The Thom isomorphism theorem asserts the existence of a class on E the restriction of which to each fiber is U_F.

***Theorem VI.4.1* (Thom isomorphism).** *There exists a unique cohomology class $U \in H^n(E, E_0; \mathbf{Z})$ the restriction of which to $H^n(F, F_0; \mathbf{Z})$ is U_F (for any fiber F). Moreover the map:*

$$\begin{array}{ccc} H^\star(E) & \longrightarrow & H^\star(E, E_0; \mathbf{Z}) \\ y & \longmapsto & y \smile U \end{array}$$

is an isomorphism.

For the proof, first look at the case of the trivial bundle, which is rather easy, then cover the base space by open subsets which trivialize the bundle, and (at least if the base space B is compact) deduce the result by induction on the number of open subsets, with the help of Mayer–Vietoris and the Five Lemma (see [**109**] for details). □

The class U is called the *Thom class* of the oriented bundle E. It is clear that a change in orientation will change the sign of U. It is also clear that U is *natural:* if $f : B' \to B$ is a continuous map, then $U(f^\star E) = f^\star U(E)$. The isomorphism $y \mapsto y \smile U$ is the *Thom isomorphism.*

VI.4.b. The Euler class of an oriented bundle, equivariant Euler class. The Thom class U allows us to define the *Euler class* of the oriented vector bundle $\pi : E \to B$. This is the unique class $e \in H^n(B; \mathbf{Z})$ such that $\pi^\star e = j^\star U$ where the maps are the natural ones

$$H^n(E, E_0; \mathbf{Z}) \xrightarrow{j^\star} H^n(E; \mathbf{Z}) \xleftarrow{\pi^\star} H^n(B; \mathbf{Z}).$$

Remark VI.4.2. Changing the orientation of E changes the sign of e. On the other hand, e is natural as is the Thom class we used to define it.

The equivariant Euler class. Recall that $E \to B$ is a G-vector bundle if it is a vector bundle, endowed with a G-action which is linear in the fibers and compatible with some G-action on the base space. This is what we need to form the vector bundle $EG \times_G E \to EG \times_G B$. As such, it has an Euler class (if E is oriented) which we shall denote $e_{\mathrm{G}}(E) \in H^\star_{\mathrm{G}}(B)$ as we intend to call it the *equivariant Euler class* of the bundle $E \to B$.

A G-vector bundle may be trivial as a vector bundle without being trivial as a G-vector bundle, as the following example shows.

Example VI.4.3. Let B be a point, so that a G-vector bundle over B is nothing other than a linear representation of the group G. Such a representation has an Euler class in $H^\star B\mathrm{G}$. For instance if $\mathrm{G} = \mathbf{S}^1$ and if $E = \mathbf{C}$ with the action $u \cdot v = u^m v$, then $E\mathbf{S}^1 \times_{\mathbf{S}^1} E$ is the bundle $\mathcal{O}(-m)$ over $B\mathbf{S}^1$ (see Appendix VI.5).

***Proposition VI.4.4* (Gysin exact sequence).** *There exists an exact sequence*

$$\longrightarrow H^i(B) \xrightarrow{\smile e} H^{i+n}(B) \xrightarrow{\pi_0^\star} H^{i+n}(E_0) \longrightarrow H^{i+1}(B) \longrightarrow .$$

Proof. Use the long cohomology exact sequence of the pair (E, E_0) and the previous isomorphisms in the diagram

$$\begin{array}{ccccccccc}
\longrightarrow & H^k(E,E_0) & \xrightarrow{j^\star} & H^k(E) & \longrightarrow & H^k(E_0) & \xrightarrow{\delta} & H^{k+1}(E,E_0) & \longrightarrow \\
& \smile U \uparrow & & \pi^\star \uparrow & \nearrow \pi_0^\star & & & \uparrow \smile U & \\
& H^{k-n}(E) & & H^k(B) & & & & H^{k-n+1}(E) & \\
& \pi^\star \uparrow & \nearrow \smile U & & & & & \uparrow \pi^\star & \\
& H^{k-n}(B) & & & & & & H^{k-n+1}(B) &
\end{array}$$

to conclude. □

The cohomology of the projective space. To prove Proposition VI.1.14, we use the Gysin exact sequence for the "tautological" line bundle

$$E = \mathcal{O}(-1) = \left\{ (\ell, x) \mid \ell \subset \mathbf{C}^{n+1}, x \in \ell \right\} \subset \mathbf{P}^n(\mathbf{C}) \times \mathbf{C}^{n+1}.$$

The complement E_0 of the zero section is the set of all pairs (ℓ, x) for $x \neq 0$. It is easily identified with $\mathbf{C}^{n+1} - \{0\}$. Its homotopy type is that of the sphere $\mathbf{S}^{2n+1}$. Let v be the Euler class of the bundle E (the class u in the statement of Proposition VI.1.14 is just $-v$). We may write the Gysin exact sequence

$$\longrightarrow H^{k+1}(\mathbf{S}^{2n+1}) \longrightarrow H^k(\mathbf{P}^n(\mathbf{C})) \xrightarrow{\smile v} H^{k+2}(\mathbf{P}^n(\mathbf{C})) \xrightarrow{\pi_0^\star}$$

$$\xrightarrow{\pi_0^\star} H^{k+2}(\mathbf{S}^{2n+1}) \longrightarrow$$

from which we immediately deduce that all the $H^{2k+1}(\mathbf{P}^n(\mathbf{C}))$ vanish, and that

$$\smile v : H^{2k}(\mathbf{P}^n(\mathbf{C})) \longrightarrow H^{2k+2}(\mathbf{P}^n(\mathbf{C}))$$

is an isomorphism for $0 \leqslant 2k \leqslant 2n-2$. □

The cohomology of a blow up. Recall that the blow up $\widetilde{\mathbf{C}}^n$ of $\mathbf{C}^n$ at 0 is also the total space of a complex line bundle over $\mathbf{P}^{n-1}(\mathbf{C})$, so that it has the same homotopy type (and the same cohomology) as this projective space. Using the Mayer–Vietoris exact sequence, one easily deduces the cohomology of any (symplectic or complex) manifold of dimension $2n$ blown-up at a point.

Proposition VI.4.5. *As a group, the cohomology of $\widetilde{W}$ is the direct sum*

$$H^\star(\widetilde{W}) = H^\star(W) \oplus \widetilde{H}^\star(\mathbf{P}^{n-1}(\mathbf{C})). \qquad \square$$

In dimension 4 (or complex dimension 2), it is easy to be more precise. As in the proposition, the cohomology groups of the blown-up manifold $\widetilde{W}$ are the same as those of W, except for the H^2, which is the direct sum

$$H^2(\widetilde{W}) \cong H^2(W) \oplus H^2(\mathbf{P}^1(\mathbf{C})) \cong H^2(W) \oplus \mathbf{Z}.$$

Moreover, the factor $\mathbf{Z}$ is generated by the cohomology class dual to the class $E \in H_2(\widetilde{W})$ of the exceptional divisor and the intersection pairing (or, equivalently, the cup product) is determined by the fact that

$$E \cdot E = -1 \text{ and } E \cdot C = 0 \text{ for all } C \in H_2(W).$$

VI.4.c. The Gysin homomorphism. Let us start with the case of an embedding. Let $i : Z \hookrightarrow W$ denote the inclusion of a codimension-m submanifold, with normal bundle ν (assumed to be oriented). We define the *Gysin homomorphism*

$$i_\star : H^k(Z) \longrightarrow H^{k+m}(W)$$

(which goes the wrong way and which does not preserve the graduation) as the composition

$$H^k(Z) \xrightarrow{\ \Phi\ } H^{k+m}(\nu, \nu_0) \xrightarrow{\text{excision}} H^{k+m}(W, W - Z) \xrightarrow{\text{restriction}} H^{k+m}(W)$$

where of course, Φ denotes the Thom isomorphism for the bundle ν.

Proposition VI.4.6. *Let $i : W \subset W$ be the inclusion of a submanifold, the normal bundle of which is denoted ν. We have*

$$i^\star i_\star 1 = e(\nu) \quad \text{and} \quad i^\star i_\star x = x \smile e(\nu).$$

Proof. The commutative diagram

$$\begin{array}{ccccc}
1 & \longmapsto & U & & \\
H^0(Z) & \longrightarrow & H^m(\nu, \nu_0) & \longrightarrow & H^m(W) \\
 & & \downarrow j^\star & & \downarrow i^\star \\
 & & H^m(\nu) & \xleftarrow{\ \pi^\star\ } & H^m(Z) \ni e(\nu)
\end{array}$$

proves the first equality. As $\pi^\star$ and $j^\star$ preserve multiplicative structures and $x \mapsto x \smile U$ is the Thom isomorphism, this also proves the second one. □

The equivariant case. Suppose the Lie group G acts on W and Z, the embedding i being equivariant. Up to finite approximations, we may consider that the inclusion

$$i : EG \times_{\mathrm{G}} Z \hookrightarrow EG \times_{\mathrm{G}} W$$

is the embedding of a submanifold. Therefore we have a Gysin homomorphism

$$i_\star : H^\star_{\mathrm{G}}(Z) \longrightarrow H^\star_{\mathrm{G}}(W).$$

Computing $i^\star i_\star 1$ gives the Euler class of the normal bundle of $EG \times_{\mathrm{G}} Z$ that is, the equivariant Euler class of ν:

$$i^\star i_\star x = x \smile e_{\mathrm{G}}(\nu).$$

Integration in the fibers. Let $\pi : E \to B$ be a locally trivial bundle with compact fiber F. Assume that B and E are manifolds and, moreover, that F is an oriented manifold of dimension m. Let ω be a k-form on E. We can associate to it a $(k-m)$-form on B, denoted $\pi_\star\omega$ and defined in the following way. We first define an m-form α on $F = \pi^{-1}(b)$ by

$$\alpha_x\left(V_1, \dots, V_m\right) = \omega_x\left(\widetilde{X}_1(x), \dots, \widetilde{X}_{k-m}(x), V_1, \dots, V_m\right)$$

in which formula $\widetilde{X}_i(x)$ is any vector in $\mathsf{T}_x E$ such that $\mathsf{T}_x\pi(\widetilde{X}_i(x)) = X_i \in \mathsf{T}_b E$ (it is clear that the result of this process is well-defined). Then we define the form $\pi_\star\omega$ by

$$\begin{cases} \text{if } k < m, & \pi_\star\omega = 0 \\ \text{otherwise}, & (\pi_\star\omega)_b\left(X_1, \dots, X_{k-m}\right) = \int_{\pi^{-1}(b)} \alpha. \end{cases}$$

We have defined this way a map

$$\pi_\star : \Omega^k(E) \longrightarrow \Omega^{k-m}(B)$$

at the level of forms.

Proposition VI.4.7. *The diagram*

$$\begin{array}{ccc} \Omega^k(E) & \xrightarrow{\ \pi_\star\ } & \Omega^{k-m}(B) \\ \downarrow d & & \downarrow d \\ \Omega^{k+1}(E) & \xrightarrow{\ \pi_\star\ } & \Omega^{k-m+1}(B) \end{array}$$

commutes.

Proof. This is a straightforward verification. □

We thus get a morphism, still denoted $\pi_\star$,

$$\pi_\star : H^k(E) \longrightarrow H^{k-m}(B),$$

of which it is not hard to make a relative version which will appear in the following example (of course, in this paragraph, we are dealing with de Rham cohomology).

Example VI.4.8. The (constant) map $\pi : \mathbf{S}^m \to \mathrm{pt}$ may be considered as a fibration with fiber $\mathbf{S}^m$. The homomorphism

$$\begin{array}{rcl} \pi_\star : H^m(\mathbf{S}^m) & \longrightarrow & H^0(\mathrm{pt}) \\ \omega & \longmapsto & \int_{\mathbf{S}^m} \omega \end{array}$$

sends the generator to 1. From the definition of the Thom class U, we deduce that if E is an oriented rank n vector bundle over B, then

$$\pi_\star : H^n(E, E_0) \longrightarrow H^0(B)$$

maps U to 1.

The reader may be frightened by the present terminology and ask (with some reason) what relationship exists between the $_\star$ considered here in the case of a fibrations and the $_\star$ defined above in the case of an embedding. She or he may remain calm: $\pi_\star$ and $i_\star$ are two aspects of a single *Gysin homomorphism* $f_\star$ defined for any *proper* map $f : V \to W$ from one manifold into another. One possible way to define $f_\star$ is precisely to decompose f as the composition of an embedding and a fibration

$$i : V \longrightarrow W \times \mathbf{S}^n \text{ for } n \text{ large and } \pi : W \times \mathbf{S}^n \longrightarrow W$$

and to put $f_\star = \pi_\star \circ i_\star$ (it is not very hard to prove that this does not depend on the choices).

Consider for instance the case of the inclusion of a *section* $i : B \hookrightarrow E$ of the bundle. We then have, for any $y \in H^\star(B)$, $\pi_\star i_\star y = y$. In fact, $i_\star$ is the composition of the horizontal maps in the diagram

$$\begin{array}{ccccc} y & \longmapsto & y \smile U & & \\ H^m(B) & \longrightarrow & H^m(\nu, \nu_0) & \longrightarrow & H^{m+n}(E) \\ & & & \searrow & \downarrow \pi_\star \\ & & & & H^n(B) \end{array}$$

but we have seen in the previous example that the Thom class is mapped to 1.

VI.5. Appendix: various notions of Euler classes

We have already met two notions of Euler classes for fixed point free $\mathbf{S}^1$-actions: in Chapter I for 3-manifolds, where we got a *number*, and in § VI.1.c where a degree-2 cohomology class was constructed. On the other hand, we have also used the Euler class of a (real) oriented vector bundle. In this appendix, we shall investigate the relations between these various notions.

VI.5.a. The case of $\mathbf{S}^1$-bundles. Because all relevant definitions can be ultimately expressed in terms of principal bundles (or free actions), it is enough to study this case. We shall thus prove:

Theorem VI.5.1. *Let V be an oriented compact 3-manifold endowed with an $\mathbf{S}^1$-action. Let B be the oriented surface which is the quotient. If $e \in H^2(B)$ is the Euler class of the $\mathbf{S}^1$-bundle $V \to B$ (in the sense of Chapter* VI*), then $\langle e, [B]\rangle \in \mathbf{Z}$ is the Euler number in the sense of Chapter* I.

Proof. We first reduce to the case where B is a sphere. Indeed, write $B = \overline{B - D_0} \cup \overline{D_0}$ so that the bundle is trivialized over each piece. Define a map $f : B \to \mathbf{S}^2$ by collapsing the complement of a collar $\overline{B - D_0}$ onto a point (see Figure 9, the small white disc is mapped to the upper hemisphere, the complement of the annulus to the South pole, the annulus itself onto the lower hemisphere).

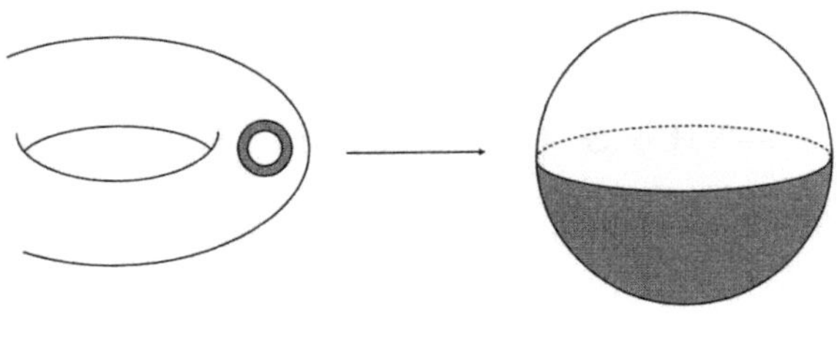

Figure 9

The map $f^\star : H^2(\mathbf{S}^2) \to H^2(B)$ is an isomorphism. Moreover, by the definition of f, the bundle $V \to B$ is exactly the pull-back by f of the bundle over the sphere, which is trivial over each hemisphere and described by the same gluing data as $V \to B$ along the equatorial circle.

We thus assume that B is a sphere $\mathbf{S}^2$. We reduce now to the case where $V \to \mathbf{S}^2$ is the Hopf bundle. In fact, if $g : \mathbf{S}^2 \to \mathbf{S}^2$ is a degree-m map, it multiplies Euler classes (in the sense of Chapter VI) by m, but in the sense of Chapter I as well: we may assume it induces the map $z \mapsto z^m$ on the equatorial circle ∂D_0. In the diagram

$$\begin{array}{ccc} V' & \longrightarrow & V \\ \downarrow & & \downarrow \\ \mathbf{S}^2 & \xrightarrow[g]{} & \mathbf{S}^2 \end{array}$$

with obvious notation $a' \mapsto ma$ and similarly $\partial\sigma' \mapsto m\partial\sigma$, but $b' \mapsto b$ as it is an orbit. We deduce that $e = me'$ (in the sense of Chapter I).

For the Hopf bundle $\mathbf{S}^3 \to \mathbf{P}^1(\mathbf{C})$, the statement is clear, as we have $e = -1$ in the sense of Chapter I and in the sense of Chapter VI as well by definition of u. □

Remark VI.5.2. We have just checked by explicit computation, and used the fact that the Hopf bundle $\mathbf{S}^3 \to \mathbf{P}^1(\mathbf{C})$ is universal among the principal $\mathbf{S}^1$-bundles over surfaces.

VI.5.b. Complex line bundles. To any principal $\mathbf{S}^1$-bundle $P \to B$ we associate (this is actually what is called an *associated bundle*) a rank-1 complex vector bundle $E \to B$ using the usual linear $\mathbf{S}^1$-action over $\mathbf{C}$

$$E = P \times_{\mathbf{S}^1} \mathbf{C}$$

where $\mathbf{S}^1$ acts on $P \times \mathbf{C}$ by $t \cdot (v, z) = (tv, \bar{t}z)$.

Example VI.5.3. Take $P = \mathbf{S}^{2n+1}$ with the action defining the Hopf fibration. The map

$$\begin{aligned} \mathbf{S}^{2n+1} \times \mathbf{C} &\longrightarrow \mathbf{P}^n(\mathbf{C}) \\ (z_0, \dots, z_n; z) &\longmapsto [z_0, \dots, z_n] \end{aligned}$$

becomes, taking quotients, the vector bundle

$$\mathbf{S}^{2n+1} \times_{\mathbf{S}^1} \mathbf{C} \longrightarrow \mathbf{P}^n(\mathbf{C}).$$

Using the injection

$$\begin{aligned} \mathbf{S}^{2n+1} \times_{\mathbf{S}^1} \mathbf{C} &\hookrightarrow \mathbf{P}^n(\mathbf{C}) \times \mathbf{C}^{n+1} \\ [z_0, \dots, z_n; z] &\longmapsto ([z_0, \dots, z_n], z_0 z, \dots, z_n z) \end{aligned}$$

we may identify the vector bundle obtained with the tautological bundle $\mathcal{O}(-1)$.

For any $m \in \mathbf{Z}$ we also know how to define a bundle called $\mathcal{O}(m)$ over $\mathbf{P}^n(\mathbf{C})$

$$\mathcal{O}(m) = \mathbf{S}^{2n+1} \times \mathbf{C}/(z_0, \dots, z_n; z) \sim (tz_0, \dots, tz_n; t^m z).$$

Conversely, given a complex line bundle E over a manifold B, the choice of a Hermitian metric on E allows us to consider its *circle* bundle $S(E)$, which is then a principal $\mathbf{S}^1$-bundle over B (thanks to the linear $\mathbf{S}^1$-action by rotations in the fibers). Of course we may recover E from $S(E)$ as above.

We now have to compare the Euler class of E as an oriented vector bundle and the Euler class of $S(E)$ as a principal $\mathbf{S}^1$-bundle. But we have actually defined the class $u \in H^2(\mathbf{P}^n(\mathbf{C}); \mathbf{Z})$ in such a way that $e(\mathcal{O}(-1)) = -u$ (see VI.4.b). By universality, we deduce that they coincide.

As a conclusion to this appendix, let us now check the assertion about multiplication by n which we used to define the Euler class in the non-principal case and in § VI.1.c.

Proposition VI.5.4. *The map $z \mapsto z^n$ induces multiplication by n on $H^2(B\mathbf{S}^1)$.*

Proof. Coming back to the previous examples, we see that we have to prove that the Euler class of $\mathcal{O}(n)$ is nu. On the other hand, as this only concerns H^2, the assertion needs only be checked on $\mathbf{P}^1(\mathbf{C})$. Consider thus the map

$$\begin{aligned} \varphi : \mathbf{P}^1(\mathbf{C}) &\longrightarrow \mathbf{P}^1(\mathbf{C}) \\ [x, y] &\longmapsto [x^n, y^n]. \end{aligned}$$

It has degree n, which means that it induces multiplication by n in the group H^2. Moreover it is (almost) obvious that it satisfies $\varphi^\star \mathcal{O}(1) \cong \mathcal{O}(n)$. By naturality we thus have

$$e(\mathcal{O}(n)) = \varphi^\star e(\mathcal{O}(1)) = ne(\mathcal{O}(1)) = nu. \qquad \square$$

Exercises

***Exercise VI.1* (Contractibility of $\mathbf{S}^\infty$).** Consider the infinite sphere

$$\mathbf{S}^\infty = \left\{x = (x_0, \dots, x_n, \dots) \mid \text{ almost all } x_i = 0 \text{ and } \sum x_i^2 = 1\right\}$$

with its subspheres

$$\mathbf{S}^n = \left\{x \in \mathbf{S}^\infty \mid \sum_{i=0}^n x_i^2 = 0\right\}$$

and their neighborhoods

$$U_n = \left\{x \in \mathbf{S}^\infty \mid \sum_{i=0}^n x_i^2 \neq 0\right\}, \quad V_n = \left\{x \in \mathbf{S}^\infty \mid \sum_{i=0}^n x_i^2 > \frac{2}{3}\right\}.$$

The aim of this exercise is to prove that $\mathbf{S}^\infty$ deforms into the simplex

$$\Delta = \{x \in \mathbf{S}^\infty \mid x_i \geqslant 0 \text{ for all } i\}.$$

Prove that Δ is contractible (so that you will have proved, eventually, that $\mathbf{S}^\infty$ is contractible as well).

Prove that $\mathbf{S}^n$ deforms to a point (of Δ) in $\mathbf{S}^{n+1}$. More precisely, check that the map

$$\begin{aligned} f_n : \mathbf{S}^n \times [0,1] &\longrightarrow \mathbf{S}^{n+1} \\ (x,t) &\longmapsto (\sqrt{1-t}x_0, \ldots, \sqrt{1-t}x_n, \sqrt{t}, 0, \ldots) \end{aligned}$$

is a deformation with the required properties.

Check that the map $g_n : U_n \times [0,1] \to U_n$ which maps (x,t) to

$$\left(\sqrt{\frac{t+(1-t)\sum_{j=0}^n x_j^2}{\sum_{j=0}^n x_j^2}}x_0, \ldots, \sqrt{\frac{t+(1-t)\sum_{j=0}^n x_j^2}{\sum_{j=0}^n x_j^2}}x_n, \sqrt{1-t}x_{n+1}, \ldots\right)$$

is continuous, deforms U_n to $\mathbf{S}^n$ and maps $\Delta \times [0,1]$ into Δ. With f_n and g_n, we get a map

$$h_n : U_n \times [0,1] \longrightarrow U_{n+1}$$

that deforms U_n to a point of Δ. Notice however that the point obtained this way depends on n. Prove that the maps[(8)]

$$\varphi_n : U_n \times [0,1] \longrightarrow U_{n+1}$$

defined by induction on n and by $\varphi_0 = h_0$,

$$\varphi_{n+1}(x,t) = \begin{cases} h_{n+1}(x,t) & \text{if } 0 \leqslant \sum_{j=0}^n x_j^2 \leqslant \frac{1}{3} \\ h_{n+1}(\varphi_n(x, 2t(3\sum_{j=0}^n x_j^2 - 1)), t) & \text{if } \frac{1}{3} \leqslant \sum_{j=0}^n x_j^2 \leqslant \frac{1}{2} \\ h_{n+1}(\varphi_n(x,t), 2t(2 - 3\sum_{j=0}^n x_j^2)) & \text{if } \frac{1}{2} \leqslant \sum_{j=0}^n x_j^2 \leqslant \frac{2}{3} \\ \varphi_n(x,t) & \text{if } \frac{2}{3} \leqslant \sum_{j=0}^n x_j^2. \end{cases}$$

are continuous deformations and satisfy

$$\varphi_{n+1}|_{V_n \times [0,1]} = \varphi_n|_{V_n \times [0,1]} \text{ and } \varphi_n(x,1) \in \Delta.$$

Deduce a deformation of $\mathbf{S}^\infty$ on Δ.

Exercise VI.2. Prove that the map

$$\begin{aligned} V_k(\mathbf{C}^N) &\longrightarrow \mathbf{S}^{2N-1} \\ (v_1, \ldots, v_k) &\longmapsto v_k \end{aligned}$$

is a locally trivial fibration with fiber $V_{k-1}(\mathbf{C}^N)$. Deduce that the natural map

$$\pi_n(V_{k-1}(\mathbf{C}^N)) \longrightarrow \pi_n(V_k(\mathbf{C}^N))$$

is an isomorphism for $n \leqslant 2N-2$ and that $V_k(\mathbf{C}^\infty)$ is weakly contractible.

---

(8) This is using essentially the same strategy and the same formulas that Dold shows in [**40**] that the spaces EG constructed by Milnor are contractible.

Exercise VI.3. We have seen that $\mathbf{P}^n(\mathbf{C})$ is a finite dimensional approximation to $B\,\mathrm{U}(1) = B\mathbf{S}^1$. Prove that the same is true of the Grassmannian $\widetilde{G}_2(\mathbf{R}^{n+1})$. Using Exercise IV.9, prove that,

$$\text{for } n+2=2k, \quad H^{2k-2}(\widetilde{G}_2(\mathbf{R}^{2k};\mathbf{Z}) = \mathbf{Z}\oplus\mathbf{Z}$$

(so that the two spaces $\mathbf{P}^{2k-2}(\mathbf{C})$ and $\widetilde{G}_2(\mathbf{R}^{2k})$ are, indeed, different).

Consider $\widetilde{G}_2(\mathbf{R}^{n+2})$ as a quadric in $\mathbf{P}^{n+1}(\mathbf{C})$ (see Exercises I.21 and III.15). Let $u \in H^2(\widetilde{G}_2(\mathbf{R}^{n+2});\mathbf{Z})$ be the image of the generator of $H^2(\mathbf{P}^{n+1}(\mathbf{C});\mathbf{Z})$, so that u is dual to the intersection of the quadric with the hyperplane $z_{n+1}=0$. Check that $\widetilde{G}_2(\mathbf{R}^{n+2})$ is simply connected and that

$$\pi_2\widetilde{G}_2(\mathbf{R}^{n+2}) = \mathbf{Z},$$

generated by a map

$$f : \mathbf{S}^2 \longmapsto \widetilde{G}_2(\mathbf{R}^{n+2})$$

which pulls back the tautological bundle to a bundle of Euler class 1. Let $v \in H^{2n-2}(\widetilde{G}_2(\mathbf{R}^{n+2});\mathbf{Z})$ be the class dual to $f_\star[\mathbf{S}^2]$. Check that

$$u^{n-1} = 2v,$$

so that, although for n odd,

$$H^\star(\widetilde{G}_2(\mathbf{R}^{n+2});\mathbf{Z}) \text{ is isomorphic with } H^\star(\mathbf{P}^n(\mathbf{C});\mathbf{Z})$$

as groups, the two cohomology rings (and hence the two spaces) are different.

Exercise VI.4. Show that, if α is an invariant k-form ($\alpha \in \Omega^k(W)^{\mathrm{G}}$), $d\alpha$ is an invariant $(k+1)$-form, in other words that there is a commutative diagram

$$\begin{array}{ccc} \Omega^k(W)^{\mathrm{G}} & \xrightarrow{\ d\ } & \Omega^{k+1}(W)^{\mathrm{G}} \\ \downarrow{\scriptstyle j} & & \downarrow{\scriptstyle j} \\ \Omega^k(W) & \xrightarrow{\ d\ } & \Omega^{k+1}(W). \end{array}$$

Assume G is a compact connected Lie group. Show that the inclusion j of invariant forms in the space of all forms is a *quasi-isomorphism* (that is, induces an isomorphism at the cohomology level).

Exercise VI.5. Let $\pi : P \to B$ be a principal $\mathbf{S}^1$-bundle, and let η be a 2-form on B such that $[\eta/2\pi]$ is the Euler class of the bundle. Show that there exists a connection form α on P such that $d\alpha = \pi^\star\eta$.

Exercise VI.6. Let (W,ω) be a compact symplectic manifold of dimension $2n$ endowed with a Hamiltonian action of a torus $\mathbf{T}^n$ (of half the dimension). Let P be the polyhedron which is the image of W under the momentum mapping μ.

(1) Let AB be an edge in P. Show that it is the image of a symplectic sphere $S \subset W$.
(2) Write $\overrightarrow{AB} = \lambda \overrightarrow{u}$ where $\overrightarrow{u} \in \mathbf{Z}^n$ is indivisible and $\lambda \in \mathbf{R}$. What can be said about the stabilizer of the points in S? What is the volume $\int_S \omega$ of the sphere S?

***Exercise VI.7* (Symplectic blow up).** Let ω_1 be the standard form on $\mathbf{C}^2$, ω_2 that on $\mathbf{P}^1(\mathbf{C})$. Consider the inclusion

$$j : \widetilde{\mathbf{C}}^2 \hookrightarrow \mathbf{C}^2 \times \mathbf{P}^1(\mathbf{C})$$

and define, for any $r > 0$, $\Omega_r = j^\star(\omega_1 + r\omega_2)$. Check that this is a symplectic form on $\widetilde{\mathbf{C}}^2$, which gives volume r to the exceptional divisor.

Let D_r be the ball of radius r in $\mathbf{C}^2$,

$$D_r = \left\{(x, y \in \mathbf{C}^2 \mid |x|^2 + |y|^2 \leqslant r\right\}$$

and let

$$\begin{aligned} f_r : \mathbf{C}^2 - D_r &\longrightarrow \mathbf{C}^2 - \{0\} \\ v &\longmapsto \left(\frac{\|v\| - r}{\|v\|}\right) v \end{aligned}$$

be the radial diffeomorphism. Identify $\mathbf{C}^2 - \{0\}$ with $\widetilde{\mathbf{C}}^2 - \pi^{-1}(0)$ *via* π. Show that

$$f_r^\star \Omega_r = \omega_1|_{\mathbf{C}^2 - D_r}.$$

Exercise VI.8. Let ζ be a primitive m-th root of 1. Consider the lens space $L^{2n+1}(m)$ which is the quotient of $\mathbf{S}^{2n+1}$ by the equivalence relation generated by $v \sim \zeta v$. Check that the natural projection map $L^{2n+1}(m) \to \mathbf{P}^n(\mathbf{C})$ is a principal $\mathbf{S}^1$-bundle. Prove that the complex line bundle associated with this principal bundle is $\mathcal{O}(-m)$.

Exercise VI.9. Prove that $\mathcal{O}(m)$ is the complex line bundle dual to $\mathcal{O}(-m)$. Prove that the cohomology class u used in Proposition VI.1.14 is $e(\mathcal{O}(1))$.

Exercise VI.10. Let $E \to B$ be a complex vector bundle over a closed surface B. Decompose

$$B = \overline{B - D_0} \cup \overline{D_0}$$

as in the proof of Theorem VI.5.1 and prove that the isomorphism class of the complex bundle is given by an element of $\pi_1 \operatorname{GL}(n; \mathbf{C})$. Recall (or check) that $\operatorname{GL}(n; \mathbf{C})$ retracts on $\operatorname{U}(n)$ (this is the Gram–Schmidt process) and that the determinant

$$\det : \operatorname{U}(n) \longrightarrow \mathbf{S}^1$$

induces an isomorphism at the level of π_1's.

Consider now the projectivized bundle $\mathbf{P}(E) \to B$. Prove that its isomorphism type is given by an element of $\pi_1 \mathbf{P}\,\mathrm{GL}(n;\mathbf{C})$. Look at the exact sequence of groups (and fibration)

$$1 \longrightarrow \mathbf{C}^\star \longrightarrow \mathrm{GL}(n;\mathbf{C}) \longrightarrow \mathbf{P}\,\mathrm{GL}(n;\mathbf{C}) \longrightarrow 1$$

and prove that $\pi_1 \mathbf{P}\,\mathrm{GL}(n;\mathbf{C}) \cong \mathbf{Z}/n$.

An illustration. Prove that the total space of the bundle

$$\mathbf{P}(\mathcal{O}(k_1) \oplus \mathcal{O}(k_2)) \longrightarrow \mathbf{P}^1(\mathbf{C})$$

is diffeomorphic (by a diffeomorphism compatible with the fibration) with the total space $\mathbf{P}(\mathcal{O}(k_1 - k_2) \oplus \mathbf{1})$.

Prove that[(9)] the Hirzebruch surface W_k of Exercises IV.4 and IV.17 is diffeomorphic with $\mathbf{P}^1(\mathbf{C}) \times \mathbf{P}^1(\mathbf{C})$ if k is even and with $\widetilde{\mathbf{P}}^2(\mathbf{C})$ if k is odd.

[(9)]This was already shown in Exercise IV.17.

CHAPTER VII

TORIC MANIFOLDS

The goal of this chapter is to present a very beautiful family of symplectic manifolds endowed with Hamiltonian torus actions, the *toric varieties*(1). Long before their symplectic aspects were understood, toric manifolds were introduced by Demazure as closures of complex torus orbits in algebraic manifolds. They have been investigated since then by numerous authors(2). These are algebraic varieties which can be defined over any field (of course we shall restrict ourselves to the field of complex numbers). The prototype is the closure of any orbit of a (complex) torus acting in a linear way on a projective space which we already met in Theorem IV.4.25. We shall present here an alternative description.

The manifolds under consideration have properties which coincide *a priori* with those which we are interested in:

- They are endowed with the action of a "big" torus (of maximal dimension: half that of the manifold) and with symplectic structures for which this action is Hamiltonian.
- They are constructed from something very close to a convex polyhedron, namely a *fan* (this is exactly what allows us to recover the combinatorics of the faces of a convex polyhedron, but *not* the size of the faces, see VII.1.a). Moreover, an invariant ample line bundle on such a manifold allows us to construct a convex polyhedron with integral vertices.
- According to the uniqueness theorem of Delzant (here Theorem IV.4.20), they are *the* compact symplectic manifolds endowed with completely integrable torus actions.

---

(1)This is to insist on the fact that we are mainly interested in the complex and smooth case that we used the word "manifold" in the title of the chapter.

(2)There are rather complete, but already a little out of date, lists of references in [**38**] and [**114**], beautifully updated in [**35, 36, 37**].

After the publication of the convexity theorem (Theorem IV.4.3), various authors hurried to remark that, in the case where the cohomology class of the symplectic form is the Euler class of a complex line bundle, the polyhedron image by the momentum mapping was actually the one that was classically associated with the line bundle under consideration (see Jurkiewicz's paper [**80**] and the proof by Khovanski and Arnold of Theorem IV.4.25 in [**8**] that we have already used).

Here, we shall describe a complex toric variety(3) as the "quotient" of some $\mathbf{C}^N$ by a complex subtorus of $(\mathbf{C}^\star)^N$. Here, the "quotient" will be the honest topological quotient of a big open subset of $\mathbf{C}^N$. According to a rather general idea of Kirwan, this quotient is identified to the symplectic reduction of some regular level of a momentum mapping on $\mathbf{C}^N$. We thus meet a description due to Delzant, from which we deduce, as in [**39**] that with each (integral or not) convex polyhedron satisfying the necessary conditions implied by Proposition IV.4.12, we are able to associate a compact symplectic manifold of dimension double that of the polyhedron, the latter being the image of the momentum mapping.

This construction gives, in a very natural way, many of the classical results on the topology of these manifolds (simple connectedness, cohomology, invariant line bundles and symplectic forms).

The ideas in this chapter come from [**85**], [**8**] and, of course, [**39**]. I have chosen to discuss toric varieties here, mainly because I wanted to relate the topological aspects of the theory I enjoyed learning in the very beautiful paper by Danilov [**38**]. I enjoyed reading [**30**] as well, so that I have used it, especially in the description of fibrations and of toric surfaces.

VII.1. Fans and toric varieties

VII.1.a. Fans. The idea one must have in mind is that a fan is what is left from a convex polyhedron when the "sizes" of its faces are forgotten.

Consider the vector space $\mathbf{R}^n$ with its integral lattice $\mathbf{Z}^n$. Let P be a convex polyhedron in the dual space $(\mathbf{R}^n)^\vee$. We are thinking of polyhedra as those we have met in Chapter IV (this is why they live in a dual space) so that we will assume that the faces of P have integral equations, namely equations of the form

$$\langle v, \varphi \rangle = a \text{ for some } v \in \mathbf{Z}^n \text{ and } a \in \mathbf{R}.$$

With any face Γ we may associate its *tangent cone* (Figure 1). Chose a point m in the (relative) interior of Γ (notation $m \in \mathring{\Gamma}$) and define

$$\sigma_\Gamma = \bigcup_{r \geqslant 0} r \cdot (P - m).$$

(3) Not in the generality needed by algebraic geometers.

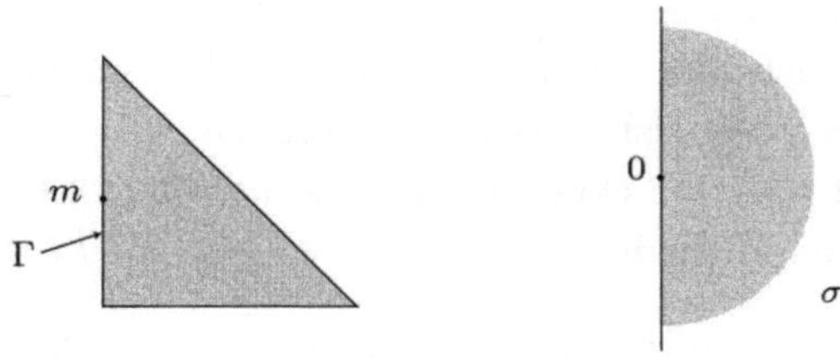

Figure 1

The result is obviously a convex cone (and, as it is easily checked, it does not depend on the choice of the point m in $\mathring{\Gamma}$). This is the tangent cone[4]. Figure 2

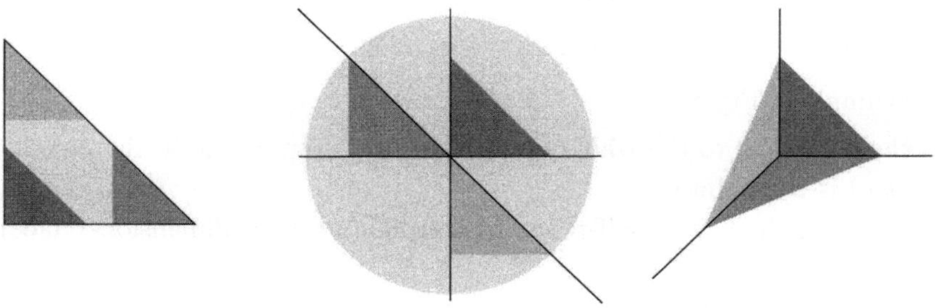

Figure 2

shows the polyhedron (on the left) and its tangent cones (in the middle). Notice that all the tangent cones have dimension n (if the interior of the polyhedron is not empty). For the example shown in Figure 2, the tangent cone

- to the dimension-2 face is the whole vector space,
- to each edge is a half-plane limited by the line through 0 parallel to the line supporting the edge (note that, for clarity, these cones have not been drawn on the picture),
- to each vertex is a cone with vertex at 0.

Lemma VII.1.1. *For any two faces* Γ, Γ', *we have*

$$\Gamma \subset \Gamma' \Longrightarrow \sigma_\Gamma \subset \sigma_{\Gamma'} \quad \textit{and} \quad \sigma_{\Gamma\cap\Gamma'} = \sigma_\Gamma \cap \sigma_{\Gamma'}.$$

Proof. The first property is due to the fact that, if the face Γ is included in the face Γ', any neighborhood of $m \in \mathring{\Gamma}$ contains a point of the relative interior of Γ'. This implies the inclusion $\sigma_{\Gamma\cap\Gamma'} \subset \sigma_\Gamma \cap \sigma_{\Gamma'}$, so that we only need to check the

[4]The word "tangent" is used here in the same sense that the tangent space to an affine space is the underlying vector space.

reverse inclusion. But, if $x \in \sigma_\Gamma \cap \sigma_{\Gamma'}$, there is an m in the relative interior of the intersection $\Gamma \cap \Gamma'$ such that the half line $[mx)$ contains points of the polyhedron. Again, as close to m as we wish, we can find points n and n' of the relative interiors of Γ and Γ' and, if they are close enough, the half-lines $[nx)$ and $[n'x)$ will also contain points of the polyhedron. □

The right part of Figure 2 shows the associated fan, that we define now. This is the family

$$\Sigma(P) = \{\sigma_\Gamma^\vee\}$$

of the *convex dual cones*, that is, the family of all the $\sigma^\vee$, where

$$\sigma^\vee = \{v \in \mathbf{R}^n \mid \varphi(v) \geqslant 0 \ \forall \varphi \in \sigma\}.$$

Notice that, as usual with duality,

$$\sigma \subset \sigma' \Longrightarrow \sigma^\vee \supset \sigma'^\vee.$$

In the example of Figure 2,

- the cone dual to the whole vector space (corresponding to the dimension-2 face) is the point 0,
- the cone dual to a half-plane (corresponding to a dimension-1 face) is a half-line,
- the cone dual to a cone corresponding to a vertex is a convex cone of dimension 2 (in the picture, the "color" of a cone is the same as that of its dual).

A *fan* in $\mathbf{R}^n$ is a family Σ of convex polyhedral cones in $\mathbf{R}^n$ all having 0 as a vertex, all generated by integral vectors, and such that

- any face of a cone in Σ is a cone in Σ,
- the intersection of two cones in Σ is a face in each of them.

Remark VII.1.2. One may replace $\mathbf{R}^n$ by $\mathbf{Q}^n$ in this definition: the important things are the vector space structure, allowing us to define cones, and (last but not least) the lattice $\mathbf{Z}^n$.

Proposition VII.1.3. *The family of the convex dual cones associated with a convex polyhedron the faces of which have integral equations is a fan. The cone associated with a codimension-k face has dimension k.*

Here the dimension of a cone is, of course, the dimension of the linear subspace it spans.

Proof. Consider a *facet* Γ, namely a codimension-1 face of the polyhedron. It has an equation of the form

$$\langle v, \varphi \rangle = a$$

for some integral vector v and real number a. The tangent convex cone is one of the half-spaces limited by the linear subspace spanned by the facet, say

$$\sigma_\Gamma = \{\varphi \in (\mathbf{R}^n)^\vee \mid \langle v, \varphi\rangle \geqslant 0\}.$$

Then the dual cone is the half-line generated by v. Now a codimension-k face is the intersection of k facets, say $\Gamma = \Gamma_1 \cap \cdots \cap \Gamma_k$, so that

$$\sigma_\Gamma = \sigma_{\Gamma_1} \cap \cdots \cap \sigma_{\Gamma_k}$$

and that $\sigma_\Gamma^\vee$ is the convex cone generated by the vectors $v_1, \ldots, v_k$ defining the facets. Hence the result. □

The union $\Sigma^{(k)}$ of all the cones of dimension k is called the *k-skeleton* of Σ. The 1-skeleton $\Sigma^{(1)}$ may (and actually will) be considered as the set of the N (primitive) vectors in $\mathbf{Z}^n$ which generate the half-lines which are the cones in $\Sigma^{(1)}$.

One of the reasons why people usually prefer to work with the fan instead of using the family of tangent convex cones is that the properties of the polyhedron are easier to read on the fan. Think for instance of the simple description of the cones of the fan in Proposition VII.1.3. Remember also how complicated the figure of the tangent cones is (even for instance in the simple case represented in Figure 2). Here comes another illustration.

The *support* of the fan Σ is the union $|\Sigma|$ of all the cones. We say that a fan is *complete* if its support is the whole space $\mathbf{R}^n$.

Proposition VII.1.4. *The convex polyhedron P is compact if and only if the fan $\Sigma(P)$ is complete.*

Proof. The convex polyhedron P is compact if and only if its projection on any line is compact, and in particular if and only if for any $v \in \mathbf{R}^n$ the map

$$\begin{aligned} (\mathbf{R}^n)^\vee &\longrightarrow \mathbf{R} \\ \varphi &\longmapsto \varphi(v) \end{aligned}$$

sends P onto a compact interval. Considering $-v$ as well, we see that P is compact if and only if

$$\forall v \in \mathbf{R}^n,\ \exists a \in \mathbf{R} \text{ such that } \forall \varphi \in P,\ \varphi(v) \geqslant a.$$

As P is a convex polyhedron, a may be considered as the value at v of a linear form m on the (relative) interior of some face of P. Thus P is compact if and only if

$$\forall v \in \mathbf{R}^n,\ \exists \Gamma \text{ face of } P \text{ and } m \in \mathring{\Gamma} \text{ such that } \forall \varphi \in P,\ \varphi(v) - m(v) \geqslant 0.$$

The last condition is equivalent to the existence for any v of a face Γ such that v is in the dual convex cone: the fan covers the whole space. $\square$

VII.1.b. The action of $\mathbf{T}_{\mathbf{C}}^N$ and its subgroups on $\mathbf{C}^N$. We denote by $\mathbf{T}^N$ the compact (real) torus of dimension N,

$$\mathbf{T}^N = \{(t_1, \ldots, t_N) \in \mathbf{C}^N \mid |t_i| = 1\} \cong (\mathbf{S}^1)^N$$

and $\mathbf{T}_{\mathbf{C}}^N$ the complex torus

$$\mathbf{T}_{\mathbf{C}}^N = \{(t_1, \ldots, t_N) \in \mathbf{C}^N \mid t_i \neq 0\} \cong (\mathbf{C}^\star)^N.$$

We make these two groups act on $\mathbf{C}^N$ as usual,

$$(t_1, \ldots, t_N) \cdot (z_1, \ldots, z_N) = (t_1 z_1, \ldots, t_N z_N).$$

Notice(5) that, restricted to $(\mathbf{C}^\star)^N$ the action is free and transitive!

Nontrivial stabilizers. Call $(e_1, \ldots, e_N)$ the canonical basis in any of the spaces

$$\mathbf{Z}^N \subset \mathbf{Q}^N \subset \mathbf{R}^N \subset \mathbf{C}^N.$$

For $I \subset \{1, \ldots, N\}$, denote

– by e_I the "coordinate" subspace generated by the e_i's for $i \in I$, that is,

$$e_I = \{z = (z_1, \ldots, z_N) \mid j \notin I \Rightarrow z_j = 0\}$$

– by $\mathbf{T}_I$ the corresponding complex torus(6)

$$T_I = \{t = (t_1, \ldots, t_N) \mid j \notin I \Rightarrow t_j = 1\}$$

– by e'_I the open cone

$$e'_I = \{z \mid j \notin I \Leftrightarrow z_j = 0\}$$

– and by $\overline{I}$ the complement of I.

... so that $z \in e_I$ if and only if its stabilizer contains $\mathbf{T}_{\overline{I}}$ and $z \in e'_I$ if and only if its stabilizer *is* $\mathbf{T}_{\overline{I}}$.

Figure 3 shows the example of e_I and e'_I for $I = \{1, 2\}$ and $I = \{1\}$, in $\mathbf{C}^3$.

(5) Silly as it may seem, this remark is the basis of this chapter.

(6) Trying to economize on notation which is already very heavy we shall denote in the same way both the complex and real torus. No confusion is to be feared.

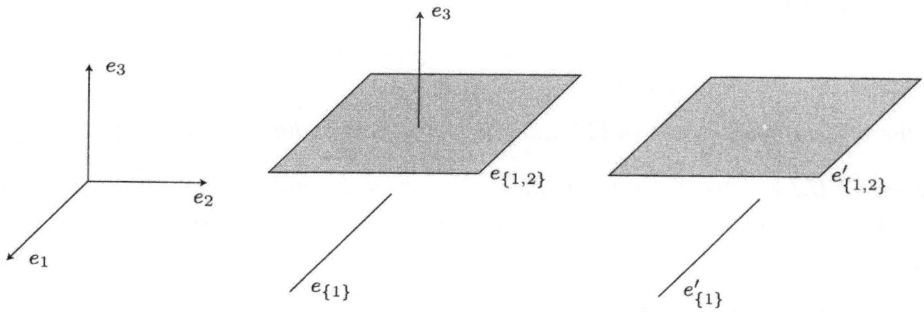

Figure 3

Subtori. Consider a linear map

$$\pi : \mathbf{Z}^N \longrightarrow \mathbf{Z}^n.$$

We shall use the same letter π (when no confusion is to be feared) for the maps $\pi \otimes \mathbf{Q}$, $\pi \otimes \mathbf{R}$ and $\pi \otimes \mathbf{C}$. Assume that $\pi \otimes \mathbf{Q}$ is surjective and call $\mathbf{K} \subset \mathbf{Z}^N$ the kernel of π. Similarly $K \subset \mathbf{T}^N$ will be the kernel of

$$\mathbf{R}^N/\mathbf{Z}^N \xrightarrow{\ \pi\ } \mathbf{R}^n/\mathbf{Z}^n.$$

The definition of $K_{\mathbf{C}}$ is analogous. All these groups are now acting on $\mathbf{C}^N$ as subgroups of $\mathbf{T}^N_{\mathbf{C}}$. Call $k = N - n$ the dimension of K.

Proposition VII.1.5. *The points in the singular orbits of $K_{\mathbf{C}}$ in $\mathbf{C}^N$ are the points of the subspaces e_I for I such that*

$$\mathbf{K} \otimes \mathbf{C} \cap e_{\overline{I}} \neq 0.$$

Proof. We know that z lies in a non-principal orbit of $K_{\mathbf{C}}$ if and only if there exists a (proper) subset I of indices in $\{1, \ldots, N\}$ such that

$$z \in e_I \text{ and } K_{\mathbf{C}} \cap T_{\overline{I}} \neq 1.$$

Linearizing the latter condition, we find that $\mathbf{K} \otimes \mathbf{C} \cap e_{\overline{I}} \neq 0$. Note that this gives no information on the exceptional orbits (case where $K_{\mathbf{C}} \cap T_{\overline{I}}$ is a finite group). □

Example VII.1.6. Put $N = n = 2$ (and $k = 0$), and let $\pi : \mathbf{Z}^2 \to \mathbf{Z}^2$ be the linear map, the matrix of which in the canonical bases is $\left(\begin{smallmatrix} 1 & 1 \\ 0 & 2 \end{smallmatrix}\right)$. Then, π itself is injective (in the notation above, $\mathbf{K} = 0$) but its avatar $\mathbf{T}^2 \to \mathbf{T}^2$ has a nontrivial kernel

$$K \cong \{(\varepsilon, \varepsilon) \mid \varepsilon = \pm 1\} \subset \mathbf{T}^2.$$

In this case, $\mathbf{K}_{\mathbf{C}}$ acts on $\mathbf{C}^2$ without singular orbit, but the point $(0, 0)$ is, in itself, an exceptional orbit.

Given the (integral) map π, we thus deduce:

Proposition VII.1.7. *The open set*

$$\mathcal{U}_\pi = \mathbf{C}^N - \bigcup_{\mathbf{K}\cap e_{\overline{I}}\neq 0} e_I$$

is the biggest open subset in $\mathbf{C}^N$ *on which* $K_\mathbf{C}$ *acts without singular orbits.* □

Remark VII.1.8. Recall that $\dim \mathbf{K} = N - n = k$, and that, if $\#\overline{I} > n$, then $\mathbf{K}\otimes\mathbf{C}\cap e_{\overline{I}} \neq 0$. Therefore, in $\mathcal{U}_\pi$, there is no coordinate subspace e_I such that $\#I < k$.

Example VII.1.9. Let $\pi : \mathbf{Z}^{n+1} \to \mathbf{Z}^n$ be the map which sends the n first vectors $e_1, \ldots, e_n$ of the canonical basis to the vectors having the same names, representing the canonical basis of $\mathbf{Z}^n$ and sending e_{n+1} to $-(e_1 + \cdots + e_n)$. The kernel $\mathbf{K}$ is the line generated by $e_1 + \cdots + e_n + e_{n+1}$ and K the diagonal torus $(t, \ldots, t)$. Here $\mathcal{U}_\pi$ is $\mathbf{C}^{n+1} - 0$. This is shown in Figure 4 in the case $n = 2$.

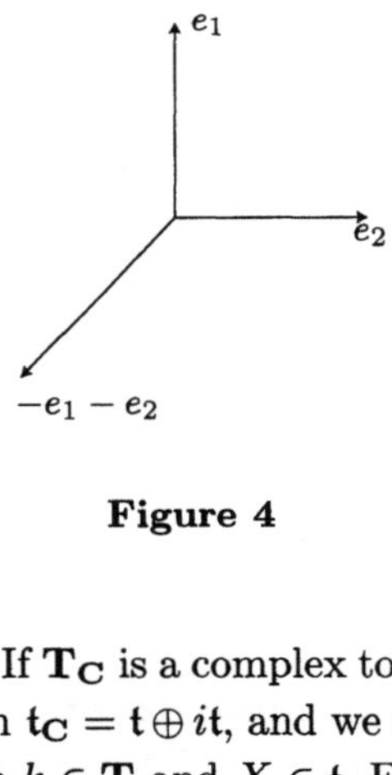

Figure 4

Real and imaginary parts. If $\mathbf{T}_\mathbf{C}$ is a complex torus and $\mathbf{T}$ the real torus which is its compact component, then $\mathfrak{t}_\mathbf{C} = \mathfrak{t} \oplus i\mathfrak{t}$, and we shall simply write the element g of $\mathbf{T}_\mathbf{C}$ as $g = k \exp iX$ where $k \in \mathbf{T}$ and $X \in \mathfrak{t}$. For example, if $\mathbf{T}_\mathbf{C} = \mathbf{C}^\star$, then $\mathbf{T} = \mathbf{S}^1$, $\mathfrak{t} = i\mathbf{R}$ and any element $z \in \mathbf{C}^\star$ may be written $z = ue^x$ where $x \in \mathbf{R}$ and $u \in \mathbf{S}^1$.

VII.1.c. Closing a fan, toric varieties. Given a fan Σ, choose a numbering of the primitive vectors generating its 1-skeleton,

$$\Sigma^{(1)} = (x_1, \ldots, x_N).$$

Consider the unique linear map

$$\pi : \mathbf{Z}^N \longrightarrow \mathbf{Z}^n$$

which sends the vectors $(e_1, \ldots, e_N)$ of the canonical basis (respectively) onto the vectors $(x_1, \ldots, x_N)$ and assume that $\pi \otimes \mathbf{Q}$ is surjective (this will be the case for instance if Σ contains a cone of dimension n). Here n is the dimension of the ambient vector space (and lattice) while N is the number of cones of dimension 1 in the fan.

Denote $\langle x_I$ the cone generated by the vectors x_i $(i \in I)$, that is,

$$\langle x_I = \{\alpha_1 x_{i_1} + \cdots + \alpha_r x_{i_r} \mid \alpha_j \geqslant 0, I = \{i_1, \ldots, i_r\}\}.$$

To the projection π, we know how to associate an open subset $\mathcal{U}_\pi$ in $\mathbf{C}^N$. Just as π, the latter depends only on the 1-skeleton of Σ. We now want to construct an open subset $\mathcal{U}_\Sigma$ of $\mathcal{U}_\pi$ which will really depend on Σ. Let us make a restriction: we will use smooth fans. A fan Σ is *smooth* if each of its cones is generated by a part of a $\mathbf{Z}$-basis of the lattice $\mathbf{Z}^n$.

Example VII.1.10. The fan associated with the standard n-simplex (as in Figure 2) is smooth, while the one shown on the right part of Figure 5 is not: as

$$\begin{vmatrix} -2 & 0 \\ -1 & 1 \end{vmatrix} \neq \pm 1,$$

the darkest cone does not satisfy the requirement. See also Exercise VII.7.

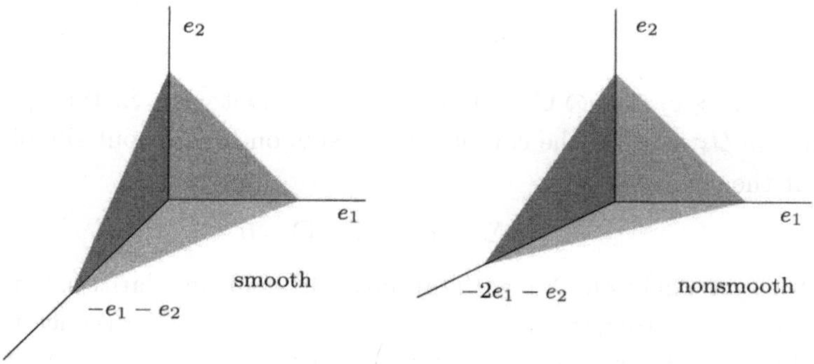

Figure 5

This notion is of course related with the primitivity property of the polyhedra we have met in Chapter IV.

Proposition VII.1.11. *The fan associated with a polyhedron is smooth if and only if the polyhedron is* primitive. □

Remark VII.1.12. As we have assumed that $\pi \otimes \mathbf{Q}$ was surjective, then

$$e_I \cap \mathbf{K} \otimes \mathbf{C} = \{0\} \Longrightarrow \#I \leqslant n$$

and therefore the smoothness condition implies that all the cones of dimension n in Σ are *simplicial*, that is, generated by n vectors.

We can now define

$$\mathcal{U}_\Sigma = \mathbf{C}^N - \bigcup_{\{I \mid \langle x_I \notin \Sigma\}} e_I.$$

This is an economical way of writing that a point z in $\mathcal{U}_\Sigma$ may have zero coordinates

$$z_{i_1} = \cdots = z_{i_r} = 0 \text{ if and only if } \langle x_{i_1}, \ldots, x_{i_r} \in \Sigma.$$

Notice that the $K_{\mathbf{C}}$-orbits in $\mathcal{U}_\Sigma$ are closed, so that the quotient $\boldsymbol{X}_\Sigma = \mathcal{U}_\Sigma / K_{\mathbf{C}}$ is Hausdorff. The quotient space $\boldsymbol{X}_\Sigma$ is called the *toric variety* associated with Σ.

Remark VII.1.13. Assume the fan Σ is smooth. Let $\langle x_I$ be one of its cones of dimension h. Up to renumbering in $\Sigma^{(1)}$, we may assume that $I = \{1, \ldots, h\}$. Then

$$\begin{aligned} e_I \cap \mathbf{K} \otimes \mathbf{C} &= \left\{ \sum_{i=1}^h \alpha_i e_i \mid \pi\left(\sum \alpha_i e_i\right) = 0 \right\} \\ &= \left\{ \sum_{i=1}^h \alpha_i e_i \mid \sum \alpha_i x_i = 0 \right\} \\ &= 0. \end{aligned}$$

Hence $\langle x_I \in \Sigma \Rightarrow e_I \cap \mathbf{K} \otimes \mathbf{C} = 0$. Thus, for a smooth fan Σ, the open set $\mathcal{U}_\Sigma$ is contained in $\mathcal{U}_\pi$, so that the complex torus acts on $\mathcal{U}_\Sigma$ without singular orbits. Notice that the condition

$$\langle x_I \in \Sigma \Rightarrow e_I \cap \mathbf{K} \otimes \mathbf{C} = 0$$

is sufficient to get a quotient $\boldsymbol{X}_\Sigma$ with (at most) orbifold singularities. Our smoothness assumption is stronger, as it gives the *smoothness* of $\boldsymbol{X}_\Sigma$ (hence the terminology). However, the weaker assumption has better properties, such as the fact that, if $\Sigma' \subset \Sigma$ and if Σ satisfies it, then Σ' satisfies it as well.

Proposition VII.1.14. *The toric variety $\boldsymbol{X}_\Sigma$ is smooth if and only if the fan Σ is smooth.*

Proof. The quotient $\boldsymbol{X}_\Sigma$ is smooth if and only if $K_{\mathbf{C}}$ has no finite stabilizer in $\mathcal{U}_\Sigma$, in other words if and only if

$$\langle x_I \in \Sigma \Rightarrow K \cap T_I = \{1\}.$$

As we assumed the linear map $\pi \otimes \mathbf{Q}$ to be surjective, this is equivalent to

$$\langle x_I \in \Sigma \text{ and } \#I = n \Rightarrow K \cap T_I = \{1\}$$

as well. But the torus K is the kernel of $\pi : \mathbf{R}^N/\mathbf{Z}^N \to \mathbf{R}^n/\mathbf{Z}^n$. In other words, if K contains a nontrivial element, this means that there exists a nonintegral vector in $\mathbf{Q}^N$ the image of which lies in $\mathbf{Z}^n$. Renumbering the x_i's if necessary, assume that $I = \{1, \ldots, n\}$. Then $K \cap T_I$ is a nontrivial finite group if and only if there exist relatively prime integers $a_1, \ldots, a_n$ and $m \geqslant 2$ such that

$$\pi\left(\sum_{i=1}^{n} \frac{a_i}{m} e_i\right) \in \mathbf{Z}^n.$$

This says that we have found a vector $\sum a_i x_i$ of $\mathbf{Z}^n$ the "coordinates" of which are relatively prime and which is divisible: thus it is not written in a $\mathbf{Z}$-basis. □

First properties of X_Σ. By definition, all the cones $\langle x_i$ of the 1-skeleton belong to the fan Σ, so that

$$\langle x_{\overline{I}} \notin \Sigma \Longrightarrow \#\overline{I} \geqslant 2 \Longrightarrow \operatorname{codim} e_I \geqslant 2.$$

To construct $\mathcal{U}_\Sigma$, we have thus taken off subspaces of complex codimension $\geqslant 2$ in $\mathbf{C}^N$. Hence $\mathcal{U}_\Sigma$ is 2-connected. The homotopy exact sequence of the fibration

$$\begin{array}{ccc} K_{\mathbf{C}} & \longrightarrow & \mathcal{U}_\Sigma \\ & & \downarrow \\ & & X_\Sigma \end{array}$$

gives that X_Σ is 1-connected and that

$$\pi_2(X_\Sigma) \longrightarrow \pi_1(K_{\mathbf{C}})$$

is an isomorphism. The latter is isomorphic with $H_2(BK_{\mathbf{C}}; \mathbf{Z})$. Dualizing, we get:

Proposition VII.1.15. *The toric variety X_Σ is simply connected. Moreover, the Euler class of the principal $K_{\mathbf{C}}$-bundle $\mathcal{U}_\Sigma \to X_\Sigma$ gives an isomorphism*

$$H^2(BK; \mathbf{Z}) \longrightarrow H^2(X_\Sigma; \mathbf{Z}).$$ □

Notice that the Euler class of this principal $K_{\mathbf{C}}$-bundle lives in $H^2(X_\Sigma; \mathbf{Z})^k$ where $k = \dim K$ is a $\mathbf{Z}$-basis of $H^2(X_\Sigma; \mathbf{Z})$. Notice also that $H^2(BK; \mathbf{Z})$ can be considered as the integral lattice in $\mathfrak{k}^\star$.

Examples. To begin with, as a zero-th example, the open set $\mathcal{U}_\Sigma$ itself is a toric variety of dimension N, associated with the fan $\widetilde{\Sigma}$ obtained when *closing* Σ, that is,

$$\widetilde{\Sigma}^{(1)} = (e_1, \ldots, e_N) \text{ and } \langle e_I \in \widetilde{\Sigma} \Longleftrightarrow \langle x_I \in \Sigma.$$

In Figure 2, the reader can imagine the fan pictured on the right as being (in three dimensions) in $\mathbf{R}^3$.

By definition, the closed fan $\widetilde{\Sigma}$ contains only cones of the first quadrant. Moreover, as soon as $N > n$ (which will always be the case if Σ is complete), it contains no cone of dimension N. In other words we then have $\mathcal{U}_\Sigma \subset \mathbf{C}^N - 0$.

(1) Let Σ be the fan consisting of the unique cone $\sigma = \{0\} \subset \mathbf{R}^n$. We are not exactly in the framework investigated above, since the 1-skeleton is empty, so that we are looking at the zero map

$$\pi : \mathbf{Z}^0 \longrightarrow \mathbf{Z}^n.$$

The kernel is zero as well, $\mathcal{U}_\Sigma = (\mathbf{C}^\star)^n$ and $\boldsymbol{X}_\Sigma$ is the complex torus as well.

(2) Take for $(x_1, \ldots, x_n)$ the canonical basis of $\mathbf{Z}^n$ and let Σ be the set of all the cones (of the first quadrant). Then $N = n$, $\pi = \mathrm{Id}$, $\mathbf{K} = 0$, $K_{\mathbf{C}} = 1$ and

$$\mathcal{U}_\pi = \mathcal{U}_\Sigma = \mathcal{U}_\Sigma / K_{\mathbf{C}} = \mathbf{C}^n.$$

The toric variety associated with the first quadrant fan is $\mathbf{C}^n$. This is shown (in the case $n = 2$) on the left part of Figure 6.

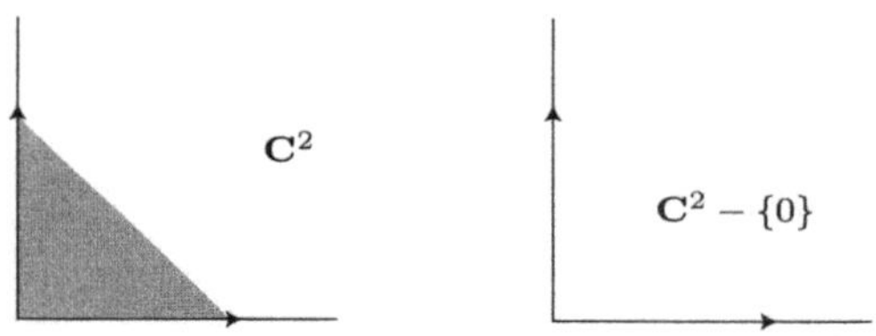

Figure 6

(3) With the same 1-skeleton as in the previous example, define now Σ as the set of all the cones in the first quadrant except the n-dimensional one. Nothing has changed except $\mathcal{U}_\Sigma$ from which we must take off $e_\varnothing = 0$. Hence $\boldsymbol{X}_\Sigma = \mathcal{U}_\Sigma = \mathbf{C}^n - 0$. This is what the right part of Figure 6 shows (for $n = 2$).

(4) Consider again the fan in Figure 2, that is the vectors $x_i = e_i$ $(1 \leqslant i \leqslant n)$ and $x_{n+1} = -(x_1 + \cdots + x_n)$ and the fan in $\mathbf{R}^n$ consisting of all dimension $\leqslant n$ cones on these vectors (at least if $n = 2$ we have already seen that it is the fan associated with the standard simplex). Here $\mathcal{U}_\Sigma = \mathbf{C}^{n+1} - 0$, the map π is that shown in Figure 4, so that $\mathbf{K}$ is the diagonal $\mathbf{Z}$, namely the set $(m, \ldots, m) \in \mathbf{Z}^{n+1}$ and therefore $K_{\mathbf{C}}$ is the diagonal $\mathbf{C}^\star$ and $\boldsymbol{X}_\Sigma$ is the complex projective space $\mathbf{P}^n(\mathbf{C})$ (once again).

(5) As on the left part of Figure 7, consider the fan in $\mathbf{R}^2$ the 1-skeleton of which is $(e_1, e_2, -e_1, -e_2)$ and the 2-skeleton of which consists of all the cones $\langle x_{i,i+1}$ (considering the numbering mod4). Looking for the I's such that $\langle x_I \notin \Sigma$ we find that

$$\mathcal{U}_\Sigma = \mathbf{C}^4 - \Big(\{z_1 = z_3 = 0\} \bigcup \{z_2 = z_4 = 0\}\Big).$$

Thus

$$\mathcal{U}_\Sigma = (\mathbf{C}^2 - 0) \times (\mathbf{C}^2 - 0).$$

The group $K_\mathbf{C}$ is $\mathbf{C}^\star \times \mathbf{C}^\star$ acting diagonally. The toric variety is the product $\mathbf{P}^1(\mathbf{C}) \times \mathbf{P}^1(\mathbf{C})$.

(6) The open subset $\mathcal{U}_\Sigma$ depends only on the combinatorics of the fan. For instance, all the fans in $\mathbf{R}^2$ the 1-skeleton of which consists of four vectors x_1, x_2, x_3, x_4 and whose 2-skeleton contains all the cones $\langle x_{i,i+1}$ give the same $\mathcal{U}_\Sigma = (\mathbf{C}^2 - 0) \times (\mathbf{C}^2 - 0)$. Of course, the torus $K_\mathbf{C}$ changes. Consider the case where $x_1 = e_1$, $x_2 = e_2$, $x_3 = -e_1$, $x_4 = -e_1 - ae_2$ for some integer a (right part of Figure 7). This time, the torus $K_\mathbf{C}$ acts by

$$(u, v) \cdot (z_1, z_2, z_3, z_4) = (u^a v z_1, u z_2, v z_3, u z_4).$$

The toric variety $\boldsymbol{X}_\Sigma$ we get this way is a Hirzebruch surface (already met in Exercise IV.4) as we shall prove in Section VII.4.

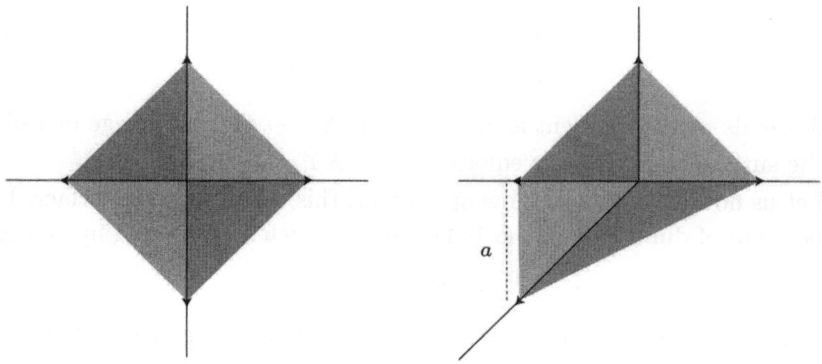

Figure 7

(7) Here is an example of a nonsmooth fan. We have shown one in Figure 5. To simplify the investigation here, concentrate on the cone which was responsible for the trouble there and consider Figure 8. Working in $\mathbf{Z}^2$, the 1-skeleton is $x_1 = e_1$, $x_2 = e_1 + 2e_2$. As the determinant of (x_1, x_2) equals 2, this is not a $\mathbf{Z}$-basis. Add now the cone generated by x_1 and x_2. Then $\mathcal{U}_\Sigma = \mathcal{U}_\pi = \mathbf{C}^2$, and $\mathbf{K} = 0$. The group K is the kernel of

$$\pi = \begin{pmatrix} 1 & 1 \\ 0 & 2 \end{pmatrix} : \mathbf{T}^2 \longrightarrow \mathbf{T}^2,$$

which we have already considered in § VII.1.b, where we saw that the group K is generated by $(-1, -1) \in \mathbf{T}^2$. The variety $\boldsymbol{X}_\Sigma$ is the quotient

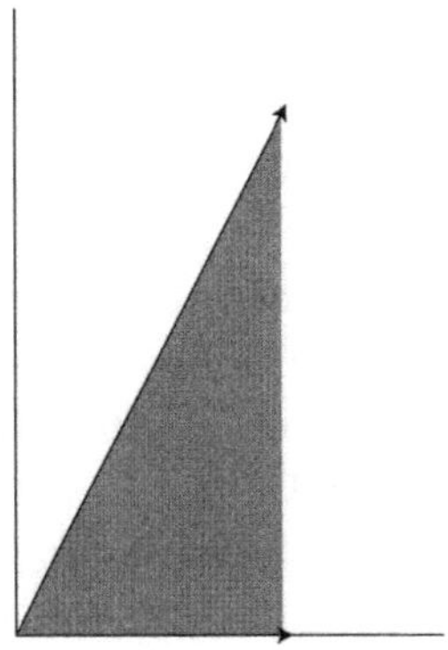

Figure 8

$\mathbf{C}^2/(x,y) \sim (-x,-y)$, it has a singularity at 0. Here is another description of this complex surface: the map

$$\begin{aligned} \mathbf{C}^2 &\longrightarrow \mathbf{C}^3 \\ (x,y) &\longmapsto (x^2,y^2,xy) \end{aligned}$$

descends to the quotient as an injection $\boldsymbol{X}_\Sigma \to \mathbf{C}^3$, the image of which is the surface satisfying the equation $Z^2 = XY$.

(8) Let us now (once again) blow up a point, this time in a toric surface. Let Σ be a fan of dimension 2, the 1-skeleton of which is generated by N vectors

$$\Sigma^{(1)} = (x_1,\ldots,x_N)$$

and the cones of dimension 2 of which are the cones generated by two consecutive vectors

$$(\langle x_{i,i+1}\rangle_{i \bmod N}.$$

The open subset $\mathcal{U}_\Sigma$ is defined by $z \notin \mathcal{U}_\Sigma \subset \mathbf{C}^N$ if and only if it has at least two zero coordinates, and, if it has only two they are not consecutive (mod N). Let Σ' be a complete fan, the 1-skeleton of which is $(x_1,\ldots,x_N,x_{N+1}=x_1+x_N)$. The map

$$\begin{aligned} \mathbf{C}^{N+1} &\xrightarrow{\ \widetilde{\sigma}\ } \mathbf{C}^N \\ (z_1,\ldots,z_{N+1}) &\longmapsto (z_1z_{N+1},z_2,\ldots,z_{N-1},z_Nz_{N+1}) \end{aligned}$$

clearly maps $\mathcal{U}_{\Sigma'}$ onto $\mathcal{U}_\Sigma$. Let $K_\mathbf{C}$ act on $\mathbf{C}^N$ by

$$u\cdot(z_1,\ldots,z_N) = (u^{b_1}z_1,\ldots,u^{b_N}z_N).$$

Then $K'_\mathbf{C} = K_\mathbf{C}\times\mathbf{C}^\star$ acts on $\mathbf{C}^{N+1}$ by

$$(u,v)\cdot(z_1,\ldots,z_{N+1}) = (u^{b_1}\overline{v}z_1,u^{b_2}z_2,\ldots,u^{b_N}\overline{v}z_N,vz_{N+1}),$$

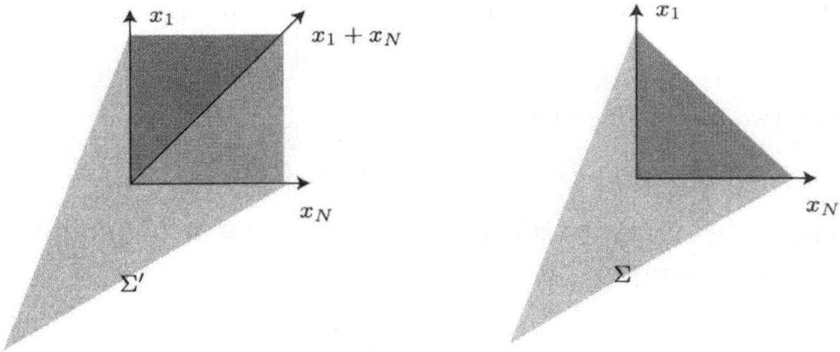

Figure 9

so that $\widetilde{\sigma}$ induces a map

$$\sigma : \boldsymbol{X}_{\Sigma'} \longrightarrow \boldsymbol{X}_{\Sigma}$$

which is the blow up of the fixed point of $Q_{\mathbf{C}}$ associated with the cone $\langle x_{N,1}$ in Σ.

The big torus. From the action of the torus $\mathbf{T}^N_{\mathbf{C}}$, the toric manifold $\boldsymbol{X}_\Sigma$ keeps an operation of a torus. Indeed, the $\mathbf{T}^N_{\mathbf{C}}$-action descends to the quotient and induces a $\mathbf{T}^N_{\mathbf{C}}/K_{\mathbf{C}}$-action on $\boldsymbol{X}_\Sigma$, which is therefore endowed with the action of a complex torus we shall denote $Q_{\mathbf{C}}$, the real torus it contains being denoted Q. We have

$$\dim_{\mathbf{C}} \boldsymbol{X}_\Sigma = \dim_{\mathbf{C}} \mathcal{U}_\Sigma - \dim_{\mathbf{C}} K_{\mathbf{C}} = N - k = \dim_{\mathbf{C}} Q_{\mathbf{C}} = n.$$

Moreover, the complex torus $Q_{\mathbf{C}}$ has an open dense orbit in $\boldsymbol{X}_\Sigma$. Each cone $\langle x_I$ in Σ defines an open subset $\boldsymbol{X}_I$ in $\boldsymbol{X}_\Sigma$. For example, the cone $\{0\} = \langle x_\varnothing$ defines $\mathcal{U}_\varnothing = (\mathbf{C}^\star)^N = \mathbf{T}^N_{\mathbf{C}}$ and $\boldsymbol{X}_\varnothing = \mathbf{T}^N_{\mathbf{C}}/K_{\mathbf{C}} = Q_{\mathbf{C}}$, which thus appears as an open dense subset in $\boldsymbol{X}_\Sigma$. The torus Q is the *big torus* of the toric manifold $\boldsymbol{X}_\Sigma$.

Morphisms. The map we have used in the last example above is a special case of a morphism of fans. A *morphism*

$$f : \Sigma \longrightarrow \Sigma'$$

is an integral linear map

$$f : \mathbf{Z}^n \longrightarrow \mathbf{Z}^{n'}$$

such that, for any cone σ of Σ, there exists a cone σ' of Σ' for which $f(\sigma) \subset \sigma'$. For instance, in the example of the blow up, the linear map f is the identity $\mathbf{Z}^2 \to \mathbf{Z}^2$, giving a morphism of fans $\Sigma' \to \Sigma$ (I apologize for the inversion in the notation).

Proposition VII.1.16. *A morphism of fans $\Sigma \to \Sigma'$ defines a map*

$$\boldsymbol{X}_\Sigma \longrightarrow \boldsymbol{X}_{\Sigma'}.$$

Remark VII.1.17. This is an analytic map (and even an algebraic map, when this makes sense).

Proof. The two fans define open subsets

$$\mathcal{U}_\Sigma \subset \mathbf{C}^N, \qquad \mathcal{U}_{\Sigma'} \subset \mathbf{C}^{N'}.$$

We map $\mathcal{U}_\Sigma$ to $\mathcal{U}_{\Sigma'}$ using f. Calling as usual (x_i) a system of primitive generators of $\Sigma^{(1)}$ and analogously (x'_i) for $\Sigma'^{(1)}$, we must have

$$f(x_i) = \sum a_i^j x'_j$$

for some $a_i^j \in \mathbf{N}$ (this is to say that the image of the cone generated by x_i is contained in a cone of Σ'). We define

$$\begin{aligned} \widetilde{f} : \mathbf{C}^N &\longrightarrow \mathbf{C}^{N'} \\ (z_1, \dots, z_N) &\longmapsto (z_1^{a_1^1} \cdots z_N^{a_N^1}, \dots, z_1^{a_1^{N'}} \cdots z_N^{a_N^{N'}}) \end{aligned}$$

and check that it maps $\mathcal{U}_\Sigma$ into $\mathcal{U}_{\Sigma'}$. Let z be a point in $\mathcal{U}_\Sigma$. Look at the coordinates of $\widetilde{f}(z)$ that are zero,

$$z'_{j_1} = \cdots = z'_{j_r} = 0,$$

say. If this is the case, there are indices $i_1, \dots, i_r$ with

$$a_{i_1}^{j_1} \neq 0, \dots, a_{i_r}^{j_r} \neq 0 \text{ and } z_{i_1} = \cdots = z_{i_r} = 0.$$

As $z \in \mathcal{U}_\Sigma$, the cone $\langle x_{i_1}, \dots, x_{i_r}$ is in Σ. Its image under f is contained in a cone σ' of Σ'. Now

$$f\left(\sum_{i=1}^r \lambda_k x_{i_k}\right) \sum_{j,k} \lambda_k a_{i_k}^j x'_j.$$

To say that all such vectors lie in σ' is to say that there is a set K of indices which contains all the indices j for which $a_{i_k}^j \neq 0$ and such that

$$\sigma' = \langle x'_K \in \Sigma'.$$

But K contains $\{j_1, \dots, j_r\}$, so that

$$\langle x'_{j_1}, \dots, x'_{j_r} \in \Sigma'$$

as well (as a face of σ'). We have proved that, if

$$z'_{j_1} = \cdots = z'_{j_r} = 0, \text{ then } \langle x'_{j_1}, \dots, x'_{j_r} \in \Sigma',$$

that is, that $\widetilde{f}(\mathcal{U}_\Sigma) \subset \mathcal{U}_{\Sigma'}$.

We must now compare the actions of $K_\mathbf{C}$ and $K'_\mathbf{C}$ on these open subsets. The formula we have used to define $\widetilde{f}$ shows that $\widetilde{f}$ is equivariant with these actions. Recall our standing assumption, namely that the projections $\pi \otimes \mathbf{R}$ and $\pi' \otimes \mathbf{R}$,

$$\mathbf{R}^N \longrightarrow \mathbf{R}^n \text{ and } \mathbf{R}^{N'} \longrightarrow \mathbf{R}^{n'}$$

are surjective. Then $\widetilde{f}$ lifts to a linear map

$$\begin{aligned} \mathbf{Z}^N &\longrightarrow \mathbf{Z}^{N'} \\ e_i &\longmapsto \sum_{j=1}^{N'} a_i^j e'_j \end{aligned}$$

and this new mapping maps $\mathbf{K}$ into $\mathbf{K}'$ and $K_{\mathbf{C}}$ into $K'_{\mathbf{C}}$. Moreover, we have

$$\widetilde{f}(u \cdot (z_1, \ldots, z_N)) = \widetilde{f}(u) \cdot \widetilde{f}(z_1, \ldots, z_N)$$

so that $\widetilde{f}$ is equivariant and defines a map

$$\boldsymbol{X}_\Sigma \longrightarrow \boldsymbol{X}_{\Sigma'},$$

which is equivariant with respect to the big torus actions. □

Examples VII.1.18

(1) The linear map $\pi : \mathbf{Z}^N \to \mathbf{Z}^n$ defines a morphism

$$\boldsymbol{X}_{\widetilde{\Sigma}} \longrightarrow \boldsymbol{X}_\Sigma$$

which is just the canonical map $\mathcal{U}_\Sigma \to \boldsymbol{X}_\Sigma$.

(2) The identity map $\mathbf{Z}^n \to \mathbf{Z}^n$ maps the fan consisting of the cone $\{0\}$ into any fan Σ. The corresponding map of toric varieties is the inclusion of $(\mathbf{C}^\star)^n$ in $\boldsymbol{X}_\Sigma$.

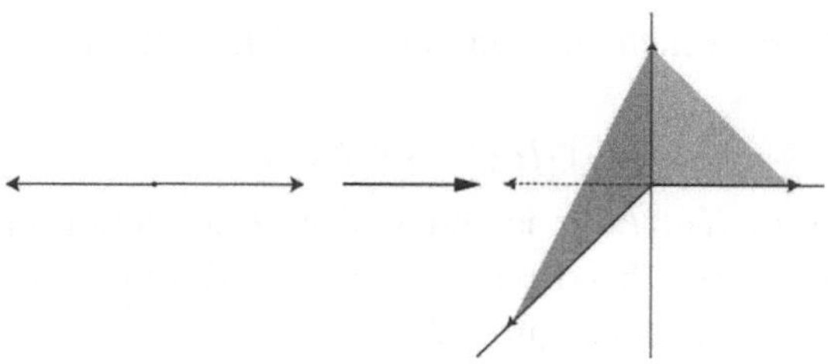

Figure 10

(3) Let Σ' be the fan in $\mathbf{R}^2$, the dimension-2 cones of which are $\langle e_1, e_2$ and $\langle e_2, -e_1 + be_2$. Assume $b < 0$. then the map

$$\begin{aligned} \mathbf{Z} &\longrightarrow \mathbf{Z}^2 \\ x &\longmapsto (x, 0) \end{aligned}$$

defines a morphism of the fan Σ consisting of the cones $\langle e_1$, $\langle -e_1$ in $\mathbf{R}$ into Σ' (Figure 10). It defines an inclusion

$$\boldsymbol{X}_\Sigma = \mathbf{P}^1(\mathbf{C}) \lhook\joinrel\longrightarrow \boldsymbol{X}_{\Sigma'}.$$

We shall come back to this example in § VII.2.b.

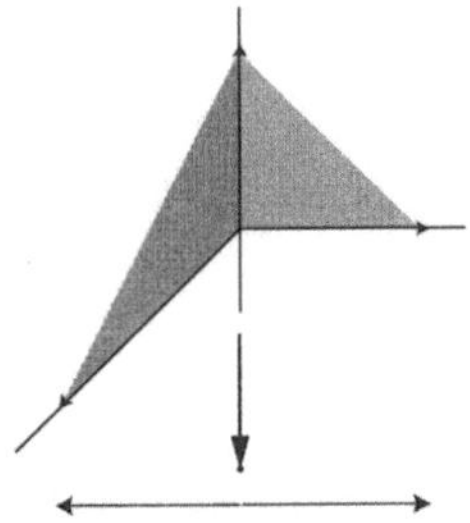

Figure 11

(4) Conversely, for any b, the fan, the dimension-2 cones of which are $\langle e_1, e_2$ and $\langle e_2, -e_1+be_2$, is mapped to the fan defining $\mathbf{P}^1(\mathbf{C})$ by a projection, the kernel of which is the central vector (Figure 11). We shall see in § VII.2.b that this is the projection of a line bundle onto $\mathbf{P}^1(\mathbf{C})$.

Fixed points and complex curves. A cone of dimension n in Σ corresponds to a fixed point of the torus $Q_{\mathbf{C}}$ in the toric manifold $\boldsymbol{X}_\Sigma$. This can be deduced from the examples of the fans shown in Figure 6 above.

Consider now a cone $\langle x_I$ of dimension $n-1$. If $\#I = n-1$, the coordinate subspace $e_{\overline{I}}$ is a subspace of $\mathbf{C}^N$, of dimension $k+1$, the set $e_{\overline{I}} \cap \mathcal{U}_\Sigma$ is an open subset (as $\langle x_I \in \Sigma$) on which the complex torus $K_{\mathbf{C}}$ still acts. We thus have the inclusion

$$e_{\overline{I}} \cap \mathcal{U}_\Sigma / K_{\mathbf{C}} \subset \mathcal{U}_\Sigma / K_{\mathbf{C}}$$

of a complex curve in X_Σ. As it is endowed with an action of the torus $Q_{\mathbf{C}}$, it is a $\mathbf{P}^1(\mathbf{C})$. There are two fixed points, corresponding to the adjacent cones of dimension n. This is shown in Figure 12.

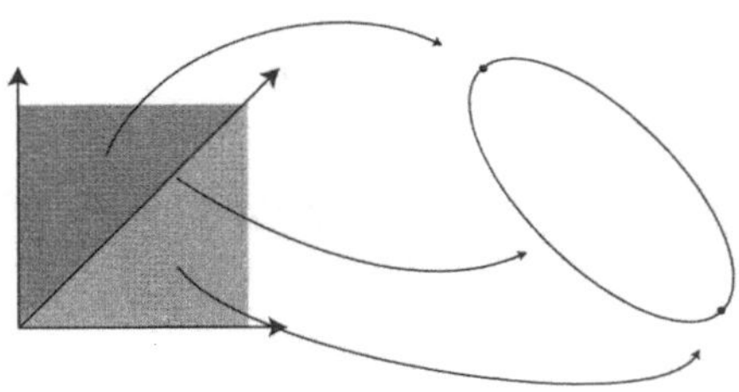

Figure 12

Charts. We first exhibit open subsets trivializing the principal bundle

$$\mathcal{U}_\Sigma \longrightarrow \boldsymbol{X}_\Sigma.$$

A convenient way to present $\mathcal{U}_\Sigma$ from the point of view of the $K_{\mathbf{C}}$-bundle $\mathcal{U}_\Sigma \to X_\Sigma$ is the following:

Lemma VII.1.19. *Let $\mathcal{U}_I$ be the open subset (complement of coordinate subspaces)*

$$\mathcal{U}_I = \left\{ z \in \mathbf{C}^N \mid z_j = 0 \Rightarrow j \in I \right\}.$$

This is the product of $\#\overline{I}$ copies of $\mathbf{C}^\star$ and of $\#I$ copies of $\mathbf{C}$ and we have

$$\mathcal{U}_\Sigma = \bigcup_{\langle x_I \in \Sigma} \mathcal{U}_I.$$

Proof. We may write $\mathcal{U}_I = e_I \times e'_{\overline{I}}$. For any $z \in \mathbf{C}^N$, call $I(z) \subset \{1, \ldots, N\}$ the set of the indices of those coordinates which are zero (thus $z \in \mathcal{U}_{I(z)}$). We have

$$z \in e_I \Longleftrightarrow \overline{I} \subset I(z)$$

and

$$\begin{aligned} z \in \mathcal{U}_\Sigma &\Longleftrightarrow (z \in e_I \Rightarrow \langle x_{\overline{I}} \in \Sigma) \\ &\Longleftrightarrow (I \subset I(z) \Rightarrow \langle x_I \in \Sigma) \\ &\Longleftrightarrow \langle x_{I(z)} \in \Sigma \end{aligned}$$

so $z \in \mathcal{U}_\Sigma \Leftrightarrow \langle x_{I(z)} \in \Sigma \Leftrightarrow \mathcal{U}_{I(z)} \subset \mathcal{U}_\Sigma$. □

In the case of a smooth fan, and thus of a *principal* $K_{\mathbf{C}}$-bundle $\mathcal{U}_\Sigma \to X_\Sigma$, the open subsets $\mathcal{U}_I$ of $\mathcal{U}_\Sigma$ give local trivializations of the bundle.

Proposition VII.1.20. *With each cone $\langle x_I$ of the smooth fan Σ is associated an open subset X_I of the toric variety X_Σ, over which the principal $K_{\mathbf{C}}$-bundle $\mathcal{U}_\Sigma \to X_\Sigma$ is trivialized.*

Proof. Choose a basis of $\mathbf{K}$. Let A be the matrix (with k columns and N rows) giving this basis in the canonical basis of $\mathbf{C}^N$. The torus $K_{\mathbf{C}}$ acts on $\mathbf{C}^N$ by

$$(u_1, \ldots, u_k) \cdot (z_1, \ldots, z_N) = (u^{m_1} z_1, \ldots, u^{m_n} z_N)$$

where u^{m_i} is the monomial in the u_j's given by the multi-exponent $m_i \in \mathbf{Z}^k$ which is the i-th line of A.

Consider this action on one of the open subsets $\mathcal{U}_I$ we are interested in. Remark first that the $\#I = n$ case is actually sufficient; as the fan is complete, any cone $\langle x_J \in \Sigma$ with $\#J < n$ is a face of a cone $\langle x_I \in \Sigma$ with $\#I = n$ and then $\mathcal{U}_J \subset \mathcal{U}_I$. Furthermore, we merely have to change the numbering of the vectors in the 1-skeleton of Σ to be able to assume that $I = \{1, \ldots, n\}$. Then

$$\mathcal{U}_I = \{(z_1, \ldots, z_N) \mid z_{n+1} \neq 0, \ldots, z_N \neq 0\}.$$

The fact that $K_{\mathbf{C}}$ then acts freely on $\mathcal{U}_I$ corresponds to the fact that the k last rows (recall that $N = n + k$) of the matrix A give an invertible (over the integers) matrix. In other words, we can change the parametrization of $K_{\mathbf{C}}$ in such a way that the action is written:

$$(v_1,\dots,v_k)\cdot(z_1,\dots,z_N)=(v^{p_1}z_1,\dots,v^{p_n}z_n,v_1z_{n+1},\dots,v_kz_N).$$

Denote (in analogy with homogeneous coordinates in the projective space) $[z_1,\dots,z_N]$ the element in X_Σ which represents the class of $(z_1,\dots,z_N)\in\mathcal{U}_\Sigma$ modulo $K_{\mathbf{C}}$. We can define a section $\sigma_I : X_I \to \mathcal{U}_I$ to the projection $\mathcal{U}_I \to X_I$ by

$$\sigma_I[z_1,\dots,z_N]=\left(\frac{z_1}{z^{p_1}},\dots,\frac{z_n}{z^{p_n}},1,\dots,1\right)$$

where z^{p_i} denotes the monomial in $z_{n+1},\dots,z_{n+k}$ associated with $p_i\in\mathbf{Z}^k$. A trivialization of the principal $K_{\mathbf{C}}$-bundle over X_I is easily deduced:

$$\begin{aligned} K_{\mathbf{C}}\times X_I &\longrightarrow \mathcal{U}_I\\ (v,[z]) &\longmapsto v\cdot\sigma_I[z]. \end{aligned}$$

□

VII.2. Symplectic reduction and convex polyhedra

In the previous section, we have associated, with every convex polyhedron, a fan. And to any fan a toric manifold. We try now to go back and to associate, to the toric manifold, a polyhedron.

Recall that a toric manifold $\boldsymbol{X}_\Sigma$ of complex dimension n is endowed with the action of a torus Q of dimension n, its big torus.

The idea (of Delzant and) of what follows is to prove that $\boldsymbol{X}_\Sigma$ can be endowed with symplectic forms for which the action of the compact torus Q is completely integrable and the image of the momentum mapping is one of the convex polyhedra leading to the fan Σ—the shape of the polyhedron determines the fan, the volumes of its faces then determine the symplectic form. This way, we will prove (this is Theorem VII.2.1 below) that all the primitive polyhedra are indeed images of momentum mappings of completely integrable torus actions(7).

Look once again at the example of $\mathbf{P}^n(\mathbf{C})$ as a quotient of $\mathbf{C}^{n+1}-\{0\}$ (this is one of the examples above). We consider the fan described above and shown in Figure 2. The closed fan $\widetilde{\Sigma}$ is just the first quadrant fan in $\mathbf{R}^{n+1}$, with all the possible cones of dimension $\leqslant n$, so that $\mathcal{U}_\Sigma$ is $\mathbf{C}^{n+1}-\{0\}$. The linear map

$$\pi:\mathbf{Z}^{n+1}\longrightarrow\mathbf{Z}^n$$

maps e_i to x_i and e_{n+1} to $-\sum x_i$, so that the torus K is the diagonal $\mathbf{C}^\star$ and $\boldsymbol{X}_\Sigma$ is indeed $\mathbf{P}^n(\mathbf{C})$. What we aim to do now is to understand $\mathbf{P}^n(\mathbf{C})$ with its reduced symplectic forms, in this context.

(7) which, together with the uniqueness theorem IV.4.20, constitute the classification theorem of Delzant [39].

For this particular toric variety, let us make a few observations:

(1) The complex space $\mathbf{C}^{n+1}$ can be replaced by the compact $\mathbf{S}^{2n+1}$, noticing that $\mathbf{S}^{2n+1} \times]0, +\infty[\cong \mathbf{C}^{n+1} - 0$.
(2) The previous decomposition is compatible with that of $\mathbf{C}^\star$ as the product $\mathbf{S}^1 \times]0, +\infty[$, and thus $\mathbf{P}^n(\mathbf{C})$ is the quotient of $\mathbf{S}^{2n+1}$ by $\mathbf{S}^1$.
(3) Moreover, all this can be done in the Hamiltonian framework: the sphere is a level manifold of a momentum mapping and the projective space is its symplectic reduction.
(4) Of the action of the torus $\mathbf{T}^{n+1}$ on $\mathbf{C}^{n+1}$, there remains a (Hamiltonian) $\mathbf{T}^n$-action on the projective space, for which the image of the momentum mapping is the standard simplex in $\mathbf{R}^n$... polyhedron, the associated fan of which is precisely the one we used to construct $\mathbf{P}^n(\mathbf{C})$ as a toric variety (see Figure 2)!

Of course, the aim of this section is to generalize the above remarks to all toric manifolds.

VII.2.a. Delzant's existence theorem.

Theorem VII.2.1. *Let P be a* primitive *polyhedron in $(\mathbf{R}^n)^\vee$. Let Σ be the fan of $\mathbf{R}^n$ associated with P. There exists, on the toric manifold $\boldsymbol{X}_\Sigma$, a symplectic form such that the action of the compact component of the big torus of $\boldsymbol{X}_\Sigma$ is Hamiltonian and P is the image of the momentum mapping.*

Together with Theorem IV.4.20, this asserts the identity between primitive polyhedra and compact connected symplectic manifolds endowed with the integrable action of a torus.

Proof of the theorem

First step. Following Delzant [**39**], we describe the convex polyhedron P in $(\mathbf{R}^n)^\vee$ as the intersection of some n-dimensional affine subspace with the first quadrant in a big space $(\mathbf{R}^N)^\vee$.

Here N will be the number of facets of the polyhedron. We will write equations of the hyperplanes spanned by these facets in the form

$$\langle x_i, \varphi \rangle = \lambda_i \quad 1 \leqslant i \leqslant N$$

where the vectors $x_i \in \mathbf{R}^n$ have been chosen integral, primitive and such that the polyhedron is

$$P = \{\varphi \in (\mathbf{R}^n)^\vee \mid \langle x_i, \varphi \rangle + \lambda_i \geqslant 0, 1 \leqslant i \leqslant N\}.$$

As above, we consider the projection

$$\pi : \mathbf{R}^N \longrightarrow \mathbf{R}^n$$

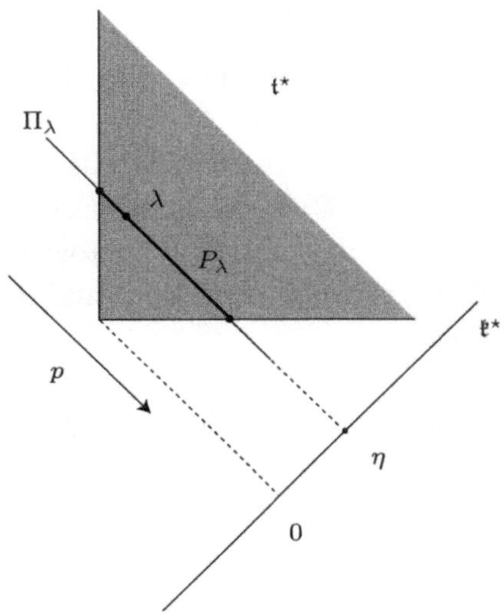

Figure 13

which sends the vectors of the canonical basis to the x_i's and the dual inclusion

$$^t\pi : (\mathbf{R}^n)^\vee \longrightarrow (\mathbf{R}^N)^\vee.$$

Consider the affine space Π_λ through $\lambda = (\lambda_1, \ldots, \lambda_N)$ parallel to the image of $(\mathbf{R}^n)^\vee$ in $(\mathbf{R}^N)^\vee$, so that P can be considered as the intersection of Π_λ with the first quadrant, that is, with the image of the momentum mapping

$$\mu : \mathbf{C}^N \longrightarrow (\mathbf{R}^N)^\vee$$

for the standard $\mathbf{T}^N$-action.

Let us look now at the torus $j : K \subset \mathbf{T}^N$ as above and at the momentum mapping

$$\mu' : \mathbf{C}^N \xrightarrow{\ \mu\ } (\mathbf{R}^N)^\vee \xrightarrow{\ p\ } \mathfrak{k}^\star.$$

Let $\eta = p(\lambda)$. The projection p is the transpose of the injection $j : \mathfrak{k} \subset \mathbf{R}^N$, namely,

$$\forall \varphi \in \mathfrak{k}^\star, \forall x \in \mathfrak{k}, \quad \langle p(\varphi), x\rangle = \langle \varphi, j(x)\rangle.$$

The kernel $p^{-1}(0)$ is thus the annihilator of $\mathfrak{k}$ in $\mathbf{R}^N$, that is, $^t\pi(\mathbf{R}^n)^\vee$, and $p^{-1}(\eta)$ is the subspace parallel to $^t\pi(\mathbf{R}^n)^\vee$ through λ, namely Π_λ, so that

$$\mu'^{-1}(\eta) = \mu^{-1}(\Pi_\lambda \cap \mu(\mathbf{C}^N)) = \mu^{-1}(P).$$

Notice that μ is obviously proper, and we have assumed P to be compact, so that $\mu'^{-1}(\eta)$ is compact.

Let us prove now that the torus K acts freely on $\mu'^{-1}(\eta)$. Fix a point $z \in \mu'^{-1}(\eta) \subset \mathbf{C}^N$. Let $I = \{i_1, \ldots, i_r\}$ be the set of indices of the coordinates of z

that are zero,

$$z_i = 0 \text{ if and only if } i \in I.$$

The intersection of the corresponding hyperplanes

$$\langle x_{i_1}, \varphi\rangle = \lambda_{i_1}, \dots, \langle x_{i_r}, \varphi\rangle = \lambda_{i_r}$$

is not empty, and its intersection with P is not empty either, since $\mu(z)$ belongs to its image in Π_λ. So that there is a vertex of the polyhedron in this intersection and, by the primitivity assumption, there exist indices $i_{r+1}, \dots, i_n$ such that $(x_{i_1}, \dots, x_{i_n})$ is an integral basis of the lattice $\mathbf{Z}^n$. The linear map

$$\pi : \mathbf{R}^N \longrightarrow \mathbf{R}^n$$

is thus (of course, we could have to renumber coordinates)

$$\begin{aligned} \mathbf{R}^n \times \mathbf{R}^k &\longrightarrow \mathbf{R}^n \\ e_i &\longmapsto e_i \quad 1 \leqslant i \leqslant n \\ e_j &\longmapsto x_j \quad n+1 \leqslant j \leqslant N. \end{aligned}$$

The kernel of π is parametrized by

$$(b_1, \dots, b_k) \longmapsto \left(-(x_{n+1}, \dots, x_N) \begin{pmatrix} b_1 \\ \vdots \\ b_k \end{pmatrix}, b_1, \dots, b_k \right) \in \mathbf{R}^N$$

and of course the torus $K = \mathbf{R}^k/\mathbf{Z}^k$ is parametrized, as a subgroup of $\mathbf{T} = \mathbf{R}^N/\mathbf{Z}^N$, by the same formulae.

Now the coordinates $z_1, \dots, z_N$ of our point z are all nonzero, so that the stabilizer of this point is trivial. Hence we have proved that the K-action on $\mu'^{-1}(\eta)$ is free. The quotient is a compact $2n$-dimensional manifold W and it is endowed with a reduced symplectic form ω_η.

Moreover the Hamiltonian action of the n-dimensional torus $Q = \mathbf{T}^N/K$ on $\mathbf{C}^N$ descends to the quotient and so does the map μ. It takes values in the subspace Π_λ and its image is, by construction, the polyhedron P.

Second step. Let us prove now that the manifold W is, indeed, the toric manifold $\boldsymbol{X}_\Sigma$ associated with the fan Σ defined by P. On the one hand, we have performed the quotient of the open set $\mathcal{U}_\Sigma$ by the complex cone $K_{\mathbf{C}}$ (think of $\mathbf{C}^N - \{0\}$ with the $\mathbf{C}^\star$-action). On the other one, we have performed the quotient of the compact set $\mu'^{-1}(\eta)$ by the compact torus K (think of $\mathbf{S}^{2N-1}$ with the $\mathbf{S}^1$-action).

We need first to remark that $\mu'^{-1}(\eta)$ is contained in $\mathcal{U}_\Sigma$, namely that, if $z \in \mu'^{-1}(\eta)$ and $z_i = 0$ for all $i \in I$, then the cone $\langle x_{\overline{I}} \in \Sigma$. But this is exactly what we have done above in the proof of the freeness of the K-action.

We follow now a general argument due to Kirwan ([**85**, Theorem 7.4]). Choose a metric on $\mathfrak{k}^\star$ and consider the function $f : \mathcal{U}_\Sigma \to \mathbf{R}$ which, to z, associates the real number $\|\mu'(z)-\eta\|^2$. The function f is nonnegative and it is minimal on $\mu'^{-1}(\eta)$. Its derivative is given by

$$\mathsf{T}_z f(X) = 2(\mu'(z)-\eta)\cdot \mathsf{T}_x\mu'(X),$$

hence it vanishes at the points of $\mu'^{-1}(\eta)$ and at the points z such that $\operatorname{Im}\mathsf{T}_z\mu' \subset (\mu'(z)-\eta)^\perp$... but we have assumed that $z \in \mathcal{U}_\Sigma$, where we know that K acts freely and thus, that μ' has maximal rank. So that our only critical points are the points of $\mu'^{-1}(\eta)$. Hence for any $z \in \mathcal{U}_\Sigma$, the gradient trajectory descending from z will reach a point in $\mu'^{-1}(\eta)$.

Let us calculate this gradient:

$$\begin{aligned}
(\operatorname{grad}_z f, X) &= 2\left((\mu'(z)-\eta), \mathsf{T}_z\mu'(X)\right)\\
&= 2\left({}^t\mathsf{T}_z\mu'(\mu'(z)-\eta), X\right)\\
&= 2\left(i_{\underline{\mu'(z)-\eta}_z}\omega, X\right)\\
&= 2\omega_z(\underline{\mu'(z)-\eta}_z, X)\\
&= -2i(\underline{\mu'(z)-\eta}_z, X).
\end{aligned}$$

Hence $\operatorname{grad}_z f = -2i\underline{\mu'(z)-\eta}_z$, where of course we have used the metric to identify $\mu'(z)-\eta$ with an element of $\mathfrak{k}$ and underlining denotes as usual the associated fundamental vector field. Thus the gradient flow consists of exponentials of elements of $i\mathfrak{k}$: this is the "imaginary part" of the torus $K_{\mathbf{C}}$ (the analogue of $]0,+\infty[$ in the torus $\mathbf{C}^\star$).

In particular, any gradient trajectory is included in a $K_{\mathbf{C}}$-orbit. From what we have said before, we derive that each $K_{\mathbf{C}}$-orbit meets $\mu'^{-1}(\eta)$. Let us now show that they meet along a unique K-orbit, that is to say that if $z \in \mu'^{-1}(\eta)$, then $K_{\mathbf{C}}\cdot z \cap \mu'^{-1}(\eta) = K\cdot z$.

Assuming that $\mu'(g\cdot z) = \eta$ for some $g \in K_{\mathbf{C}}$ we shall show that we then have $g\cdot z = k\cdot z$ for some $k\in K$. As $\mu'^{-1}(\eta)$ is K-invariant, we may assume that $g = \exp iX$ for some $X \in \mathfrak{k}$. We thus want to check that $g\cdot z = z$.

Consider the function $h : \mathbf{R} \to \mathbf{R}$ defined by $h(t) = \mu'((\exp itX)\cdot z)\cdot X$. We have $h(0) = h(1) = \eta\cdot X$ and therefore h' must vanish between 0 and 1: $\exists\, t_0 \in [0,1]$ such that $h'(t_0) = 0$. But

$$\begin{aligned}
h'(t) &= \mathsf{T}_y\mu'(i\underline{X}_y)\cdot X\\
&= \omega_y(i\underline{X}_y, \underline{X}_y) = \pm\left\|\underline{X}_y\right\|^2
\end{aligned}$$

putting $y = \exp itX$ to save notation. For $y_0 = \exp it_0X$, we thus have $\underline{X}_{y_0} = 0$ and therefore $\exp itX$ fixes y_0 and z. □

Notice that the fan (namely the u_i's) determines the manifold, but this is the polyhedron (the λ_i's) which determines its symplectic form.

Compactness of $\boldsymbol{X}_\Sigma$. In the case of a fan defined by a polyhedron P, we have expressed $\boldsymbol{X}_\Sigma$ as $\mu'^{-1}(\eta)/K$ where K was a compact group. For $\boldsymbol{X}_\Sigma$ to be compact, it is thus necessary (and sufficient) that $\mu'^{-1}(\eta)$ is.

On the other hand, $\mu'^{-1}(\eta) = \mu^{-1}(p^{-1}(\eta))$ and it is clear that μ is proper. We need thus only know when $p^{-1}(\eta) \cap \operatorname{Im}(\mu)$ is compact, but this is precisely the polyhedron under consideration. We have therefore proved that the toric manifold $\boldsymbol{X}_\Sigma$ is compact if and only if one of the polyhedra defining the fan Σ is compact. Taking Proposition VII.1.4 into account, we thus also have proven:

Proposition VII.2.2. *The toric manifold $\boldsymbol{X}_\Sigma$ associated with a convex polyhedron is compact if and only if the fan Σ is complete.* □

Examples VII.2.3

(1) Consider the triangle ABC in Figure 14. It is defined by the inequalities

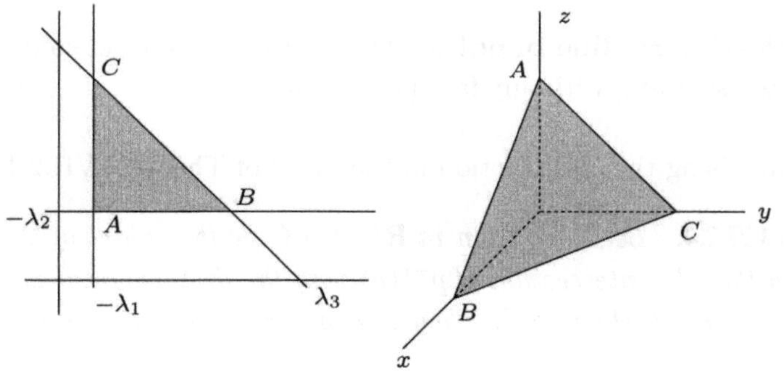

Figure 14

$$x_1 + \lambda_1 \geqslant 0, \quad x_2 + \lambda_2 \geqslant 0, \quad -(x_1 + x_2) + \lambda_3 \geqslant 0.$$

These inequalities define indeed a triangle if and only if $\lambda_1 + \lambda_2 + \lambda_3 > 0$. The linear map π maps e_i to x_i for $i = 1, 2$ and e_3 to $-x_1 - x_2$. Thus the image of the transpose ${}^t\pi$ is the plane $x + y + z = 0$, while Π_λ is the plane

$$x + y + z = \lambda_1 + \lambda_2 + \lambda_3$$

which does intersect the positive quadrant when $\lambda_1 + \lambda_2 + \lambda_3 > 0$.

(2) Consider now the trapezium in Figure 15. It is defined by the inequalities

$$x_1 + \lambda_1 \geqslant 0, \quad x_2 + \lambda_2 \geqslant 0, \quad -(x_1 + x_2) + \lambda_3 \geqslant 0, \quad -x_2 + \lambda_4 \geqslant 0,$$

which define a trapezium if and only if

$$\lambda_1 + \lambda_2 + \lambda_3 > 0, \quad \lambda_2 + \lambda_4 > 0, \quad \lambda_1 + \lambda_3 > \lambda_4.$$

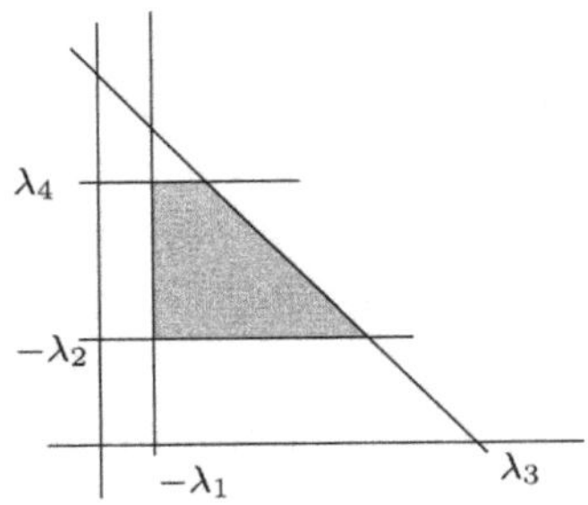

Figure 15

The affine plane $\Pi_\lambda \subset \mathbf{R}^4$ has equations

$$x+y+z+t=\lambda_1+\lambda_2+\lambda_3, \quad y+t=\lambda_2+\lambda_4.$$

VII.2.b. Polyhedra, line bundles. The next question is, of course, whether it is possible to associate, with any fan, polyhedra.

Polyhedra. Using the construction in the proof of Theorem VII.2.1, we prove:

Proposition VII.2.4. *Let Σ be a fan in $\mathbf{R}^n$. Let C be the subset of $\mathfrak{k}^\star$ consisting of the η's such that the intersection of $p^{-1}(\eta)$ with the first quadrant is a polyhedron, the associated fan of which is Σ. Then C is an open convex cone in $\mathfrak{k}^\star$.*

Proof. Consider a cone σ of dimension n in the fan Σ. Call $I(\sigma)$ the set of indices $(i_1,\ldots,i_n)$ such that the corresponding cones of the 1-skeleton generate σ. Given $\lambda \in \mathbf{R}^N$, define a linear form $\varphi_{\sigma,\lambda}$ by

$$\varphi_{\sigma,\lambda}(x_i)=-\lambda_i \quad \forall\, i \in I(\sigma).$$

This defines a linear form on $\mathbf{R}^n$, since the x_i's with $i \in I(\sigma)$ constitute a basis of (the lattice in) $\mathbf{R}^n$. Hence we have a finite set of points

$$\{\varphi_{\sigma,\lambda}\}_{\sigma\in\Sigma^{(n)}} \subset (\mathbf{R}^n)^\vee.$$

According to the Krein–Milman theorem, the convex set

$$P_\lambda=\{\varphi\in(\mathbf{R}^n)^\vee \mid \varphi(x_i)+\lambda_i \geqslant 0, 1\leqslant i\leqslant N\}$$

is the convex hull of the set of $\varphi_{\sigma,\lambda}$ if and only if for all i and for all σ,

$$\varphi_{\sigma,\lambda}+\lambda_i \geqslant 0.$$

Hence the set of λ's such that P_λ is the convex hull of the $\varphi_{\sigma,\lambda}$ is a closed convex cone. □

The boundary of the cone C consists of the λ's for which there exist an n-cone σ and a vector x_i not in σ (that is, $i \notin I(\sigma)$) such that $\varphi_{\sigma,\lambda}(x_i) = -\lambda_i$: through the vertex corresponding to σ, there are too many supporting hyperplanes. Here are two examples of this situation, continuations of Examples VII.2.3.

Examples VII.2.5

(1) In the case of the triangle, the map $p : (\mathbf{R}^3)^\vee \to \mathfrak{k}^\star$ is the projection $(x, y, z) \mapsto x + y + z$. A real number is in the image of the set of $(\lambda_1, \lambda_2, \lambda_3)$ such that $\lambda_1 + \lambda_2 + \lambda_3 > 0$ if and only if it is positive. The cone $C \subset \mathfrak{k}^\star$ is the positive half-line in this case.

(2) In the case of the trapezium, the dimension of $\mathfrak{k}^\star$ is 2, the map $p : (\mathbf{R}^4)^\vee \to \mathfrak{k}^\star$ sends (x, y, z, t) to $(u, v) = (x + y + z, y + t)$, so that the inequalities

$$\lambda_1 + \lambda_2 + \lambda_3 > 0, \quad \lambda_2 + \lambda_4 > 0, \quad \lambda_1 + \lambda_3 > \lambda_4$$

define the cone

$$C = \left\{(u, v) \in \mathbf{R}^2 \mid u > 0, v > 0 \text{ and } u > v\right\}.$$

This is suggested in Figure 16. Notice that, when the fat dotted line moves up, the polyhedron will change its shape (we shall be on the boundary of the cone C), becoming a triangle, then stop to be convex (we shall be outside C), the inequalities defining the empty set.

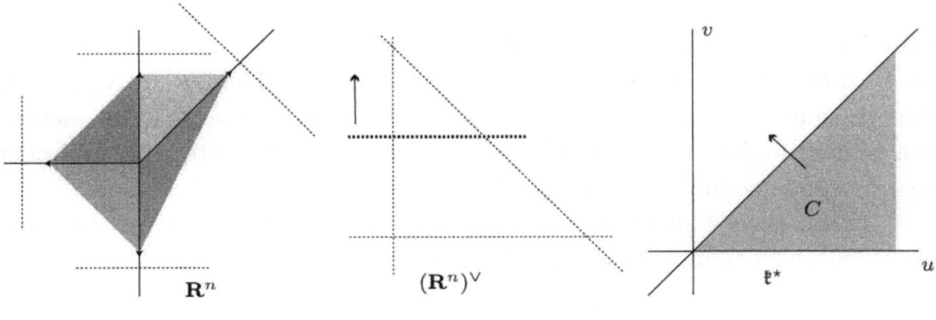

Figure 16

In order that the interior of the cone C be nonempty, there must exist $\lambda = (\lambda_1, \ldots, \lambda_N)$ such that the $\#\Sigma^{(n)}$ points $\varphi_{\sigma,\lambda}$ of $(\mathbf{R}^n)^\vee$ are the vertices of their convex hull. This is not the case, for instance, for the fan described in Figure 17. This is a very classical example (see [**38**]). The fan consists of the convex cones on the triangles shown (this is a triangulation of a tetrahedron with a vertex cut off), the origin (vertex of the cones) is in the interior of the polyhedron.

Let $g : \mathbf{R}^n = |\Sigma| \to \mathbf{R}$ be a convex function such that

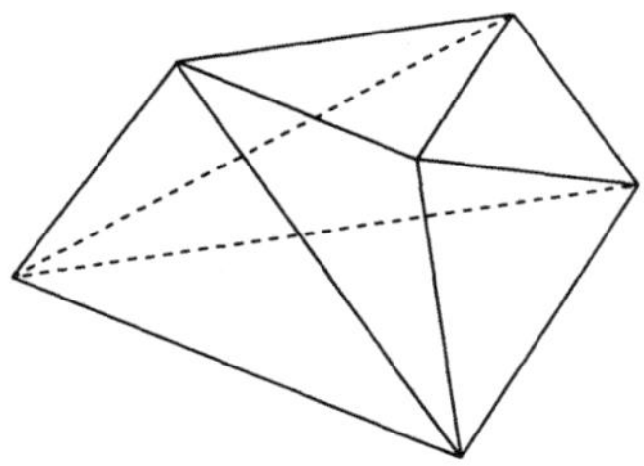

Figure 17

– it is linear on each n cone of Σ, but
– distinct cones correspond to distinct linear forms.

Such a function g is usually said to be *strictly convex* (as g is piecewise linear, it is not strictly convex in the usual sense, however, this terminology is very convenient in the present context, so that we will use it). The problem with the fan evoked in Figure 17 is that a convex function on such a fan cannot be strictly convex.

Corollary VII.2.6. *Let $g : \mathbf{R}^n \to \mathbf{R}$ be a strictly convex map, with respect to a smooth complete fan Σ. Then the convex set*

$$P = \{\varphi \in (\mathbf{R}^n)^\vee \mid \varphi(x_i) \geqslant g(x_i)\}$$

is a convex compact polyhedron, the fan associated to which is Σ. □

Line bundles. It turns out (see [**38, 30**]) that integral strictly convex functions on Σ correspond to *ample*, and even to *very ample* line bundles on $\boldsymbol{X}_\Sigma$. Recall that these are the line bundles which have enough sections so that the sections can be used to define an embedding of the toric variety $\boldsymbol{X}_\Sigma$ into a big projective space... and, by pull back, a Kähler form on it.

This is completely coherent with the fact that the strictly convex functions are those which allow us to define a (reduced) symplectic form on $\boldsymbol{X}_\Sigma$.

In this paragraph, we consider all the complex line bundles $E \to \boldsymbol{X}_\Sigma$ the total space of which are themselves toric manifolds.

Let $g : \mathbf{R}^n \to \mathbf{R}$ be a continuous map satisfying the two properties:

(1) This is the restriction of a linear map g_σ on each n-cone $\sigma \in \Sigma^{(n)}$.
(2) It takes integral values on $|\Sigma| \cap \mathbf{Z}^n$.

Remark VII.2.7. This is equivalent to requiring that g is the projection of an integral *linear* map, still denoted $g : \mathbf{R}^N \to \mathbf{R}$. We shall consider g sometimes as one and sometimes as the other.

With such a map, we associate a fan Σ_g in $\mathbf{R}^{n+1} = \mathbf{R}^n \times \mathbf{R}$ which is more or less the graph of g on Σ:

(1) Put $x'_i = (x_i, g(x_i))$ for $1 \leqslant i \leqslant N$ and $x'_{N+1} = (0,1)$; this defines the 1-skeleton $\Sigma_g^{(1)}$ from $\Sigma^{(1)}$.
(2) More generally, accept a cone $\langle x'_I$ as a member of Σ_g when (and only when)

$$\begin{cases} N+1 \notin I \text{ and } \langle x_I \in \Sigma \\ I = J \cup \{N+1\} \text{ and } \langle x_J \in \Sigma. \end{cases}$$

Notice that the projection $\mathbf{Z}^{n+1} = \mathbf{Z}^n \times \mathbf{Z} \to \mathbf{Z}^n$ defines a morphism from the fan Σ_g to the fan Σ and thus also a map $\boldsymbol{X}_{\Sigma_g} \to \boldsymbol{X}_\Sigma$.

Proposition VII.2.8. *The toric variety $\boldsymbol{X}_{\Sigma_g}$ is smooth if and only if $\boldsymbol{X}_\Sigma$ is. It is endowed with a natural map*

$$\boldsymbol{X}_{\Sigma_g} \longrightarrow \boldsymbol{X}_\Sigma$$

which makes it a complex line bundle over $\boldsymbol{X}_\Sigma$.

Proof. The former assertion follows immediately from Proposition VII.1.14, since Σ_g is obviously complete if and only if Σ is. To prove the latter, consider carefully the open subset $\mathcal{U}_{\Sigma_g}$ of $\mathbf{C}^{N+1}$:

$$\mathcal{U}_{\Sigma_g} = \mathbf{C}^{N+1} - \bigcup_{\langle x'_{\overline{I}} \notin \Sigma_g} e_I$$

and notice that

$$\mathcal{U}_{\Sigma_g} = \mathcal{U}_\Sigma \times \mathbf{C}.$$

Indeed,

$$\mathcal{U}_{\Sigma_g} = \mathbf{C}^{N+1} - \bigcup_{\langle x'_I \notin \Sigma_g} f_{\overline{I}},$$

where we have called $f = (e_1, \ldots, e_N, f_{N+1})$ the canonical basis. Now

$$\langle x'_I \in \Sigma_g \Longleftrightarrow \begin{cases} I \subset \{1, \ldots, N\} & \text{and } \langle x_I \in \Sigma \Leftrightarrow f_{\overline{I}} = e_{\overline{I}} \times \mathbf{C} \\ I = J \cup \{N+1\} & \text{and } \langle x_J \in \Sigma \Leftrightarrow f_{\overline{I}} = e_{\overline{J}} \times 0 \end{cases}$$

and therefore

$$\begin{aligned} \mathcal{U}_{\Sigma_g} &= \mathbf{C}^{N+1} - \textstyle\bigcup_{\langle x_I \in \Sigma} (e_{\overline{I}} \times 0 \bigcup e_{\overline{I}} \times \mathbf{C}) \\ &= \mathbf{C}^{N+1} - \textstyle\bigcup_{\langle x_I \in \Sigma} e_{\overline{I}} \times \mathbf{C} \\ &= \mathcal{U}_\Sigma \times \mathbf{C}. \end{aligned}$$

Let us now construct the complex torus K' associated with the fan Σ_g:

$$0 \longrightarrow \mathbf{K}' \longrightarrow \mathbf{Z}^N \times \mathbf{Z} \xrightarrow{\ \pi'\ } \mathbf{Z}^n \times \mathbf{Z}$$

where $\pi'(a,b) = (\pi(a), g(a) + b)$, that is,

$$\mathbf{K}' = \{(a,b) \mid \pi(a) = 0 \text{ and } b = -g(a)\}$$

and the map

$$\begin{array}{ccc} \mathbf{K} & \longrightarrow & \mathbf{K}' \\ a & \longmapsto & (a, -g(a)) \end{array}$$

is an isomorphism. The complex torus $K'_{\mathbf{C}} \subset T_{\mathbf{C}}^{N+1} = T_{\mathbf{C}}^N \times \mathbf{C}^\star$ is the isomorphic image of $K_{\mathbf{C}}$ by the restriction of

$$\begin{array}{ccc} \mathbf{T}_{\mathbf{C}}^N & \longrightarrow & T_{\mathbf{C}}^N \times \mathbf{C}^\star \\ t & \longmapsto & (t, \widetilde{g}(t)^{-1}) \end{array}$$

where $\widetilde{g}$ is the multiplicative avatar of the linear map g. In other words, the $K'_{\mathbf{C}}$-action on $\mathcal{U}_{\Sigma_g}$ may be identified with that of $K_{\mathbf{C}}$ on $\mathcal{U}_\Sigma \times \mathbf{C}$ by

$$u \cdot (x, y) = (u \cdot x, g(u)^{-1} y).$$

Taking quotients of both sides gives the announced line bundle:

$$\begin{array}{ccc} \mathcal{U}_{\Sigma_g} = \mathcal{U}_\Sigma \times \mathbf{C} & \longrightarrow & \mathcal{U}_\Sigma \\ K'_{\mathbf{C}} \big\downarrow & & \big\downarrow K_{\mathbf{C}} \\ X_{\Sigma_g} & \longrightarrow & X_\Sigma. \end{array}$$

□

Example VII.2.9. In the case where Σ is the fan describing $\mathbf{P}^n(\mathbf{C})$ (that of Figure 2), if we define

$$g : \mathbf{Z}^{n+1} \longrightarrow \mathbf{Z}$$

by $g(e_i) = 0$ for $1 \leqslant i \leqslant n$ and $g(e_{n+1}) = m$, we get the bundle $\mathcal{O}(-m)$ (see § VI.5). For $n = 1$, Figure 18 shows the bundles $\mathcal{O}(-1)$ and $\mathcal{O}(1)$ over $\mathbf{P}^1(\mathbf{C})$. Looking back

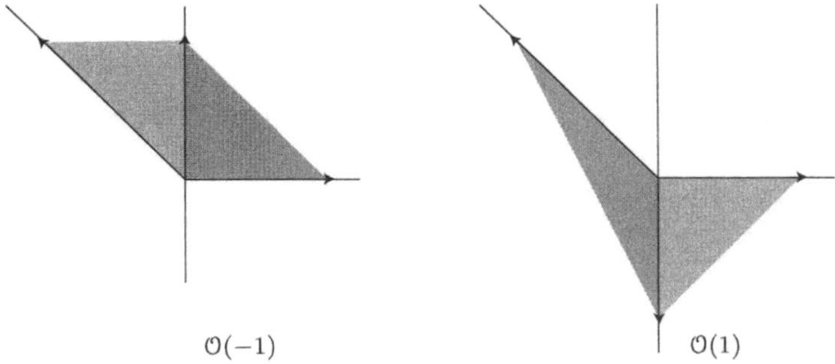

Figure 18

at the example illustrated by Figure 10 (Examples VII.1.18), notice that there is a map $\mathbf{P}^1(\mathbf{C}) \to \mathcal{O}(m)$ induced by a morphism of fans if and only if $m \geqslant 0$... and this is indeed the condition under which the bundle $\mathcal{O}(m)$ has a holomorphic section.

Remark VII.2.10. If φ is already a linear map on $\mathbf{Z}^n$, the bundle obtained is trivial. Indeed, the bundle associated with $\varphi = 0$ is trivial, since the torus $K'_{\mathbf{C}}$ does not act on the factor $\mathbf{C}$ in this case. Then, for $\varphi \in (\mathbf{Z}^n)^\vee$, define $\widetilde{\varphi} : \mathbf{Z}^n \times \mathbf{Z} \to \mathbf{Z}^n \times \mathbf{Z}$ by $(a, b) \mapsto (a, b + \varphi(a))$ and remark that the map $\widetilde{\varphi}$ defines an isomorphism from Σ_0 onto Σ_φ and thus an isomorphism from $\boldsymbol{X}_{\Sigma_0}$ onto $\boldsymbol{X}_{\Sigma_g}$ over $\boldsymbol{X}_\Sigma$.

We may thus consider that we have defined a map $g \mapsto \boldsymbol{X}_{\Sigma_g}$ from the set of all linear maps $\mathbf{Z}^N \to \mathbf{Z}$ (modulo those $\mathbf{Z}^n \to \mathbf{Z}$) into the set of isomorphism classes of Q-invariant complex line bundles over $\boldsymbol{X}_\Sigma$.

The space $(\mathbf{Z}^N)^\vee/(\mathbf{Z}^n)^\vee$ is just $\mathbf{K}^\vee$. As to the set of line bundles, and although we did not require any analyticity, we might call it $\operatorname{Pic}_Q(\boldsymbol{X}_\Sigma)$.

The map in question is a morphism of groups; if g_1, $g_2 \in (\mathbf{Z}^N)^\vee$, the group $K_{\mathbf{C}}$ acts on $\mathcal{U}_\Sigma \times \mathbf{C}$

- using g_1 by $u \cdot_1 (x, y) = (u \cdot x, g_1(u)^{-1} y)$,
- using g_2 by $u \cdot_2 (x, y) = (u \cdot x, g_2(u)^{-1} y)$,
- and using $g_1 + g_2$ by

$$u \cdot (x, y) = (u \cdot x, (g_1(u) g_2(u))^{-1} y),$$

  which correspond to the tensor product of line bundles (this is where the group structure on $\operatorname{Pic}_Q(\boldsymbol{X}_\Sigma)$ comes from).

Composing with the Euler class, we get a group morphism

$$\mathbf{K}^\vee \longrightarrow \operatorname{Pic}_Q(\boldsymbol{X}_\Sigma) \longrightarrow H^2(\boldsymbol{X}_\Sigma; \mathbf{Z})$$

which is just the isomorphism of Proposition VII.1.15.

VII.2.c. Polyhedra as projections. In the construction of Delzant above, we have used the fact that any convex polyhedron is the intersection of the "first quadrant" cone with some affine subspace. A different viewpoint is to consider the polyhedron as the projection of a big simplex. I describe this idea now and show that this leads to a description of toric manifolds which we have already met (in § IV.4.g) in connection with the proof of Kushnirenko's theorem.

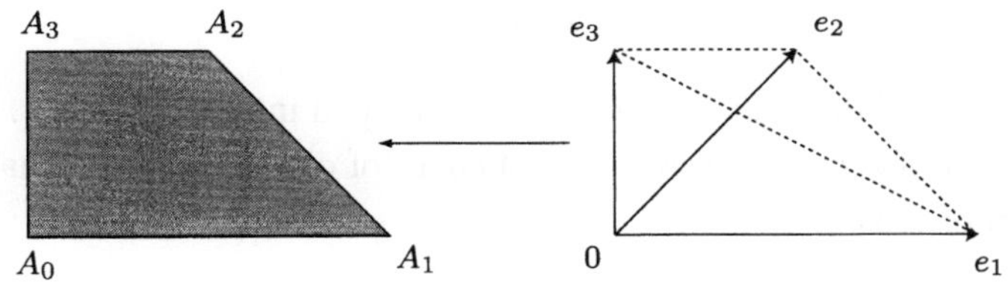

Figure 19

To simplify the exposition, let us assume here that P is a compact convex polyhedron *with integral vertices* in $(\mathbf{R}^n)^\vee$. Let m be the number of its vertices, denoted $A_0, \ldots, A_{m-1}$ and let Δ^{m-1} be the standard simplex, the vertices of which are $(0, e_1, \ldots, e_{m-1})$ in $(\mathbf{R}^{m-1})^\vee$. Notice that $m-1 \geqslant n$. Let

$$f^\vee : (\mathbf{Z}^{m-1})^\vee \longrightarrow (\mathbf{Z}^n)^\vee$$

be the integral affine map defined by $f^\vee(0) = A_0$, $f^\vee(e_i) = A_i$ for $1 \leqslant i \leqslant m-1$. It maps the simplex Δ^{m-1} to the convex hull of the A_i's, namely onto P (see Figure 19). Choose the origin of $(\mathbf{Z}^n)^\vee$ at A_0 and consider the transpose of $f^\vee$, still an integral map

$$f : \mathbf{R}^n \longrightarrow \mathbf{R}^{m-1}$$

and the map it induces at the level of tori, still denoted f,

$$f : \mathbf{T}^n = \mathbf{R}^n/\mathbf{Z}^n \longrightarrow \mathbf{R}^{m-1}/\mathbf{Z}^{m-1} = \mathbf{T}^{m-1}.$$

We know that Δ^{m-1} is the image of the momentum mapping

$$\mu : \mathbf{P}^{m-1}(\mathbf{C}) \longrightarrow (\mathbf{R}^{m-1})^\vee$$

associated with the $\mathbf{T}^{m-1}$-action, so that P is the image of the same $\mathbf{P}^{m-1}(\mathbf{C})$ by the momentum mapping

$$\mu' : \mathbf{P}^{m-1}(\mathbf{C}) \longrightarrow (\mathbf{R}^n)^\vee$$

associated with the $\mathbf{T}^n$-action. Let us now consider the fan $\Sigma = \Sigma_P$ defined by the polyhedron P.

Lemma VII.2.11. *The integral map f defines a morphism of fans from Σ_P to the fan Σ_Δ associated with the simplex Δ^{m-1}.*

Proof. Recall from Proposition VII.1.3 that the vertices of the polyhedron correspond to the cones of maximal dimension in the fan (and in particular, m is the number $\#\Sigma^{(n)}$ of dimension-n cones). Let A be a vertex of P and s a vertex of Δ such that $f^\vee(s) = A$. By convexity, we have also the inclusion

$$f^\vee(\tau_s) \subset \tau_A$$

(the τ's are the tangent cones). Dualizing, we get

$$f(\tau_A^\vee) \subset \tau_s^\vee.$$

Hence the cones of dimension n of Σ are all mapped into cones of Σ_Δ. As the fan Σ is complete, all its cones are subcones of cones of dimension n, so that f is indeed a morphism of fans. □

Hence, f defines a map

$$\boldsymbol{X}_\Sigma \longrightarrow \mathbf{P}^{m-1}(\mathbf{C})$$

and our polyhedron is the image of $\boldsymbol{X}_\Sigma$ under μ'. Notice that, if P is primitive, Σ is a smooth fan. This is an alternative way to construct a symplectic manifold the image of which is the polyhedron P (at least if the latter has integral vertices). Notice that, by construction, $\mathbf{T}^n$ is the big torus of X_Σ.

I also want to check that the construction of the "manifold" $\overline{V}_S$ in §IV.4.g gives, indeed, a toric manifold X_Σ. Recall that $\overline{V}_S$ was constructed as the closure of the orbit of the point $[1,\ldots,1] \in \mathbf{P}^{m-1}(\mathbf{C})$ under the action of a complex torus. This construction is the same as we explained just above: starting from $\alpha_0,\ldots,\alpha_{m-1} \in (\mathbf{Z}^n)^\vee$, we have constructed

$$\begin{aligned} f^\vee : (\mathbf{Z}^{m-1})^\vee &\longrightarrow (\mathbf{Z}^n)^\vee \\ e_i &\longmapsto \alpha_i - \alpha_0, \end{aligned}$$

then we have looked at the dual map f at the torus level, namely at

$$\begin{aligned} f : \mathbf{T}^n &\longrightarrow \mathbf{T}^{m-1} \\ u &\longmapsto (u^{\alpha_1-\alpha_0},\ldots,u^{\alpha_{m-1}-\alpha_0}) \end{aligned}$$

and eventually we have looked at its action on $\mathbf{P}^{m-1}(\mathbf{C})$ as a subgroup of $\mathbf{T}^{m-1}$, namely, as

$$\begin{aligned} u\cdot[z_0,\ldots,z_{m-1}] &= [z_0, u^{\alpha_1-\alpha_0}z_1,\ldots,u^{\alpha_{m-1}-\alpha_0}z_{m-1}] \\ &= [u^{\alpha_0}z_0,\ldots,u^{\alpha_{m-1}}z_{m-1}]. \end{aligned}$$

Then we have defined $\overline{V}_S$ as the closure of the orbit of the point $[1,\ldots,1]$ under the complexified torus $\mathbf{T}^n_{\mathbf{C}}$. According to our previous constructions, this closure must be the toric manifold X_Σ. Notice that there is no reason why the convex hull $\widehat{S}$ of the points $\alpha_0,\ldots,\alpha_{m-1}$ should be a *primitive* polyhedron, so that our fan Σ and our toric "manifold" X_Σ can perfectly well not be smooth.

VII.3. Cohomology of X_Σ

VII.3.a. Euler classes of line bundles and reduced symplectic forms.

We have an isomorphism

$$H^2(BK;\mathbf{Z}) \longrightarrow H^2(X_\Sigma;\mathbf{Z})$$

given by the Euler class of the principal $K_{\mathbf{C}}$-bundle, and $H^2(BK;\mathbf{Z})$ can be considered as the integral lattice in $\mathfrak{k}^\star$. Now, to any complex line bundle on X_Σ is associated an element of $H^2(X_\Sigma;\mathbf{Z})$, while, with any polyhedron P, we have associated a symplectic form ω_η and, in particular, a cohomology class in $H^2(X_\Sigma;\mathbf{R})$. Recall from Proposition VII.2.4 that $C \subset \mathfrak{k}^\star$ denotes the open cone of elements η corresponding to polyhedra.

Proposition VII.3.1. *The Euler class of the complex line bundle defined by the strictly convex integral function $g \in C \subset \mathfrak{k}^\star$ is the cohomology class of the reduced symplectic form associated with the integral polyhedron corresponding with g.*

***Corollary VII.3.2* (Delzant [39]).** *The map*

$$\begin{aligned} C &\longrightarrow H^2(\boldsymbol{X}_\Sigma;\mathbf{R}) \\ \eta &\longmapsto [\omega_\eta] \end{aligned}$$

extends to a linear isomorphism

$$\mathfrak{k}^\star \longrightarrow H^2(\boldsymbol{X}_\Sigma;\mathbf{R}),$$

the restriction to the integral lattice of which coincides with the isomorphism

$$H^2(BK;\mathbf{Z}) \longrightarrow H^2(\boldsymbol{X}_\Sigma;\mathbf{R})$$

given by the Euler class of the principal bundle $\mathcal{U}_\Sigma \to \boldsymbol{X}_\Sigma$.

Proof of the proposition. Once a basis of $\mathbf{K}$ is chosen, we may write the torus $K_{\mathbf{C}}$ as a product of k factors $\mathbf{C}^\star$. We know that a $\mathbf{Z}$-basis of $H^2(\boldsymbol{X}_\Sigma;\mathbf{Z})$ is constituted with the k components of the Euler class of the principal $K_{\mathbf{C}}$-bundle $\mathcal{U}_\Sigma \to \boldsymbol{X}_\Sigma$.

Let g be an integral element in $\mathfrak{t}^\star$ and $\eta = g|_{\mathfrak{k}} \in \mathfrak{k}^\star$ be its projection, such that $\eta \in \mu'(\mathcal{U}_\Sigma)$. Call σ $(= dz/z)$ the volume form on $\mathbf{S}^1$. Consider the principal $\mathbf{S}^1$-bundle which is the circle bundle of $X_{\Sigma_g} \to X_\Sigma$:

$$\begin{array}{ccc} \mu'^{-1}(\eta)\times\mathbf{S}^1 & \longrightarrow & \mu'^{-1}(\eta) \\ \big\downarrow & & \big\downarrow \\ \boldsymbol{X}_{\Sigma_g} & \longrightarrow & \boldsymbol{X}_\Sigma \end{array}$$

(the vertical arrows are the quotients by the dimension-k compact tori K', K).

On $\mu'^{-1}(\eta)\times\mathbf{S}^1$, $\alpha\oplus\sigma$ is a connection 1-form for the trivial bundle represented by the upper horizontal arrow. The torus K' acts on $\mu'^{-1}(\eta)\times\mathbf{S}^1$ with the fundamental vector fields $(\underline{Y}, -\eta(Y))$. In particular,

$$\forall Y\in\mathfrak{k},\ i_{\underline{Y}}(\alpha\oplus\sigma) = i_{\underline{Y}}\alpha - \eta(Y) = 0$$

and therefore $\alpha\oplus\sigma$ descends to the quotient and gives a 1-form of connection on X_{Σ_g}. Its *curvature form*, the cohomology class of which is the Euler class of $X_{\Sigma_g} \to X_\Sigma$, is the reduced symplectic form ω_η. □

Proof of the corollary. The convex open cone $\mu'(\mathcal{U}_\Sigma)$ generates $\mathfrak{k}^\star$ by definition. Moreover, the map $\eta \mapsto [\omega_\eta]$ is affine according to the Duistermaat–Heckman theorem (see Theorem VI.2.3), thus there is no difficulty to extend it as an affine map

$$\mathfrak{k}^\star \longrightarrow H^2(X_\Sigma;\mathbf{R}).$$

It suffices now to apply the previous proposition together with Theorem VI.2.3 to see that the "slope" of this map is given by a $\mathbf{Z}$-basis of $H^2(\boldsymbol{X}_\Sigma;\mathbf{Z})$. It therefore cannot avoid being a diffeomorphism. □

Corollary VII.3.3. *Assume Σ is smooth and complete. Let u_i be the Euler class of the complex line bundle over $\boldsymbol{X}_\Sigma$ which is associated with $e_i^\star \in (\mathbf{Z}^N)^\vee$. The group $H^2(\boldsymbol{X}_\Sigma; \mathbf{Z})$ is the commutative group generated by $(u_1, \ldots, u_N)$, with the relations*

$$\left(\sum_{i=1}^{N} g(e_i)u_i = 0\right)_{g\in(\mathbf{Z}^n)^\vee}.$$

Proof. We have an explicit description of $\mathbf{K}^\star$ and of the above isomorphism:

$$0 \longrightarrow (\mathbf{Z}^n)^\vee \longrightarrow (\mathbf{Z}^N)^\vee \longrightarrow H^2(X_\Sigma; \mathbf{Z}) \longrightarrow 0. \qquad \square$$

VII.3.b. The cohomology ring. Assume now that the toric manifold $\boldsymbol{X}_\Sigma$ is described by a polyhedron $P \subset \mathfrak{q}^\star$: we fix a Q-invariant symplectic form on $\boldsymbol{X}_\Sigma$ and look at the corresponding momentum mapping

$$\mu : \boldsymbol{X}_\Sigma \longrightarrow \mathfrak{q}^\star.$$

Choosing a vector $X \in \mathfrak{q}$, we get a Morse function

$$\begin{aligned} \mu_X : \boldsymbol{X}_\Sigma &\longrightarrow \mathbf{R} \\ x &\longmapsto \langle \mu(x), X \rangle \rangle. \end{aligned}$$

As in Chapter IV, the critical points of μ_X correspond to the vertices of the polyhedron (and maybe one or two edges). Choosing X generically, all the critical points are isolated. Any vertex can be given an index (relative to X). This is the number of edges starting from this vertex with a directing vector φ such that $\varphi(X) < 0$ (we call these edges *descending edges*). For instance, in Figure 20, the vertex A has index 0, B has index 1, C has index 2 and D has index 3.

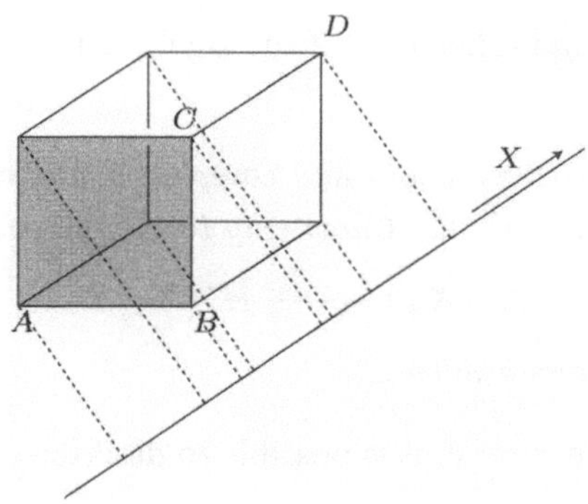

Figure 20

Recall that our polyhedron is primitive and in particular simple, so that the number of descending edges of any vertex is the dimension of the face these edges generate. For instance, in Figure 20, the shaded face corresponds to the negative bundle at the critical point corresponding to the vertex C.

Proposition VII.3.4. *Let M be a vertex of the polyhedron P, corresponding to the critical point m of μ_X. The index of m as a critical point of μ_X is twice the index of M relative to X as a vertex of the polyhedron.* □

Hence, the integral homology of $\boldsymbol{X}_\Sigma$ consists of classes represented by the faces of the polyhedron. Recall that the faces of dimension k of P correspond to the cones of codimension k of Σ and to submanifolds of $\boldsymbol{X}_\Sigma$ of complex dimension k. Moreover the intersections of these submanifolds are given by

$$\boldsymbol{X}_{\sigma \cap \tau} = \boldsymbol{X}_\sigma \cap \boldsymbol{X}_\tau$$

so that we get all the cohomology classes from the degree-2 ones and we have proved:

Theorem VII.3.5. *Assume Σ is the smooth and complete fan corresponding to a polyhedron P. The group $H_\star(\boldsymbol{X}_\Sigma; \mathbf{Z})$ is a free Abelian group of rank the number a_n of cones of dimension n in Σ, this number being the Euler characteristic of $\boldsymbol{X}_\Sigma$ as well.*

The cohomology ring $H^\star(\boldsymbol{X}_\Sigma; \mathbf{Z})$ is generated by its classes of degree 2. □

Corollary VII.3.6. *If Σ is the smooth and complete fan corresponding to a polyhedron P, the composition*

$$H^\star(B\mathbf{T}; \mathbf{Z}) \longrightarrow H^\star(BK; \mathbf{Z}) \longrightarrow H^\star(\boldsymbol{X}_\Sigma; \mathbf{Z})$$

is a surjective ring homomorphism. □

Notice that, if we had taken care of the algebraicity questions, we would have also proved:

Corollary VII.3.7. *If Σ is the smooth and complete fan corresponding to a polyhedron P, the natural map from the Chow ring to the cohomology*

$$\mathbf{A}^\star(\boldsymbol{X}_\Sigma) \longrightarrow H^\star(\boldsymbol{X}_\Sigma; \mathbf{Z})$$

is a surjective ring homomorphism. □

Being slightly more careful, it is possible to describe completely the cohomology ring.

Theorem VII.3.8. *The cohomology ring $H^\star(\boldsymbol{X}_\Sigma; \mathbf{Z})$ is isomorphic with the quotient $\mathbf{Z}[u_1, \ldots, u_N]/(\mathfrak{I} + \mathfrak{J})$, where $\mathfrak{I}$ is the ideal generated by the linear relations $(\sum g(e_i)u_i = 0)_{g \in (\mathbf{Z}^n)^\star}$ and $\mathfrak{J}$ that generated by the monomials u^I such that $\langle x_I \notin \Sigma$.*

Example VII.3.9. Consider the fan Σ defining the complex projective space $\mathbf{P}^n(\mathbf{C})$ (see Example (4), page 236). The linear map $\mathbf{Z}^{n+1} \to \mathbf{Z}^n$ maps the first n vectors of the canonical basis to $e_1, \ldots, e_n$ and the last to $-\sum e_i$, so that the dual map $(\mathbf{Z}^n)^\vee \to (\mathbf{Z}^{n+1})^\vee$ is $e_i^\star \mapsto e_i^\star - e_{n+1}^\star$. Hence $H^2(\mathbf{P}^n(\mathbf{C}); \mathbf{Z})$ is isomorphic to $\mathbf{Z}$ (as expected!), generated by any of the u_i (corresponding to the 1-skeleton). Now, to get the cohomology ring, we must look at the ideal $\mathfrak{J}$. namely we must look at the monomials $u_1^{i_1} \cdots u_m^{i_m}$ for $\langle x_{i_1} \ldots x_{i_m} \notin \Sigma$. The only such cone is $\langle x_1 \ldots x_{n+1}$, so that $\mathfrak{J}$ is generated by $u_1 \cdots u_{n+1}$ and Theorem VII.3.8 asserts, in this case, that

$$\begin{aligned} H^\star(\mathbf{P}^n(\mathbf{C}); \mathbf{Z}) &\cong \mathbf{Z}[u_1, \ldots, u_{n+1}]/(u_1 + \cdots + u_{n+1}, u_1 \cdots u_{n+1}) \\ &\cong \mathbf{Z}[u]/u^{n+1} \end{aligned}$$

as expected.

Remark VII.3.10. Theorem VII.3.8 is due, in principle, to Jurkiewicz [**79**] in the projective case and to Danilov [**38**] in the general case. However, the proof we give here relies on a previous work by Ehlers [**45**], who was smart enough to imagine and construct a cellular decomposition, the very same cellular decomposition we have here (associated to the Morse function μ_X) without even needing a Morse function!

(Half-)Proof of Theorem VII.3.8. Recall that we have a natural ring homomorphism

$$H^\star(B\mathbf{T}; \mathbf{Z}) \cong \mathbf{Z}[u_1, \ldots, u_N] \longrightarrow H^\star(\boldsymbol{X}_\Sigma; \mathbf{Z})$$

which maps u_i to the class corresponding to the cone $\langle x_i$.

Of course, the ideal $\mathfrak{I}$ is contained in the kernel and the ideal $\mathfrak{J}$ as well, since $\langle x_I \notin \Sigma$ means exactly that the facets corresponding to the i_j's in I do not intersect. Hence we have a surjective ring homomorphism

$$\mathbf{Z}[u_1, \ldots, u_N]/(\mathfrak{I} + \mathfrak{J}) \longrightarrow H^\star(\boldsymbol{X}_\Sigma; \mathbf{Z}).$$

To prove the theorem, the only thing which is left to do is to check that $\mathbf{Z}[u_1, \ldots, u_N]/\mathfrak{I} + \mathfrak{J}$ is a free Abelian group of the same rank as $H^\star(\boldsymbol{X}_\Sigma; \mathbf{Z})$, namely of rank $\#\Sigma^{(n)}$. This is a pure algebra question, which is solved in [**38**]. □

Remark VII.3.11. The statement of Theorem VII.3.8 can be given an equivariant cohomology interpretation. According to Remark VI.3.17, $H_Q^\star X_\Sigma$ is the free $H^\star BQ$-module generated by $H^\star X_\Sigma$. It is not hard to check that $E\mathbf{T} \times_Q X_\Sigma$ and $E\mathbf{T} \times_T \mathcal{U}_\Sigma$ have the same homotopy type, so that $H_Q^\star(\boldsymbol{X}_\Sigma, \mathbf{Z})$ is an $H^\star(B\mathbf{T})$-module. Thus we have ring homomorphisms

$$\begin{array}{ccccc} H^\star(B\mathbf{T}) & \longrightarrow & H_Q^\star(\boldsymbol{X}_\Sigma) & \longrightarrow & H^\star(\boldsymbol{X}_\Sigma) \\ \mathbf{Z}[u_1, \ldots, u_N] & \longrightarrow & \mathbf{Z}[u_1, \ldots, u_N]/\mathfrak{J} & \longrightarrow & \mathbf{Z}[u_1, \ldots, u_N]/(\mathfrak{I} + \mathfrak{J}) \end{array}$$

another interpretation of the two ideals $\mathfrak{I}$ and $\mathfrak{J}$.

Remark VII.3.12. The class $u_1 + \cdots + u_N$ is the first Chern class of the toric manifold X_Σ (this is proved in Exercise VII.14), which is thus the dual class to the divisor represented by the sum of all the facets of the polyhedron.

VII.4. Complex toric surfaces

In this section, we describe all the compact complex toric surfaces. A reason to do so is that we shall need the result in § VIII.2.c. The description is very simple:

***Theorem VII.4.1* (Oda [113]).** *Any compact complex toric surface is obtained from* $\mathbf{P}^2(\mathbf{C})$ *or from a Hirzebruch surface by a finite sequence of blowing ups.*

The fans. In dimension 2, a fan is well-defined once we know its 1-skeleton, and thus is given by a family of N primitive vectors $(x_1, \ldots, x_N)$ of $\mathbf{Z}^2$ (indices are considered as if in $\mathbf{Z}/N$) such that all the (x_{i-1}, x_i) are $\mathbf{Z}$-bases with the same orientation. The cones of dimension 2 are then the $\langle x_{i-1,i}$. Writing the direct basis (x_{i+1}, x_i) in terms of the direct basis (x_i, x_{i-1}), notice that

$$x_{i+1} = -x_{i-1} + m_i x_i$$

for some integer m_i. To give a fan is thus equivalent to giving the integers $(m_1, \ldots, m_N)$. Every cone of dimension 1 defines a $\mathbf{P}^1(\mathbf{C})$ in our toric surface. The integer m_i is the Euler class of the normal bundle to the $\mathbf{P}^1(\mathbf{C})$ corresponding to the cone generated by x_i.

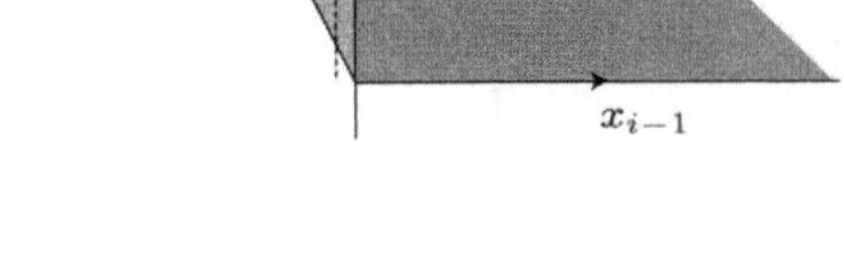

Figure 21

Fans with $N = 3$. We have three vectors x_0, x_1 and x_2. Writing

$$x_{i+1} = -x_{i-1} + m_i x_i,$$

we get

$$\begin{aligned} x_0 &= -x_1 + m_2 x_2 \\ &= -x_1 + m_2(-x_0 + m_1 x_1) \\ &= (m_1 m_2 - 1)x_1 - m_2 x_0. \end{aligned}$$

Thus $m_2 = -1$ and $m_1 = -1$ as well. There is only one fan with three cones in its 1-skeleton, this is the fan shown in Figure 2 page 227, and the toric surface is the projective plane.

Fans with $N = 4$. As above, the four vectors x_0, x_1, x_2 and x_3 must satisfy

$$\begin{aligned} x_3 &= -x_1 + m_2 x_2 \\ &= -x_1 + m_2(-x_0 + m_1 x_1) \\ &= -m_2 x_0 + (m_1 m_2 - 1)x_1 \end{aligned}$$

and the determinant of x_3 and x_0 (in this order) is equal to 1, that is,

$$-m_1 m_2 + 1 = 1,$$

so that m_1 or m_2 is zero. The 1-skeleton of our fan is

$$\begin{cases} x_2 = -x_0 \\ x_3 = -x_1 + m_2 x_2 \end{cases} \quad \text{or} \quad \begin{cases} x_2 = -x_0 + m_1 x_1 \\ x_3 = -x_1 \end{cases}$$

and the toric surface is a Hirzebruch surface.

The general case. To prove Theorem VII.4.1, it is thus sufficient to prove the next lemma.

Lemma VII.4.2. *For $N \geqslant 5$, at least one of the m_i's is equal to* 1.

Proof. We first prove that there are, in Σ^1, two opposite vectors. This is based on the following sublemma.

Sublemma VII.4.3. *Assume (u, u') and (v, v') are two direct bases of $\mathbf{Z}^2$. Then, if v is in the interior of the cone generated by u and u' and v' in the cone generated by u' and $-u$, then $v' = u'$ or $v' = -u$.*

This sublemma forbids situations shown in Figure 22 (the situation depicted on the right is the sublemma applied to the bases $(u', -u)$ and (v, v')).

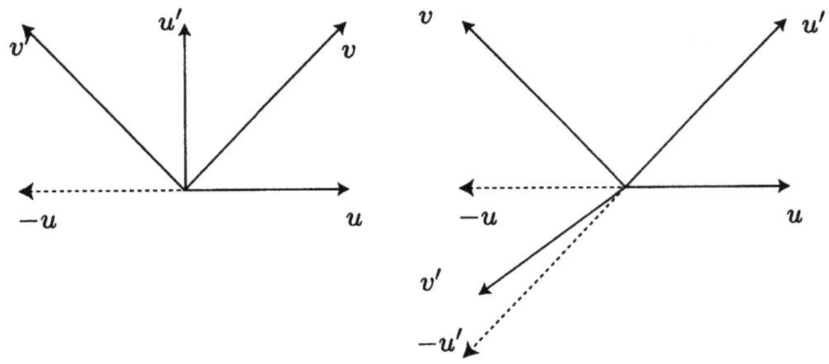

Figure 22

Proof of the sublemma. We write

$$\begin{cases} v = au + a'u' \\ v' = -bu + b'u' \end{cases}$$

for some integers a, $a' \geqslant 1$, b and $b' \geqslant 0$. The determinant of (v, v') is $ab' + ba' = 1$, so that b or b' must be zero. $\square$

Let us proceed with the proof of the lemma. Assume that there are no indices i and j such that $x_i + x_j = 0$. We can number the vectors x_i's so that, starting from x_1, more than half of them appear before $-x_1$ (and $-x_1 \neq x_i$ for all i). Let x_j be the last vector before $-x_1$. We have $j > 2$ and

$$x_{j+1} \neq -x_1 \text{ and } -x_2.$$

Applying the sublemma to the bases $(u, u') = (x_2, -x_1)$ and (x_j, x_{j+1}), we deduce that x_{j+1} cannot lie between $-x_1$ and $-x_2$. By convexity, it must be strictly between $-x_2$ and $-x_j$. This is shown in the right part of Figure 23. The left part

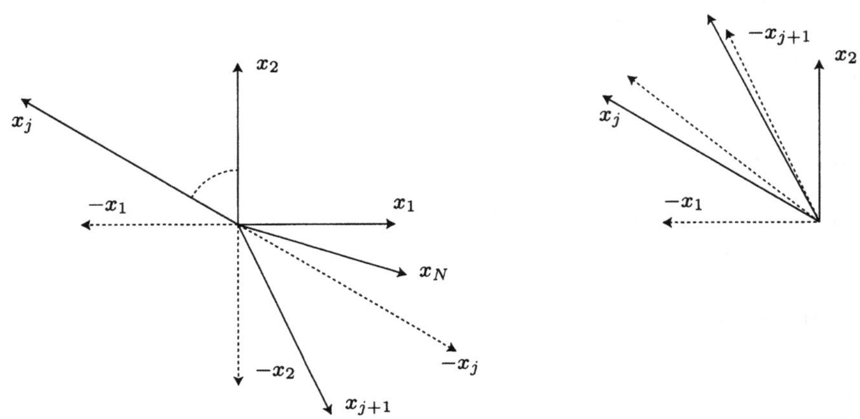

Figure 23

of the same figure shows what happens between x_2 and $-x_1$, where we find not only the vectors $x_2, \ldots, x_j$ but also $-x_{j+1}, \ldots, -x_N$ and $-x_1$ (these are the dotted vectors), N distinct vectors. Now, there is certainly a plain vector between two dotted ones, a contradiction with our sublemma.

Hence, there must be two opposite vectors among the generators of the 1-skeleton of the fan. Renumber the vectors so that $x_i = -x_0$ (for $i \geqslant 3$). Now express all the vectors x_j's in the basis (x_0, x_1):

$$x_j = -b_j x_0 + b'_j x_1$$

and define

$$c_i = b_i + b'_i.$$

We have $c_i = 1$ since $b_i = 1$ and $b'_i = 0$. Moreover, since (x_1, x_2) is a direct basis, we must have $b_2 = 1$, and we can assume that $b'_2 \geqslant 2$: if this is 0, then $x_2 = -x_0$ which is not, and if this is 1, $x_2 = -x_0 + x_1$ and this is what we wanted to find. Hence we can assume that $c_2 \geqslant 3$. As we also have $c_i = 1$, there is an index $j < i$ such that

$$\begin{cases} c_{j-1} \leqslant c_j \\ c_{j+1} < c_j. \end{cases}$$

The claim is now that, for this j, $m_j = 1$. Indeed, we have

$$\begin{cases} -b_j b'_{j+1} + b_{j+1} b'_j = 1 \\ b_j b'_{j-1} - b_{j-1} b'_j = 1, \end{cases}$$

which implies both that

$$b_j(b'_{j-1} + b'_{j+1}) - b'_j(b_{j-1} + b_{j+1}) = 0$$

and that b_j and b'_j are relatively prime. Hence there exists an integer λ such that

$$\begin{cases} b'_{j-1} + b'_{j+1} = \lambda b'_j \\ b_{j-1} + b_{j+1} = \lambda b_j. \end{cases}$$

Now, let us use the property of c_j defining the index j:

$$2c_j > c_{j-1} + c_{j+1} = b'_{j-1} + b_{j-1} + b'_{j+1} + b_{j+1} = \lambda(b_j + b'_j) = \lambda c_j.$$

Hence $\lambda = 1$,

$$\begin{cases} b_j = b_{j-1} + b_{j+1} \\ b'_j = b'_{j-1} + b'_{j+1} \end{cases}$$

and $m_j = 1$. □

Of course, one can also describe the fan of a toric surface by a weighted graph. This is done in [**113**] and [**30**] and this was repeated in [**14**]. The vertices of the graph correspond to the 2-dimensional cones of the fan, the edges to the 1-dimensional cones and there is a weight on each vertex, the corresponding m_i. For instance, Figure 24 shows the graphs corresponding, respectively to $\mathbf{P}^2(\mathbf{C})$, $\mathbf{P}^1(\mathbf{C})\times\mathbf{P}^1(\mathbf{C})$, $\widetilde{\mathbf{P}}^2(\mathbf{C})$ and, more generally, to the Hirzebruch surface W_k. Although it depends on the choice of a symplectic form, I prefer the combinatorial

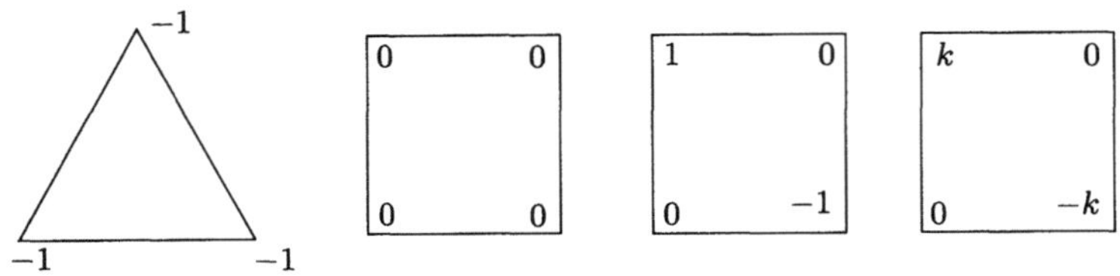

Figure 24

picture given by a polygon associated with the fan as in § VII.1.a. The same four pictures become the pictures in Figure 25. Notice that the Euler class m_i of the

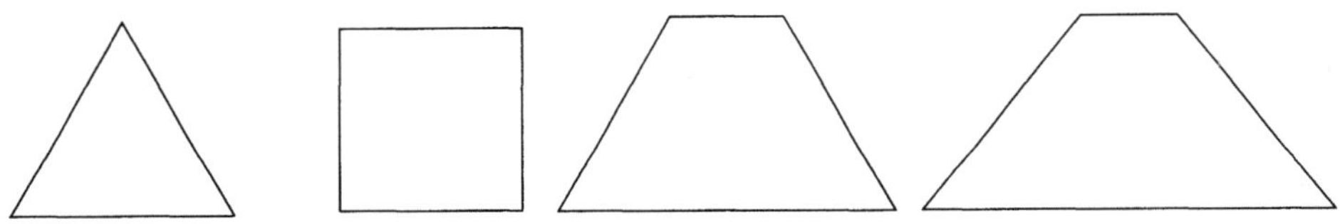

Figure 25

normal bundle to a $\mathbf{P}^1(\mathbf{C})$ corresponding to the cone generated by x_i, which is a weight on the graph, is just a slope in the polygon: two consecutive edges of our primitive polygon are generated by a basis of $\mathbf{Z}^2$, write the generator of the next edge in this basis, you get a relative slope, which is just m_i.

Exercises

Exercise VII.1. Let P be the rectangle $a \leqslant x \leqslant b$, $c \leqslant y \leqslant d$ in $\mathbf{R}^2$. Determine the tangent cones to P and their duals.

Exercise VII.2. Let Σ and Σ' be two fans. Prove that the cones $\sigma\times\sigma'$, where $\sigma\in\Sigma$, $\sigma'\in\Sigma'$, form a fan $\Sigma\times\Sigma'$. Prove that

$$\boldsymbol{X}_{\Sigma\times\Sigma'} = \boldsymbol{X}_{\Sigma}\times\boldsymbol{X}_{\Sigma'}.$$

Exercise VII.3. Let P be a convex polyhedron with integral vertices in $\mathbf{R}^n$. Assume that 0 is an interior point of P. For any face F of P, let τ_F be the cone over this face

$$\tau_F = \{rx \mid x \in F, r \geqslant 0\}.$$

Prove that the τ_F's constitute a complete fan.

Exercise VII.4. Let P be a convex polyhedron in $(\mathbf{R}^n)^\vee$. Let

$$P^\circ = \{v \in \mathbf{R}^n \mid g(v) \geqslant -1 \text{ for all } g \in P\}.$$

Prove that P° is a convex polyhedron—called the polar polyhedron of P. Assume P is rational (has vertices in $\mathbf{Q}^n$). Prove that P° is rational too. What is the polar polyhedron of the octahedron P with vertices at $(\pm 1, 0, 0)$, $(0, \pm 1, 0)$ and $(0, 0, \pm 1)$?

Exercise VII.5. What are the toric varieties associated with the fans shown in Figure 26?

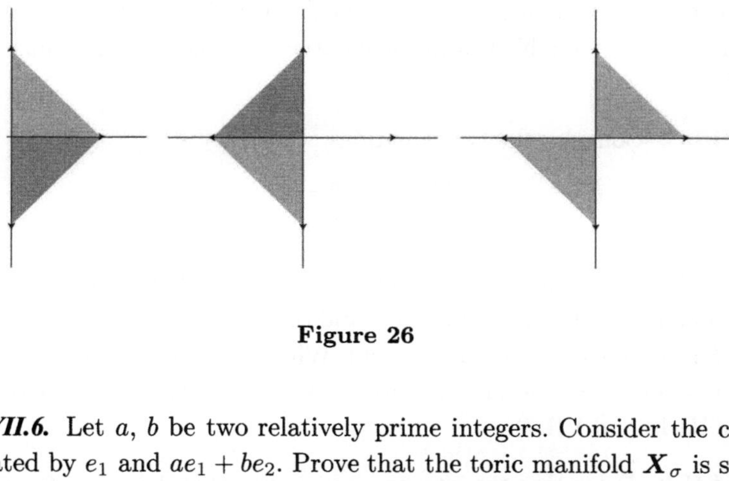

Figure 26

Exercise VII.6. Let a, b be two relatively prime integers. Consider the cone σ in $\mathbf{R}^2$ generated by e_1 and $ae_1 + be_2$. Prove that the toric manifold $\boldsymbol{X}_\sigma$ is smooth if and only if $b = \pm 1$. In the general case, prove that $\mathcal{U}_\Sigma = \mathbf{C}^2$ and that the group K is the cyclic subgroup of $\mathbf{T}^2$ generated by

$$\left(\exp\left(-\frac{2i\pi a}{b}\right), \exp\left(\frac{2i\pi}{b}\right)\right)$$

and that $\boldsymbol{X}_\sigma$ can be identified with the surface of equation

$$Z^b = XY^a \text{ in } \mathbf{C}^3.$$

Consider now the map

$$\begin{aligned} f : \mathbf{Z}^2 &\longrightarrow \mathbf{Z}^2 \\ (x, y) &\longmapsto (x, by). \end{aligned}$$

Check that f maps the cone

$$\sigma' = \{\lambda e_1 + \mu(a e_1 + e_2) \mid \lambda, \mu \geqslant 0\}$$

to the cone σ and that it defines morphisms

$$\sigma' \longrightarrow \sigma \text{ and } \boldsymbol{X}_{\sigma'} \xrightarrow{\widetilde{f}} \boldsymbol{X}_{\sigma}.$$

Identify $\boldsymbol{X}_{\sigma'}$ and show that $\widetilde{f}$ is a "desingularization" of $\boldsymbol{X}_{\sigma}$.

Exercise VII.7. Let Σ be the complete fan, the 1-skeleton of which is generated by the vectors e_1, e_2 and $-ke_1 - \ell e_2$ in $\mathbf{R}^2$. Prove that $\boldsymbol{X}_\Sigma$ is a weighted projective space (see Exercise III.18).

Exercise VII.8. Let

$$\sum_{i=1}^{k} c_i z^i = 0$$

be an equation in the complex variable z. Determine the number of its nonzero solutions for general values of the c_i's and compare to what Kushnirenko's theorem predicts (this example is mentioned in the introduction to this book, page 4). Determine V_S and $\overline{V}_S$ (this is a rational curve) and the corresponding toric manifold $\boldsymbol{X}_\Sigma$.

Exercise VII.9. Consider the two equations

$$\begin{cases} \sum_{i=0}^{k} a_i z_1^i = 0 \\ \sum_{j=0}^{\ell} b_j z_2^j = 0 \end{cases}$$

in the complex variables z_1 and z_2. Determine the number of nonzero solutions of this system for general values of the a_i's and the b_j's. What are S and $\widehat{S}$? Compare to what Kushnirenko's theorem predicts. What is the fan Σ associated with $\widehat{S}$? When is it smooth? Prove that, in general, $\boldsymbol{X}_\Sigma$ is a weighted projective space (see Exercises III.18 and VII.7).

Exercise VII.10. Prove that the integers $m_1, \ldots, m_N$ that define the fan of a smooth compact toric surface are related by

$$\begin{pmatrix} m_1 & 1 \\ -1 & 0 \end{pmatrix} \begin{pmatrix} m_2 & 1 \\ -1 & 0 \end{pmatrix} \cdots \begin{pmatrix} m_N & 1 \\ -1 & 0 \end{pmatrix} = \begin{pmatrix} 1 & 0 \\ 0 & 1 \end{pmatrix}.$$

Prove that they must also satisfy the equation

$$m_1 + \cdots + m_N = 3N - 12$$

(hint: this is a consequence of Theorem VII.4.1). Conversely, prove that any sequence of integers satisfying the two relations above comes from a smooth compact toric surface.

Exercise VII.11. Consider the fan describing $\widetilde{P}^2(\mathbf{C})$ (right part of Figure 7 page 237, with $a = 1$) and use it to compute the cohomology of $\widetilde{P}^2(\mathbf{C})$, *via* Theorem VII.3.8 (check your result, looking back at page 215).

Exercise VII.12. Let $\boldsymbol{X}_\Sigma$ be the toric surface associated with a smooth complete fan Σ. Number the primitive vectors of the 1-skeleton modulo N, in such a way that the cones of dimension 2 are the $\langle x_i x_{i+1}$. Recall that then,

$$x_{i+1} = -x_{i-1} + m_i x_i$$

for some integer m_i. We consider the ideals $\mathfrak{I}$ and $\mathfrak{J}$ as in Theorem VII.3.8.

Description of the ideal $\mathfrak{I}$ (relations in $H^2(\boldsymbol{X}_\Sigma, \mathbf{Z})$). Write the projection map $\mathbf{Z}^N \to \mathbf{Z}^2$, using the basis (x_{i-1}, x_i) of $\mathbf{Z}^2$. Check that the ideal $\mathfrak{I}$ is generated by two elements of the form

$$\begin{cases} v_1 = a_1 u_1 + \cdots + a_{i-2} u_{i-2} + u_{i-1} - u_{i+1} + a_{i+2} u_{i+2} + \cdots + a_N u_N \\ v_2 = b_1 u_1 + \cdots + b_{i-2} u_{i-2} + u_i + m_i u_{i+1} + b_{i+2} u_{i+2} + \cdots + b_N u_N. \end{cases}$$

Description of the ideal $\mathfrak{J}$ (multiplicative structure of $H^\star(\boldsymbol{X}_\Sigma; \mathbf{Z})$). Check that $\mathfrak{J}$ is generated by the monomials $u_m u_n$ for nonconsecutive (modulo N) integers m and n. What does the fact that $u_i v_1 = 0$ mean? And the fact that $u_i v_2 = 0$?

***Exercise VII.13* (The Betti numbers of $\boldsymbol{X}_\Sigma$).** Consider, as in § VII.3.b, a polytope P of dimension n, the associated fan of which is smooth and complete, and a vector X, with respect to which the vertices of P have indices. Let a_k be the number of faces of dimension k of P and let β_k be the number of its vertices of index k. Counting the faces of dimension k, prove that

$$a_i = \beta_i + \binom{i+1}{i} \beta_{i+1} + \cdots = \sum_{k=0}^{n} \binom{k}{i} \beta_k.$$

If $a(t) = \sum_{i=0}^n a_i t^i$ and $\beta(t) = \sum_{i=0}^n \beta_i t^i$, check that $\beta(t+1) = a(t)$ and deduce that

$$\beta_k = \sum_{i=k}^{n} (-1)^{i+k} \binom{i}{k} a_i.$$

Prove that the Betti numbers $b_{2k} = \operatorname{rk} H^{2k}(\boldsymbol{X}_\Sigma; \mathbf{Z})$ satisfy the formula

$$b_{2k} = \sum_{i=k}^{n} (-1)^{i+k} \binom{i}{k} \# \Sigma^{(n-i)}.$$

Check that the Euler characteristic $\chi(\boldsymbol{X}_\Sigma)$ is a_n, as stated in Theorem VII.3.5.

***Exercise VII.14* (Tangent bundle and Chern classes of $\boldsymbol{X}_\Sigma$).** We consider a polyhedron, the associated fan $\Sigma \subset \mathbf{R}^n$, assumed to be smooth and complete, and the corresponding fibration

$$\pi : \mathcal{U}_\Sigma \longrightarrow \boldsymbol{X}_\Sigma.$$

Using standard coordinates $(z_1, \ldots, z_N)$ on $\mathcal{U}_\Sigma \subset \mathbf{C}^N$, we define N vector fields $v_1, \ldots, v_N$ on $\boldsymbol{X}_\Sigma$ by

$$v_i = \pi_\star \frac{\partial}{\partial z_i} \text{ for } 1 \leqslant i \leqslant N.$$

What are the relations between the v_i's? Recall that the linear form $e_i^\star \in (\mathbf{Z})^\vee$ defines a line bundle L_i on $\boldsymbol{X}_\Sigma$, the first Chern class of which is u_i (notation of Corollary VII.3.3). Prove that the formula(8)

$$\varphi((\sigma_1, \ldots, \sigma_N)\pi(z)) = \pi_\star \left(\sum \sigma_i(\pi(z)) \frac{\partial}{\partial z_i}\right)$$

(the σ_i's are sections of the L_i's) defines a surjective morphism of vector bundles over $\boldsymbol{X}_\Sigma$,

$$\varphi : L_1 \oplus \cdots \oplus L_N \longrightarrow T\boldsymbol{X}_\Sigma,$$

the kernel of which is the trivial bundle $(\mathbf{K} \otimes \mathbf{C}) \times \boldsymbol{X}_\Sigma$. Deduce that the total Chern class of $\boldsymbol{X}_\Sigma$ is

$$c(\boldsymbol{X}_\Sigma) = \prod_{i=1}^{N} (1 + u_i).$$

In particular, as announced in Remark VII.3.12, the first Chern class is

$$c_1(\boldsymbol{X}_\Sigma) = u_1 + \cdots + u_N.$$

Check that the formula also gives that

$$\langle c_n(\boldsymbol{X}_\Sigma, [\boldsymbol{X}_\Sigma] \rangle = \chi(\boldsymbol{X}_\Sigma) = a_n$$

as determined in Theorem VII.3.5 and Exercise VII.13.

(8)This proof is a simple generalization of the usual description of the tangent bundle to the complex projective space, see [**60**, p. 409].

CHAPTER VIII

HAMILTONIAN CIRCLE ACTIONS ON MANIFOLDS OF DIMENSION 4

In this chapter, we will investigate symplectic 4-manifolds endowed with Hamiltonian circle actions. Together with what was done in the previous chapters, this will give us a rather complete description of all the compact symplectic manifolds endowed with a Hamiltonian group action.

To investigate Hamiltonian group actions on 4-manifolds, we need only to focus on the cases of the groups $\mathbf{S}^1$, $\mathbf{T}^2$, SO(3) and SU(2). There are of course symplectic and even Hamiltonian actions of bigger groups on 4-manifolds. For instance, in the same way as SO(3) acts on the sphere $\mathbf{S}^2$ (or SU(2) on $\mathbf{P}^1(\mathbf{C})$), SO(4) acts on $\mathbf{S}^2 \times \mathbf{S}^2$ (see Exercise I.20). However, if the group acting is semi-simple, we know that the action will be Hamiltonian and we will be very happy if the momentum mapping is submersive somewhere. But we must then have, for a maximal torus $\mathbf{T}$ of our group G,

$$\dim \mathrm{G} + \dim \mathbf{T} \leqslant 4,$$

so that we are left with the groups SO(3) and SU(2). For tori, the same argument leaves us the groups $\mathbf{S}^1$ and $\mathbf{T}^2$.

We have already mentioned the classification theorem of Iglesias [**70**] in § IV.5. This theorem gives a complete symplectic classification of manifolds of dimension 4 endowed with an SO(3)-action (this result was quoted here in Exercise IV.18). The case of SU(2)-actions, which is completely analogous (although simpler) was explained in § IV.5. A few years after Iglesias' theorem came the general result of Delzant [**39**] (here Theorems IV.4.20 and VII.2.1) on $\mathbf{T}^n$-actions on $2n$-dimensional manifolds, including the case of 4-manifolds endowed with Hamiltonian $\mathbf{T}^2$-actions.

The case of $\mathbf{S}^1$-actions, to which this chapter is devoted, is somewhat different, as the inequality above is strict. On the topological side, all the $\mathbf{S}^1$-actions on 4-manifolds were classified by Fintushel [**48**] a long time ago[(1)]. Of course the symplectic (and even more: Hamiltonian) character of the actions

[(1)] There are of course a lot of 4-manifolds with $\mathbf{S}^1$-action which do not admit an invariant symplectic form; to be convinced of this, the reader needs only to look at the "list" of $\mathbf{S}^1$-manifolds of dimension 4 given by Fintushel [**48**] and Exercises VIII.12 and VIII.13.

under consideration here changes the landscape, mainly because, as we have already said, one can use Morse theory to study the fixed points. Free (principal) actions were investigated in detail by Bouyakoub [**26**]. It is reasonable to think that fixed point free actions may be handled in *grosso modo* the same pattern. Of course these actions do not have the ghost of a chance to be Hamiltonian (all the manifolds we will consider are compact). A very beautiful application of the fiber connectedness theorem (Corollary IV.3.2) and of other tricks was given by McDuff [**103**]: in dimension 4, an $\mathbf{S}^1$-action is Hamiltonian if and only if it has fixed points. These methods allowed the author to give a list of those 4-manifolds which may be given a symplectic form invariant by an $\mathbf{S}^1$-action with at least one fixed point [**13**]; a somewhat longer and less geometrical proof of these results independently appeared in [**1**]. Since then, Karshon [**83**] gave a *symplectic* classification theorem.

We will present these results in this chapter.

VIII.1. Symplectic $\mathbf{S}^1$-actions, generalities

The first question is whether a given symplectic $\mathbf{S}^1$-action is, or is not, Hamiltonian.

VIII.1.a. Characterization of those circle actions that are Hamiltonian. In the paper [**51**] which we already met in Chapter IV, Frankel also showed, using Hodge theory, that, on a *Kähler* manifold[(2)], a circle action preserving the Kähler form is Hamiltonian if and only if it has a fixed point. The same is true for symplectic actions on 4-manifolds, according to a beautiful theorem of McDuff.

***Theorem VIII.1.1* (McDuff [103]).** *Let (W,ω) be a compact symplectic manifold of dimension 4, endowed with an $\mathbf{S}^1$-action which preserves the symplectic form. Then,*

- *either the action has no fixed points and W is the total space of a bundle over $\mathbf{S}^1$,*
- *or it has at least one fixed point, and then it is Hamiltonian.*

Remark VIII.1.2. The fact that, if there are no fixed points, the manifold should be fibered over the circle is part—at least when the action is free—of a theorem of Bouyakoub [**26**].

Proof. Let us first list the steps in the proof. We will show that:

(1) It suffices to investigate the case where the cohomology class of the form $[\omega]$ is integral.

(2) see the definition in § II.2.b.

(2) In this case, there exists a map h, taking values in $\mathbf{S}^1$, which generalizes the momentum mapping and has all the properties of a periodic Hamiltonian (from the Morse theoretical viewpoint).
(3) Assuming the action is not Hamiltonian, one deduces, by an argument similar to Theorem IV.3.1, that h has no local extrema.
(4) By a nice argument (*a la* Duistermaat–Heckman with singularities, as in § VI.2.c) using Euler classes of Seifert manifolds (§ I.3.d), one shows that there is no critical point at all.

First step. The set of all invariant symplectic forms η such that the 1-form $i_X\eta$ is exact is a closed subset in the set of all invariant symplectic forms. On the other hand, as close to ω as one wants, there is an invariant symplectic form the cohomology class of which is rational. Hence we get the existence of an invariant rational symplectic form ω', such that, if $i_X\omega$ is not exact, then $i_X\omega'$ is not either. As ω' is rational, we just have to multiply it by a large enough integer N to give it an integral cohomology class. Of course, it stays symplectic, and $i_X(N\omega')$ is as exact as $i_X\omega'$ is. Hence there exists a Hamiltonian for ω if and only if there exists one for $N\omega'$. Rename $N\omega'$ as ω. □

Second step. As the cohomology class of ω is integral, that of the closed 1-form $i_X\omega$ is integral as well: if C is any 1-cycle, then $\varphi(C)$, got by pushing C with the $\mathbf{S}^1$-action, is a 2-cycle, and we have

$$\int_C i_X\omega = \int_{\varphi(C)} \omega \in \mathbf{Z}$$

(for the agnostics, this is proved in Exercise VIII.1). For any x_0 (fixed) in W, the values of

$$h(x) = \int_{x_0}^{x} i_X\omega$$

are thus well-defined modulo $\mathbf{Z}$, and this defines a map $h : W \to \mathbf{R}/\mathbf{Z} = \mathbf{S}^1$ which satisfies

$$h^\star dt = i_X\omega.$$

The map h has the very same properties as a periodic Hamiltonian: restricted to any $W - h^{-1}(t_0)$, this *is* a periodic Hamiltonian. In particular, Frankel's theorem (Theorem IV.2.3) applies in this (slightly more general) situation. □

Third step. Suppose that the action is *not* Hamiltonian. Then h is surjective. From Frankel's theorem, one derives, as in the proof of Theorem IV.3.1, that the number of connected components of the fibers of h is constant: it may only change by critical submanifolds of index 1 or $n-1\ldots$ which do not appear here, because all have even indices. If h is surjective, there is no local extremum. □

Fourth step. Let us now use the fact that our manifold has dimension 4. We enumerate the possibilities for the critical submanifolds, their dimensions and indices. The critical submanifolds are symplectic, hence even dimensional, so that they are either surfaces or isolated points. The index of any critical surface, being even, may only be 0 or 2, so the critical surfaces are extrema, but we have just proved that there were no extrema. We are left with isolated critical points, the indices of which cannot be 0 or 4... in short: the map h has, at most, isolated critical points of index 2.

Near such a point, the action may be linearized as usual:

$$u \cdot (x, y) = (u^p x, u^{-q} y)$$

where $u \in \mathbf{S}^1$, $(x, y) \in \mathbf{C}^2$, and p and q are nonnegative integers which we shall assume to be relatively prime (reducing to the case of an effective action).

Consider two regular values a and b of h between which there is only one critical value c. The fibers V_a and V_b are 3-dimensional submanifolds of W, each endowed with an action of $\mathbf{S}^1$ without fixed points, since the vector field X does not vanish on the regular levels. Hence, V_a and V_b are Seifert manifolds, exactly as in the case of a genuine Hamiltonian.

The trick to end the proof is the following: one shows that the Euler class of V_b is strictly larger than that of V_a, hence, if some critical points actually appear, we get a nondecreasing function... defined on the circle, which is absurd: after a while, one has to come back to V_a. □

To end the proof of Theorem VIII.1.1, we just have to prove the next lemma, which could be seen as an orbifold Euler class version of what was done in § VI.2.c (see more precisely Example VI.2.10). This can be considered as a wall crossing formula.

Lemma VIII.1.3. *Let z be an isolated critical point of a periodic Hamiltonian h on a symplectic manifold of dimension 4. Let a and b be regular values so that $h(z)$ is the only critical value between a and b and let V_a, V_b be the corresponding level. If the $\mathbf{S}^1$-action, linearized at z, has the form*

$$u \cdot (x, y) = (u^p x, u^{-q} y),$$

then

$$e(V_a) - e(V_b) = -\frac{1}{pq}.$$

Proof. We study the surgery which allows us to go from V_a to V_b. Notice first that there exists on our manifold an invariant metric and an invariant calibrated almost complex structure so that, in a neighborhood of z, the $\mathbf{S}^1$-action is, indeed, that given by the linearization.

Consider then a neighborhood of z of the form $D_x^2 \times D_y^2 \subset \mathbf{C}^2$ (product of two discs). Our level V_a is

$$V_a \cap (D_x^2 \times D_y^2) \sim D_x^2 \times \mathbf{S}_y^1.$$

It has an exceptional fiber of type $\mathbf{Z}/q$ fiber for $x = 0$ (we shall consider this fiber as exceptional even if $q = 1$). Similarly for V_b:

$$V_b \cap (D_x^2 \times D_y^2) \sim \mathbf{S}_x^1 \times D_y^2$$

where now the exceptional (at least if $p \geqslant 2$) fiber of type $\mathbf{Z}/p$ is given by $y = 0$. Hence, the surgery when crossing the critical value $h(z)$ replaces an exceptional orbit of type $\mathbf{Z}/q$ by an exceptional orbit of type $\mathbf{Z}/p$. Let now u and v be the smallest positive integers such that

$$pv - qu = 1.$$

Then by definition, (see § I.3.c), the Seifert invariants of the orbit $x = 0$ in V_a are the numbers (q, v), those of the orbit $y = 0$ in V_b are (p, u). The other possibly exceptional orbits are the same in V_b and V_a, and, calculating as in Exercise I.13:

$$e(V_a) - e(V_b) = -\sum \frac{\beta_i}{m_i} - \frac{v}{q} - \left(-\sum \frac{\beta_i}{m_i} - \frac{u}{p}\right) = -\frac{1}{pq}. \qquad \square$$

And this is the end of the proof of Theorem VIII.1.1. □

VIII.1.b. Symplectic reduction of the regular levels. Let now H be a periodic Hamiltonian on a manifold of dimension 4. We know that the regular levels of H are Seifert manifolds V_a and that enough is kept over from the smooth structure of V_a after forming the quotient to enable us to say that the quotient B_a is a symplectic orbifold (see § III.2.f).

The previous section gave us a hint of what was happening to V_a when going through a critical value of H. We shall now concentrate on the adventures of B_a during the same operation. We still use an invariant metric and an invariant calibrated almost complex structure.

What happens near an extremum reached on a surface B. Call L the normal bundle of B in W, that we consider as a complex line bundle over B. Near each point z of B, both the action and the symplectic form are conjugate to the linear ones

$$\mathsf{T}_z W = \mathsf{T}_z B \oplus L_z$$
$$u \cdot (x, y) = (x, u^m y)$$

where m equals ± 1 because of the effectiveness of the action (the sign depending on the kind of extremum we are dealing with). The regular levels V_a (for a close to the critical value) are the circle bundles of L, the action is principal, and the quotient can be identified with B. Hence, all the reduced levels close to the extremum are the same surface as the extremum itself. This is shown, schematically, in Figure 1.

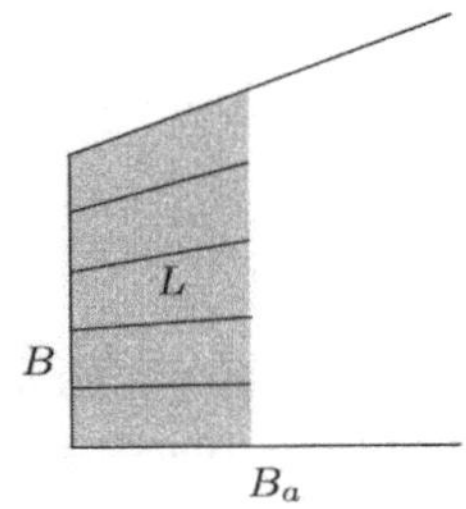

Figure 1

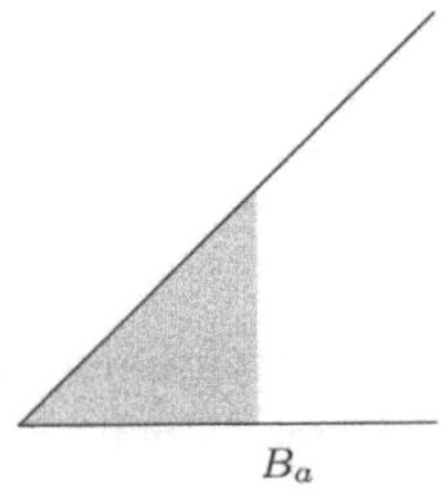

Figure 2

What happens near an extremum reached at an isolated point. With the very same method, we find ourselves in $\mathbf{C}^2$, with the action

$$u \cdot (x, y) = (u^m x, u^n y)$$

where m and n have the same sign and are relatively prime. The Hamiltonian may be written

$$H = \frac{1}{2}(m |x|^2 + n |y|^2).$$

The levels close to the critical one are ellipsoids (topologically, spheres $\mathbf{S}^3$). If m or $n \geqslant 2$, there are one or two exceptional orbits. Considering

$$\begin{array}{ccc} V_a & \longrightarrow & \mathbf{P}^1(\mathbf{C}) \\ (x, y) & \longmapsto & [x^n, y^m], \end{array}$$

one sees that the reduced surface of these regular levels is a sphere $\mathbf{S}^2$. Hence, all the reduced levels close to an extremum reached at an isolated point are 2-spheres. This is shown in Figure 2.

What happens when going through a critical value. If we are speaking of going through it, the critical level must correspond to some critical point of index 2. Another very beautiful remark due to McDuff is that the topology of the quotient surface does *not* change:

***Lemma VIII.1.4* (McDuff [103]).** *Let a and b be two regular values of H. There exists a smooth map*

$$\pi : H^{-1}([a, b]) \longrightarrow B_b$$

such that:

– *The restriction $\pi|_{V_b} : V_b \to B_b$ is the quotient map.*

– *For any regular value t of H, there exists a homeomorphism φ_t making the diagram*

$$\begin{array}{ccc} V_t & \hookrightarrow & H^{-1}([a,b]) \\ \downarrow & & \downarrow \pi \\ B_t & \xrightarrow[\varphi_t]{} & B_b \end{array}$$

commute.

Proof. The lemma is obviously true when there is no critical value between a and b, because we merely need to push using the gradient flow of H to get a map

$$\pi : H^{-1}([a,b]) \xrightarrow{\text{push}} V_b \xrightarrow{\text{proj}} B_b$$

with the desired properties.

Then, of course, it is enough to prove the lemma in the case where there is only one critical value between a and b. For this, we shall try to show, as above, that the topology of V_a does not change (!) and to understand where we are wrong.

Let us thus try to prove that it is possible to retract $H^{-1}([a,b])$ on V_b by pushing along the gradient flow. If we push by the gradient up to the level b, any

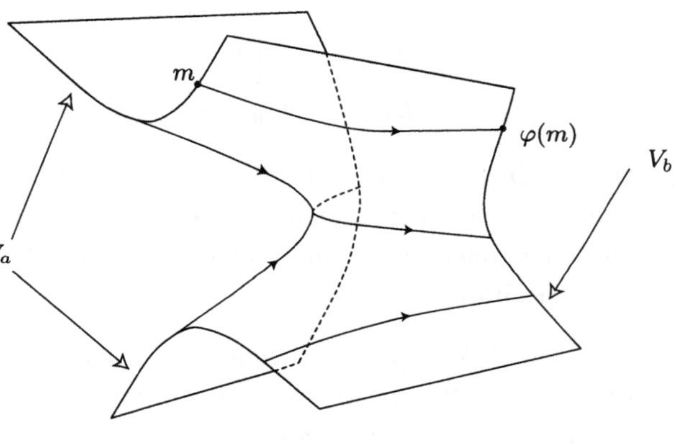

Figure 3

point of $H^{-1}([a,b])$ has an image $\varphi(m)$ which is well-defined in $V_b\ldots$ except those which fell into the hole, more precisely those of the stable manifold of a critical point at the level under consideration.

Consider as usual a neighborhood of the critical point, identified with a neighborhood of 0 in $\mathbf{C}^2$. If the action is written $(u^p x, u^{-q} y)$, the points of the y-axis are those which are forced to stay on the critical level. Associate with them the "points of the x-axis which are on V_b". Of course, this does not define a map into the level $V_b\ldots$ but notice that all these points lie in the same $\mathbf{S}^1$-orbit in V_b, so that we have a perfectly well-defined map into the quotient B_b, which has the required properties. □

Remark VIII.1.5. We have just rephrased what we have already understood in § VI.2.c, in the simple case of a manifold of dimension 4 (except that we have not assumed the action to be semi-free). Notice that the topology of the reduced levels does not change here... because we know it should change by a blowing up, but this is trivial in dimension 2.

VIII.1.c. First applications. If the maximum of H is reached on a surface $B_{\max}$, nothing prevents us, in the previous proof, from following the gradient until the end. If both the minimum and the maximum of H are reached on a surface, we get in the very same way a smooth map

$$\pi : W \longrightarrow B_{\max}$$

and homeomorphisms

$$\varphi_t : B_t \longrightarrow B_{\max}$$

(as long as t is not the image of a critical point of index 2) making the diagram

$$\begin{array}{ccc} V_t & \hookrightarrow & W \\ \downarrow & & \downarrow \pi \\ B_t & \xrightarrow[\varphi_t]{} & B_{\max} \end{array}$$

commute. In particular, $B_{\min}$ and $B_{\max}$ are two copies of the same surface.

If one of the extrema is reached at an isolated fixed point, then the other one is reached either on a sphere, or at an isolated critical point. Let us formulate these remarks as a proposition.

Proposition VIII.1.6. *Let H be a periodic Hamiltonian on a compact symplectic manifold of dimension 4. If the maximum and minimum of H are reached on surfaces, the two surfaces are diffeomorphic to each other. If one of the extrema is reached at an isolated point, the other one is reached, either at an isolated point or on a sphere.* □

Assume now that the $\mathbf{S}^1$-action is *semi-free* (that is, without any exceptional orbit) and that both the maximum and the minimum are reached at an isolated point. Near the minimum, the action can be written as

$$u \cdot (x, y) = (ux, uy)$$

(because it is semi-free). In particular, the symplectic reduction of a given nearby level is just the Hopf map

$$\mathbf{S}^3 \longrightarrow \mathbf{P}^1(\mathbf{C})$$

and so has Euler class -1 (see Example I.3.3). There must be also a critical point of index 2, otherwise, with a Morse function with only two critical points, W would be a sphere[(3)] $\mathbf{S}^4$, which would prevent it from being symplectic. Still because of the semi-freedom of the action, at the first critical point of index 2, the action must linearize as

$$u \cdot (x, y) = (ux, \overline{u}y).$$

According to Lemma VIII.1.3 (or to its more elaborate version, in this semi-free case, Theorem VI.2.9), the Euler class is zero. But now, there must be still another critical point, as otherwise we would have, near the maximum, an action of $\mathbf{S}^1$ on the sphere $\mathbf{S}^3$ with Euler class 0. Hence we have proved:

Proposition VIII.1.7. *Let W be a compact symplectic manifold of dimension 4 endowed with a semi-free Hamiltonian circle action. The action has at least four fixed points. If moreover, the fixed points are isolated, there are exactly four of them.* □

Notice that the last assertion had already been proved in § VI.3.d (see Figure 8 in Chapter VI).

VIII.2. Periodic Hamiltonians on 4-dimensional manifolds

VIII.2.a. What happens when there are only two critical values? When there are only two critical values, the minimum and the maximum, they can be reached

- either on two copies of the same surface B,
- or at an isolated point and on a sphere $\mathbf{S}^2$.

Notice that it is impossible that both are reached at an isolated point, since the 4-sphere $\mathbf{S}^4$ is still not symplectic.

Recall that our symplectic manifold is endowed with an invariant metric and a calibrated almost complex structure.

Consider first the case where one of the critical values is reached at an isolated point and the other on a sphere. Then all the regular levels are $\mathbf{S}^3$'s. The action near the isolated critical point can be written

$$u \cdot (x, y) = (u^m x, u^n y).$$

On the other side, the levels are principal bundles over $\mathbf{S}^2$, thus we must have $m = n = \pm 1$, and the principal bundles over $\mathbf{S}^2$ are the Hopf bundle.

According to the *ad hoc* version of Darboux's Theorem (equivariant form of Corollary II.1.12, see Remark II.1.13), we know that a tubular neighborhood of

(3) See, for instance [**108**] for this classical result due to Reeb.

this sphere is equivariantly and symplectically isomorphic to a neighborhood of

$$\mathbf{P}^1(\mathbf{C}) \subset \mathbf{P}^2(\mathbf{C})$$

embedded as, say, the line $x = 0$, with the symplectic form which gives it the same volume as our sphere has, say, ω_a, and the $\mathbf{S}^1$-action

$$u \cdot [x, y, z] = [x, uy, uz].$$

At the other end, a neighborhood of the point at which the minimum is reached is isomorphic with a neighborhood of the other fixed point $[1, 0, 0]$ of this $\mathbf{S}^1$-action on $\mathbf{P}^1(\mathbf{C})$. Following the gradient (minus the gradient would be more accurate here) gives an equivariant diffeomorphism preserving the Hamiltonians, from our manifold to $\mathbf{P}^2(\mathbf{C})$. Our manifold W is thus $\mathbf{P}^2(\mathbf{C})$ and the action that of $\mathbf{S}^1$ by $u \cdot (x, y) = [x, uy, uz]$. This is shown in Figure 4. The same proof, in any dimension, would apply to give a result due to Delzant (see Exercise VIII.2):

***Proposition VIII.2.1* (Delzant [39]).** *A compact connected symplectic manifold endowed with a periodic Hamiltonian with only two critical values, one of which is reached at an isolated point, is a complex projective space.* □

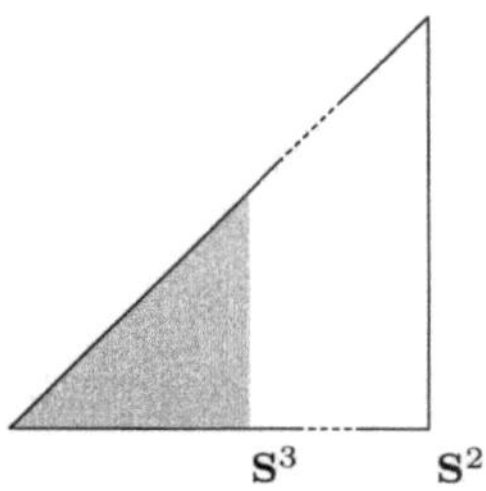

Figure 4

Remark VIII.2.2. It is true, but a little harder to prove, that there is indeed an isomorphism from W to $\mathbf{P}^2(\mathbf{C})$ (resp. $\mathbf{P}^n(\mathbf{C})$), in the sense that the symplectic structures also are the same. See [**39**].

The other possibility is that the two critical values are reached on two copies of the same surface B. Each regular level is the circle bundle of the normal bundle L of $B_{\min}$ in W. The manifold W is thus a compactification of the vector bundle L, obtained by adding to it the critical surface $B_{\max}$. As above, let us construct a model space, namely a symplectic manifold with a Hamiltonian $\mathbf{S}^1$-action and the same structure of fixed points.

Roughly speaking, if we follow the gradient, we will add a point to each fiber of L.

In order to add a point at infinity to a complex line E, the easiest thing we can do is to projectivize the plane $E \oplus \mathbf{C}$. Here we do this fiberwise, that is, we consider the projectivized bundle

$$\mathbf{P}(L \oplus \mathbf{1}) = \{(z, \ell) \mid z \in B, \ell \subset L_z \oplus \mathbf{C}\}$$

where $\mathbf{1}$ denotes the trivial complex line bundle over B, and, for any $z \in B$, L_z is the fiber of L at z. The map

$$(z, \ell) \longmapsto z$$

defines a projection onto B, the fiber of which is the projective line $\mathbf{P}^1(\mathbf{C})$. Moreover, there are two inclusions (sections of the bundle) of B in $\mathbf{P}(L \oplus \mathbf{1})$:

- the "zero section" $z \mapsto (z, 0 \oplus \mathbf{C})$,
- and the "section at infinity" $z \mapsto (z, L_z \oplus 0)$.

***Example VIII.2.3* (Hirzebruch surfaces).** In the case where the basis surface B is a sphere, $B = \mathbf{P}^1(\mathbf{C})$, the manifold $\mathbf{P}(\mathcal{O}(k) \oplus \mathbf{1})$ is a Hirzebruch surface (these 4-manifolds are actually complex surfaces). We have already met these examples in Exercise IV.4, where they appeared endowed with $\mathbf{T}^2$-actions, and in § IV.5.a, where they were endowed with SU(2)-actions. Notice that the image of the momentum mapping, a trapezium as shown in Figure 11 in Chapter IV, which we reproduce in Figure 6, rotated by a right angle, resembles very much the situation in Figure 5 here. This is not by pure chance (of course...), the projection onto

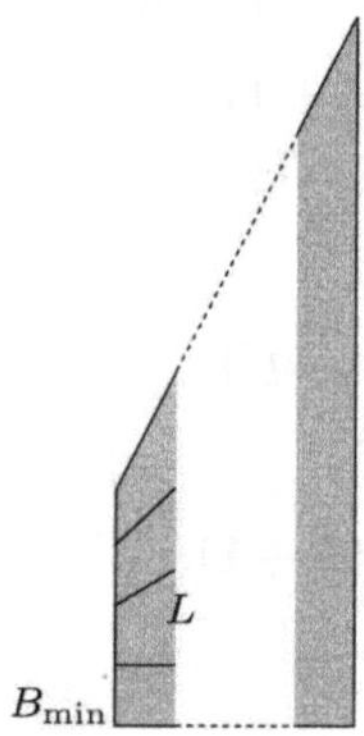

Figure 5

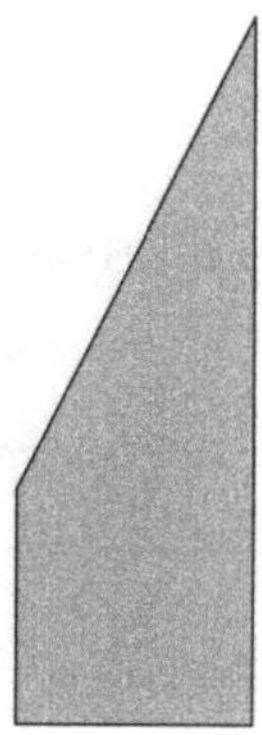

Figure 6

the horizontal in Figure 6 is indeed the Hamiltonian for the action of a subcircle of $\mathbf{T}^2$ that has a unique maximum and a unique minimum, both reached along a copy of $\mathbf{P}^1(\mathbf{C})$.

Let us come back to the general case and write (a, b) when a is a vector in L_z and b a complex number, so that $(a, b) \in L_z \oplus \mathbf{C}$. Then it is natural to write $[a, b]$—using the same notation as for honest homogeneous coordinates—for the complex line generated by (a, b), an element of $\mathbf{P}(L_z \oplus \mathbf{1})$. The zero section corresponds to $a = 0$ and the section at infinity to $b = 0$.

There is also a natural $\mathbf{S}^1$-action on $\mathbf{P}(L \oplus \mathbf{1})$, that can be written

$$u \cdot (z, [a, b]) = (z, [ua, b])$$

(in our "homogeneous" coordinates). Both the zero section and the section at infinity are submanifolds consisting of fixed points, so that the situation is close to the one we have on the symplectic manifold we are trying to model. However, we must still construct a *symplectic form* on the manifold we have just constructed. There are a lot of different possible ways to do so. We showed one in § IV.5.a when the basis B is a sphere (case of Hirzebruch surfaces). The one I give here is reminiscent of the construction of the equivariant symplectic form we have met in Proposition VI.2.1 and was suggested to me by Iglesias.

Let us endow L with a Hermitian metric, so that we can consider its unit circle bundle

$$\pi : S(L) \longrightarrow B,$$

a principal $\mathbf{S}^1$-bundle, the circle acting by rotations, or complex multiplication, in the fibers. The construction is based on the following remark.

Lemma VIII.2.4. *Let $S(L) \to B$ be the unit circle bundle (associated with any Hermitian metric) of the complex line bundle L. The mapping*

$$\begin{aligned} \widetilde{\varphi} : S(L) \times \mathbf{P}^1(\mathbf{C}) &\longrightarrow \mathbf{P}(L \oplus \mathbf{1}) \\ ((z, v), [x, y]) &\longmapsto (z, [xv, y]) \end{aligned}$$

is invariant by the $\mathbf{S}^1$-action

$$u \cdot ((z, v), [x, y]) = ((z, uv), [x, uy])$$

and induces a diffeomorphism

$$\varphi : S(L) \times_{\mathbf{S}^1} \mathbf{P}^1(\mathbf{C}) \longrightarrow \mathbf{P}(L \oplus \mathbf{1}).$$

Proof. Obviously,

$$\begin{aligned} \widetilde{\varphi}(u \cdot ((z, v), [x, y]) &= \widetilde{\varphi}((z, uv), [x, uy]) \\ &= (z, [xv, uy]) \\ &= (z, [xv, y]) \end{aligned}$$

so that $\widetilde{\varphi}$ descends to a map φ which is differentiable and one-to-one. □

Let us proceed to the construction of the symplectic form. To begin with, we construct a closed 2-form on $S(L) \times \mathbf{P}^1(\mathbf{C})$. We first consider the principal bundle

$$\pi : S(L) \longrightarrow B.$$

Notice that, if it is trivial, this means that L and $\mathbf{P}(L \oplus \mathbf{1})$ are trivial bundles,

$$\mathbf{P}(L \oplus \mathbf{1}) \cong B \times \mathbf{P}^1(\mathbf{C})$$

so that it is easy to construct a symplectic form on $\mathbf{P}(L \oplus \mathbf{1})$, the product of a symplectic form on B and of one on $\mathbf{P}^1(\mathbf{C})$.

Otherwise, let $e \in \mathbf{Z}$ be its Euler class. Fix a volume form η on the oriented surface B. There is a unique real number λ such that

$$e = \lambda \int_B \eta.$$

Notice that λ is zero exactly when the bundles L and $S(L)$ are trivializable, so that we are assuming that $\lambda \neq 0$.

We have seen in Chapter VI (more precisely in Exercise VI.5) that there exists an invariant 1-form α on $S(L)$ such that

– if X is the fundamental vector field of the $\mathbf{S}^1$-action, $i_X\alpha \equiv 1$,

– $d\alpha = \pi^\star \dfrac{\lambda\eta}{2\pi}$.

Let us consider now $\mathbf{P}^1(\mathbf{C})$ with the $\mathbf{S}^1$-action $u \cdot [x, y] = [x, uy]$ and the usual symplectic form ω_0. Let us choose a momentum mapping (a Hamiltonian) H for this action and put:

$$\widetilde{\omega} = d(H\alpha) + \omega_0.$$

This is a closed 2-form on $S(L) \times \mathbf{P}^1(\mathbf{C})$. Let us investigate its kernel. Consider a point $((z, v), [x, y]) \in S(L) \times \mathbf{P}^1(\mathbf{C})$, that we denote (ζ, ϖ) for simplicity, and two tangent vectors

$$(Z, P), (Z', P') \in \mathsf{T}_\zeta S(L) \times \mathsf{T}_\varpi \mathbf{P}^1(\mathbf{C}).$$

Then (Z, P) is in the kernel of $\widetilde{\omega}$ if and only if

$$\forall (Z', P'), \quad (d(H\alpha) + \omega_0)((Z, P), (Z', P')) = 0.$$

But we have

$$\begin{aligned}(d(H\alpha) + \omega_0)((Z, P), (Z', P')) &= dH(P)\alpha(Z') - dH(P')\alpha(Z) \\ &\quad + H(\varpi)d\alpha(Z, Z') + \omega_0(P, P') \\ &= (-\alpha(Z)dH + i_P\omega_0)(P') \\ &\quad + (dH(P)\alpha + H(\varpi)i_Z d\alpha)(Z').\end{aligned}$$

Hence (Z, P) is in $\operatorname{Ker}\widetilde{\omega}$ if and only if

$$\begin{cases} -\alpha(Z)dH + i_P\omega_0 = 0 \\ dH(P)\alpha + H(\varpi)i_Z d\alpha = 0. \end{cases}$$

Call X the fundamental vector field of the $\mathbf{S}^1$-action (on anything). The first equation gives

$$i_P\omega_0 = i_{\alpha(Z)X}\omega_0,$$

which is equivalent to $P = \alpha(Z)X$. As for the second equation, it gives, as $dH(X) = 0$,

$$H(\varpi) i_Z d\alpha = 0.$$

If we were careful enough to have chosen a function H which does not vanish on $\mathbf{P}^1(\mathbf{C})$ (this is of course possible by adding a constant to it), the second equation just gives

$$i_Z d\alpha = 0.$$

From the definition of α (starting from a *volume* form, in particular a nondegenerate form η), we know that the kernel of $d\alpha$ is generated by the vector field X hence $Z = rX$ and $P = \alpha(Z)X = rX$ for some real number r.

The vector field (X, X) is the fundamental vector field of the diagonal $\mathbf{S}^1$-action on $S(L) \times \mathbf{P}^1(\mathbf{C})$, so that we are led to consider this action. Using our convention, (X, X) should be called X. We have

$$i_X \widetilde{\omega} = 0 \text{ and } \mathcal{L}_X \widetilde{\omega} = 0.$$

This is proved exactly as the (formally) identical assertion in Proposition VI.2.1.

One deduces that $\widetilde{\omega}$ descends to the quotient as a *symplectic* form ω on $S(L) \times_{\mathbf{S}^1} \mathbf{P}^1(\mathbf{C})$, and thus, according to Lemma VIII.2.4, on $\mathbf{P}(L \oplus \mathbf{1})$. Moreover, the symplectic form ω so defined is invariant by the $\mathbf{S}^1$-action we are interested in, namely

$$u \cdot (z, [a, b]) = (z, [ua, b]).$$

Indeed, this action comes from

$$u \cdot ((z, v), [x, y]) = (z, uv, [x, y]) \text{ on } S(L) \times \mathbf{P}^1(\mathbf{C})$$

(rotations in the fibers), for which $\widetilde{\omega}$ itself is invariant.

Now, we have our model space. The consideration of tubular neighborhoods of $B_{\min}$ and $B_{\max}$ in W leads to the same conclusion as before. To summarize:

Proposition VIII.2.5. *Let W be a compact connected symplectic manifold of dimension 4 endowed with a periodic Hamiltonian H with only two critical values. Then W is equivariantly*[(4)] *diffeomorphic, by a diffeomorphism which preserves the Hamiltonians,*

– *either to $\mathbf{P}^2(\mathbf{C})$ with some multiple of the standard symplectic form and*

$$H([x, y, z] = \frac{r}{2} \frac{|x|^2}{|x|^2 + |y|^2 + |z|^2} + h_0,$$

(4) and symplectically, but we have not proved it...

– *or to the total space of a projectivized bundle* $\mathbf{P}(L \oplus \mathbf{1})$ *over an oriented closed surface* B*, with a symplectic form as defined above and Hamiltonian*

$$H(z, [a,b]) = \frac{r}{2} \frac{\|a\|^2}{\|a\|^2 + |b|^2} + h_0$$

for some real numbers r *and* h_0. □

Remark VIII.2.6. The manifolds $\mathbf{P}(L \oplus \mathbf{1})$ are called *symplectic ruled 4-manifolds*, by analogy with the ruled surfaces of algebraic geometry, since they are unions of (complex projective) lines. Symplectic ruled 4-manifolds are the subject of beautiful work by McDuff [**104**].

VIII.2.b. Critical points of index 2. To begin with, let us prove that there are other 4-manifolds endowed with Hamiltonian circle actions than the ones listed in Proposition VIII.2.5: it is easy to create examples with many critical points of index 2, simply by blowing up fixed points.

Blowing up fixed points. Let us consider now a Hamiltonian circle action (on a manifold of dimension 4) and a fixed point of the action. Using the technique explained in § VI.2.c, let us blow up this fixed point. Near this fixed point, the action can be written:

$$u \cdot (x, y) = (u^p x, u^{-q} y)$$

for some integers p, q. Notice that, as we are interested in effective actions, we must assume that p and q are relatively prime. I do not make any assumptions on the signs of p and q (that is, on the index of the critical points) yet—of course, for a genuine point of index 2, both will be positive. The $\mathbf{S}^1$-action extends to the blown up manifold by

$$u \cdot ((x, y), [a, b]) = ((u^p x, u^{-q} y), [u^p a, u^{-q} b]).$$

Consider now the fixed points on the blown up manifold.

– Outside the exceptional divisor (namely the points such that $x = y = 0$), the situation has not changed.
– In the case where the blown up fixed point was a point of a fixed surface, assuming that $p = 1$ and $q = 0$ (to preserve the effectiveness of the Hamiltonian action), the *strict transform* of the fixed surface (that is, of the submanifold $x = 0$) is a fixed surface as well (this is the submanifold consisting of the points $((0, y), [0, 1])$). We have added a new fixed point, the point $((0, 0), [1, 0])$, near which, in local coordinates (x, v) (such that $v = b/a$ and $y = vx$), the action is written

$$u \cdot (x, v) = (ux, \overline{u}v).$$

We have thus created a new fixed point, this is a critical point of *index* 2.

– If $p > 0$ and $q > 0$, the blown up fixed point was an isolated critical point of index 2 of our Hamiltonian. Let us call this a fixed point *of type* (p, q). To blow it up replaces it with two fixed points: the point $((0,0),[0,1])$, a fixed point of type $(p+q, q)$, and the point $((0,0),[1,0])$, that has type $(p, p+q)$. These two critical points are connected by a sphere, the exceptional divisor of our blowing up. Notice that, in the standard metric of $\mathbf{C}^2$ we are using near the critical point, this is a *gradient sphere*, namely the closure of the stable or unstable manifold of some critical point.
– If p and q have opposite signs, then the blown up fixed point was an isolated extremum of the Hamiltonian. Changing both the signs of p and q if necessary, the action may be written:

$$u \cdot (x, y) = (u^m x, u^n y)$$

where m and n are positive and relatively prime (with this choice of signs, the fixed point is a minimum). The $\mathbf{S}^1$-action on the blown up manifold is:

$$u \cdot ((x, y), [a, b]) = ((u^m x, u^n y), [u^m a, u^n b]),$$

so that:
• When $m = n = 1$, the whole exceptional divisor consists of fixed points; we have replaced an isolated fixed point with a fixed $\mathbf{P}^1(\mathbf{C})$.
• Otherwise, one may assume that $m > n > 0$. Then the point $((0,0),[0,1])$ is a minimum, near which the Hamiltonian action is written

$$u \cdot (v, y) = (u^{m-n} v, u^n y),$$

and the point $((0,0),[1,0])$ is an isolated critical point of index 2 and type $(n, m-n)$.

Figure 7 shows the two fixed points, connected by the gradient sphere, in the blown up manifold.

Gradient spheres and exceptional divisors. The blowing up of a critical point of index 2 gives us the following situation:

– Before blowing up (left of Figure 7) the gradient manifolds are the y-axis, stable manifold of the fixed point, the points of which have stabilizer $\mathbf{Z}/q$, and the x-axis, unstable manifold of the fixed point, the points of which have stabilizer $\mathbf{Z}/p$.
– After having blown up (right of Figure 7) we have two critical points, the u-axis (on the x side) and the v-axis (on the y side) are the two discs which form the exceptional divisor. The latter is actually the gradient manifold connecting the two fixed points.

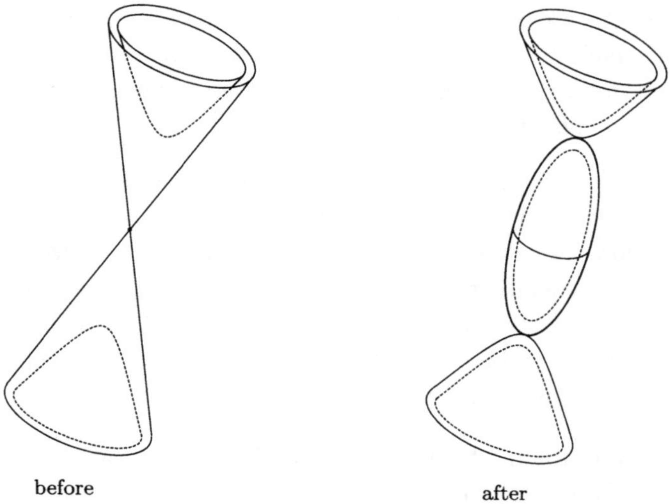

Figure 7

Notice that, the invariant metric (and, with it, the almost complex structure) having been chosen, we are in a situation similar to the Kähler situation investigated to prove Theorem IV.4.23. That is, the $\mathbf{S}^1$-action extends to a $\mathbf{C}^\star$-action. The gradient vector field is JX_H, so that the $\mathbf{C}^\star$-orbits are either fixed points or symplectic surfaces in our manifold W.

Of course, a nontrivial $\mathbf{C}^\star$-orbit is a cylinder, more precisely, it is symplectically diffeomorphic with $\mathbf{S}^1 \times]a, b[$, for some interval $]a, b[$ (and symplectic form $d\theta \wedge dx$), the momentum mapping being

$$(u, x) \longmapsto x.$$

Moreover, since the nonconstant trajectories of the gradient have limit points which are critical points of H, the closures of the $\mathbf{C}^\star$-orbits must be (at least topologically) spheres. These are the gradient spheres.

Example VIII.2.7. The exceptional divisor in the blowing up process described above is a gradient sphere.

Being defined by the gradient of H, in particular by the Riemannian metric we have chosen, the gradient spheres depend, *a priori*, on this metric. Notice however, that some of them can be defined in terms of the $\mathbf{S}^1$-action only, so that they do not, *a posteriori*, depend on any choice. These are the spheres with nontrivial cyclic stabilizer, as asserted by the next proposition.

Proposition VIII.2.8. *Let W be a compact symplectic manifold of dimension 4 endowed with a Hamiltonian $\mathbf{S}^1$-action. For any integer $k \geqslant 2$, the set $W_{\mathbf{Z}/k}$ of* *points with stabilizer* $\mathbf{Z}/k$ *is a manifold, the closure of any connected component of which is a symplectic* 2*-sphere on which* $\mathbf{S}^1$ *acts with two fixed points.*

Moreover, for any Riemannian metric compatible with the $\mathbf{S}^1$*-action, all the* $\mathbf{Z}/k$*-spheres are gradient spheres.*

Proof. The fact that $W_{\mathbf{Z}/k}$ is a manifold is a special case of Proposition I.2.2. Let us chose any invariant metric, together with a calibrated complex structure J, so that the $\mathbf{S}^1$-action extends to a $\mathbf{C}^\star$-action. This $\mathbf{C}^\star$-action preserves $W_{\mathbf{Z}/k}$, so that its connected components are $\mathbf{C}^\star$-invariant. The $\mathbf{C}^\star$-orbit of any point of $W_{\mathbf{Z}/k}$ is thus a connected component of $W_{\mathbf{Z}/k}$. We have seen that the closure of such a component is a 2-sphere, on which the $\mathbf{S}^1$-action has two fixed points... so that this closure is also a gradient sphere. □

Symplectic blow down. Assume now that Σ is an invariant symplectic sphere the Euler class of the normal bundle of which is -1. Assume its symplectic volume (area) is a positive number πr^2. Then, an equivariant neighborhood of Σ in W is isomorphic with an equivariant tubular neighborhood of the exceptional divisor in $\widetilde{\mathbf{C}}^2$. Thus we can remove this tubular neighborhood and replace it by a ball of radius r, getting a new symplectic manifold, with an $\mathbf{S}^1$-action. This is the equivariant symplectic blow down.

VIII.2.c. The list of examples. We have now introduced everything we need to understand the topology of the manifolds under consideration: they will be organized around the gradient spheres.

In § VIII.2.a, we have constructed examples of compact symplectic manifolds endowed with periodic Hamiltonians: all the $\mathbf{P}(L \oplus \mathbf{1})$ (with two surfaces as the set of fixed points), and $\mathbf{P}^2(\mathbf{C})$ with an isolated fixed point and a fixed sphere.

On $\mathbf{P}^2(\mathbf{C})$, we know that there are plenty of other periodic Hamiltonians, for example the Morse function

$$H([x,y,z]) = \frac{1}{2}\frac{m\,|x|^2 + n\,|y|^2}{|x|^2 + |y|^2 + |z|^2}.$$

This is a Hamiltonian for the $\mathbf{S}^1$-action

$$u \cdot [x,y,z] = [u^m x, u^n y, z],$$

which is effective as soon as m and n are relatively prime, and which has three fixed points when m and n are distinct and nonzero.

Similarly, when the base space B of the bundle L is a sphere, we may combine an $\mathbf{S}^1$-action on B and the action in the fibers to get more complicated actions on $\mathbf{P}(L \oplus \mathbf{1})$. This is the case of the Hirzebruch surfaces, considered here as

$$W_k = \left\{([a,b],[x,y,z]) \in \mathbf{P}^1(\mathbf{C}) \times \mathbf{P}^2(\mathbf{C}) \mid a^k y = b^k x\right\}$$

with the induced symplectic form and the $\mathbf{S}^1$-action

$$u \cdot ([a, b], [x, y, z]) = ([u^m a, b][u^{mk} x, y, u^n z]).$$

Notice that the actions we have just described are restrictions, to some subgroup $\mathbf{S}^1 \subset \mathbf{T}^2$ of $\mathbf{T}^2$-actions on $\mathbf{P}^2(\mathbf{C})$ or W_k (see Exercise IV.4).

Now that we have got all these *basic examples*, we are able to state the first result we want to prove.

Theorem VIII.2.9. *Let W be a compact symplectic manifold of dimension 4, endowed with a Hamiltonian $\mathbf{S}^1$-action. Then W can be obtained by a sequence of blow ups from*

- *either a manifold of the form $\mathbf{P}(L \oplus \mathbf{1})$, with two fixed surfaces and no other fixed point*
- *or a projective plane or a Hirzebruch surface with the restriction to some subgroup $\mathbf{S}^1$ of a Hamiltonian $\mathbf{T}^2$-action.*

If this statement gives a list of all the 4-manifolds that can be endowed with both a symplectic form and a Hamiltonian $\mathbf{S}^1$-action, this is not really a classification theorem, as

- the list might be (and actually is) redundant,
- there is no uniqueness assertion, especially on the symplectic structure, but not even on the momentum mapping.

It must thus be completed by a uniqueness result (due to Karshon), which I will explain later on (this is Theorem VIII.2.14). However, it gives an extensive list of all the compact symplectic 4-manifolds that can be endowed with a periodic Hamiltonian.

The proof given here is inspired by my original proof [**13, 14**] together with that of Karshon [**83**], and, mainly, by the observation made by McDuff that we need only to blow down spheres.

The proof of Theorem VIII.2.9. We start with two lemmas, that describe the relations between the various gradient spheres.

The submanifolds on which the extrema are reached and the gradient manifolds as well, are (almost complex) surfaces with normal bundles which describe their neighborhoods in W.

Lemma VIII.2.10. *Assume $L \to \mathbf{S}^2$ is a complex line bundle of Euler class e, endowed with an $\mathbf{S}^1$-action such that*

- *the principal stabilizer of the zero section is $\mathbf{Z}/k$,*
- *the $\mathbf{S}^1$-action on the zero section has two fixed points at which the linear action on the fibers of L has weights m and n.*

Then the four integers k, n and m are related by

$$n = m - ke.$$

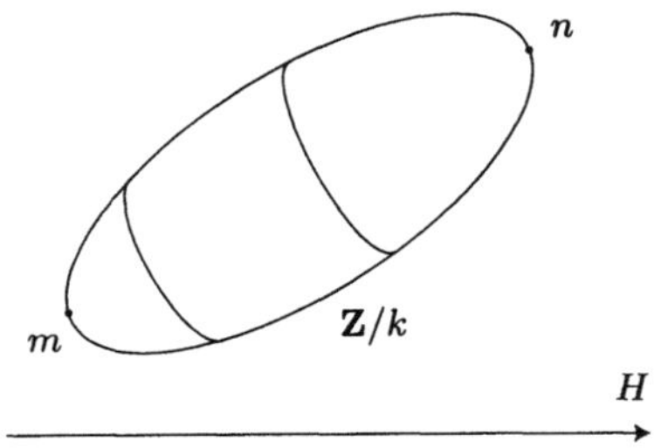

Figure 8

Proof. We decompose, as usual, the 2-sphere in two hemispheres,

– $\mathbf{S}^2_- \times \mathbf{C}$, on which $\mathbf{S}^1$ acts by

$$u \cdot (v, z) = (u^k v, u^m z)$$

– and $\mathbf{S}^2_+ \times \mathbf{C}$, with

$$u \cdot (w, z) = (\overline{u}^k w, u^n z), \qquad w = \frac{1}{v}.$$

The line bundle is obtained by gluing $\mathbf{S}^2_- \times \mathbf{C}$ and $\mathbf{S}^2_+ \times \mathbf{C}$ along their common boundary ($|v| = 1$) by

$$\varphi(v, z) = (\overline{w}, v^{-e} z).$$

In order that the $\mathbf{S}^1$-action extend, we must have

$$\varphi(u \cdot (v, z)) = u \cdot \varphi(v, z),$$

namely

$$(\overline{u}^k w, u^{-ke+m} v^{-e} z) = (\overline{u}^k \overline{v}, u^n v^{-e} z),$$

that is, $n = -ke + m$. □

Let us consider now a chain of gradient spheres.

Lemma VIII.2.11. *Let $C = \Sigma_1 \cup \cdots \cup \Sigma_r$ be a chain of gradient spheres, such that the south pole p_{i-1} (minimum of H) of Σ_i is the north pole (maximum of H) of Σ_{i-1}. Let k_i denote the order of the principal stabilizer on Σ_i. Then k_i and k_{i+1} are relatively prime and $-\dfrac{k_{i-1} + k_{i+1}}{k_i}$ is the Euler class of the normal bundle to Σ_i (in particular, this Euler class is a negative integer).*

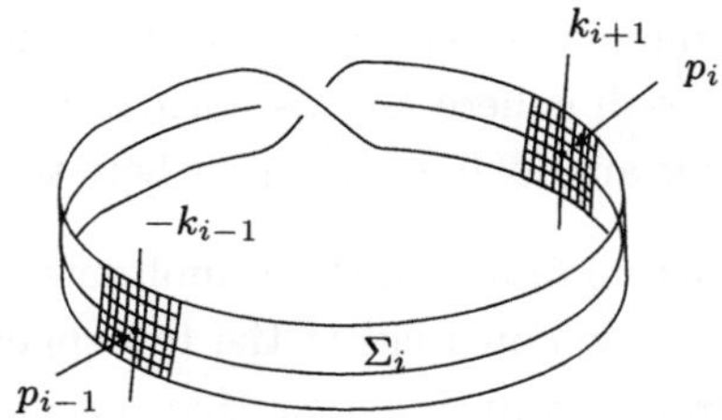

Figure 9

Proof. The first assertion is a consequence of the effectiveness of the action. The second is a consequence of Lemma VIII.2.10. Look at Figure 9: the cyclic group $\mathbf{Z}/k_i$ is the principal stabilizer of Σ_i but $-k_{i-1}$ is the weight of the action on the fiber at its south pole p_{i-1} while k_{i+1} is that of the action on the fiber at the north pole p_i. Thus, Lemma VIII.2.10 gives, for the Euler class e_i,

$$e_i = \frac{m_i - n_i}{k_i} = \frac{-k_{i-1} - k_{i+1}}{k_i}. \qquad \square$$

Let us now proceed to the proof of Theorem VIII.2.9 when both the maximum and the minimum are reached on a surface. We look at a chain of gradient spheres. It is enough to prove that such a chain contains a sphere with normal Euler class (or self-intersection) -1. Then we can blow it down and prove the theorem by induction on the number of critical points of index 2.

Denote by $k_1, \ldots, k_r$ the orders of the stabilizers of the spheres in our chain. For instance, k_1 and k_r are the weights of the $\mathbf{S}^1$-action on the neighborhood of the fixed surfaces, so that, by effectiveness of this action, $k_1 = k_r = 1$.

If we assume that all the weights k_i's are equal to 1, then, according to Lemma VIII.2.11, the last sphere has normal Euler class

$$\frac{-(1+0)}{1} = -1$$

and we are done. Assume thus that all the k_i's are not equal to 1. Let $k_s = \max\{k_i\}$. We have

$$k_s \geqslant k_{s-1}, \quad k_s \geqslant k_{s+1}, \text{ and } k_s > 1 \text{ for } s \neq 1, r.$$

As two consecutive k_i's must be relatively prime, we have actually strict inequalities

$$k_s > k_{s-1} \text{ and } k_s > k_{s+1}$$

so that

$$\frac{k_{s-1} + k_{s+1}}{k_s} < \frac{2k_s}{k_s} = 2.$$

But we know that this number should be a positive integer, so that it must be equal to 1. But then the s-th sphere Σ_s has normal Euler class -1, according to Lemma VIII.2.11. Thus we are done in the case where there are two fixed surfaces.

Assume now that the minimum of the Hamiltonian is reached at an isolated point. As above, we can blow down either the top sphere of a chain of gradient spheres or, if some isotropy group is larger than its neighbors, the corresponding sphere. Hence we can assume that all the chains of gradients spheres in our manifold consist of spheres with increasing isotropy group orders.

If there are indeed critical points of index 2 in our manifold, the $\mathbf{S}^1$-action near the minimum must have the form

$$u \cdot (z_1, z_2) = (u^{m_1} z_1, u^{m_2} z_2)$$

for some positive integers m_1 and m_2 that are not both equal to 1. Assume that $m_2 = 1$. This means that, either there is only one gradient sphere (with isotropy group $\mathbf{Z}/m_1$ starting from the minimum, or that there is another one, but without any critical point of index 2 on it. Thus, we have at most two chains of gradient spheres in our manifold.

Proposition VIII.2.12. *Let W be a compact symplectic manifold of dimension 4 endowed with an effective Hamiltonian $\mathbf{S}^1$-action such that*

- *an extremum of the Hamiltonian is an isolated fixed point,*
- *for some invariant metric, there are no more than two nontrivial chains of gradient spheres.*

Then, the $\mathbf{S}^1$-action extends to an effective Hamiltonian $\mathbf{T}^2$-action.

Assuming the proposition true for the moment, we deduce that our original manifold was obtained by a sequence of blowing ups starting from a toric surface. According to Theorem VII.4.1, the latter is itself obtained from $\mathbf{P}^2(\mathbf{C})$ or from a Hirzebruch surface by a sequence of blowing ups. So that the theorem is proved. □

Proof of the proposition. From the numerical data along the chains of gradient spheres, we construct a fan. Denote as above $k_1, \ldots, k_r$, $k'_1, \ldots, k'_s$ the orders of the isotropy groups along the two chains of gradient spheres. Recall that, having maybe blown down a few spheres, we can assume that the two sequences are strictly increasing, two consecutive weights are relatively prime and so are k_1 and k'_1, and k_r, k'_s as well.

With this data, we construct a fan. We start at the minimum, that we assume to be reached at an isolated point. Near this point, the $\mathbf{S}^1$-action reads

$$u \cdot (z_1, z_2) = (u^{k_1} z_1, u^{k'_1} z_2).$$

Bézout gives us two integers b_1 and b'_1 such that

$$k_1 b'_1 + k'_1 b_1 = 1.$$

We consider the two vectors

$$x_1 = \begin{pmatrix} k_1 \\ b_1 \end{pmatrix} \text{ and } x'_1 = \begin{pmatrix} -k'_1 \\ b'_1 \end{pmatrix},$$

which constitute an integral (direct) basis of $\mathbf{Z}^2$ by construction. Notice that there is a choice here, as we should expect: there is no reason why there should be only one torus action extending our circle action.

By induction, we construct vectors

$$x_i = \begin{pmatrix} k_i \\ b_i \end{pmatrix} \text{ and } x'_j = \begin{pmatrix} -k'_j \\ b'_j \end{pmatrix} \text{ for } 1 \leqslant i \leqslant r, 1 \leqslant i \leqslant s$$

such that (x_{i-1}, x_i) and (x'_i, x'_{i-1}) are (direct) integral bases of $\mathbf{Z}^2$. The claim now is that these vectors constitute the 1-skeleton of a fan Σ and that our manifold is equivariantly diffeomorphic with the toric manifold $\boldsymbol{X}_\Sigma$, endowed with the action of a circle subgroup of its big torus. Notice that all the x_i's are on the right side and all the x'_j's on the left side of the y-axis, so that we actually have a fan. The only problem is what happens between x_r and x'_s.

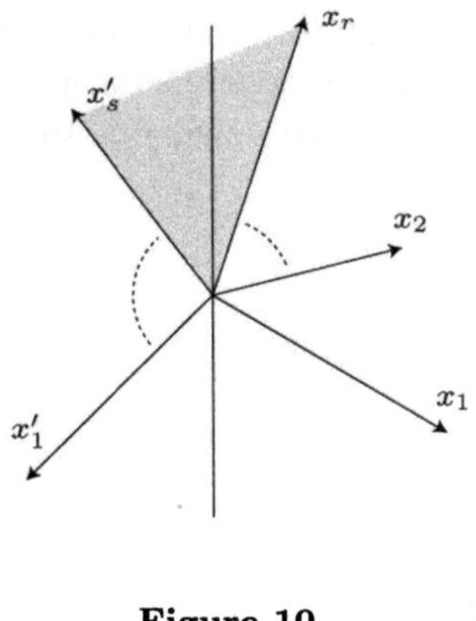

Figure 10

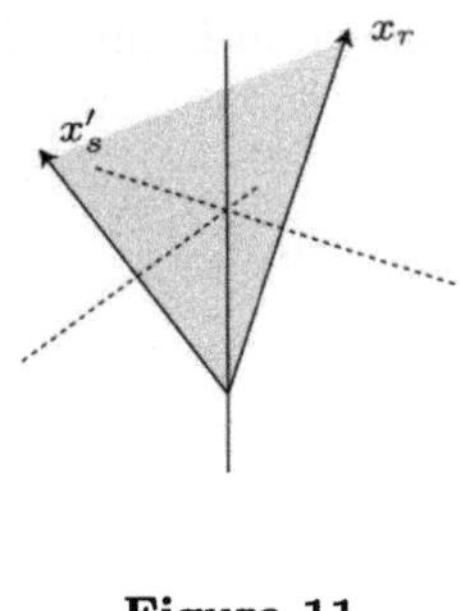

Figure 11

There are two possibilities:

- either the maximum is reached on a sphere,
- or it is reached at an isolated point.

In the first case, we add the vector $x_{r+1} = x'_{s+1}$ to our collection of vectors. The fan got this way is smooth, since we know that in this case, the stabilizers near the maximum are trivial, so that $k_r = k'_s = 1$ and

$$\left|x_r\ x_{r+1}\right| = \begin{vmatrix} k_r & 0 \\ b_r & 1 \end{vmatrix} = k_r = 1, \quad \left|x'_{s+1}\ x'_s\right| = \begin{vmatrix} 0 & -k'_s \\ 1 & b'_s \end{vmatrix} = k'_s = 1.$$

The second case is a little more intricate. We want to prove that

$$\left|x_r\ x_{s+1}\right| = \begin{vmatrix} k_r & -k'_s \\ b_r & b'_s \end{vmatrix} = 1.$$

We conclude with the help of an argument due to Karshon [**83**]. We look at the volume function on our manifold, as in § VI.2.b. Near the maximum, we know that its slope is $-\dfrac{1}{k_r k'_s}$. On the other hand, viewed from our chains of gradient spheres, this should be also be the difference of the "slopes" of the lines orthogonal to x_r and x'_s, namely

$$-\frac{1}{k_r k'_s} = \frac{b'_s}{k'_s} + \frac{b_r}{k_r} = \frac{b'_s k_r + b_r k'_s}{k'_s k_r},$$

so that $b'_s k_r + b_r k'_s = -1$, as we wanted to prove.

Now we have a toric manifold $\boldsymbol{X}_\Sigma$, which is diffeomorphic with our manifold W, with an $\mathbf{S}^1$-equivariant diffeomorphism, by construction. It is, as usual, easy to construct a symplectic form on $\boldsymbol{X}_\Sigma$ which looks like the one we have on W. From the fan Σ, we deduce a polygon, just assigning to each edge the length given by the symplectic volume of the corresponding gradient sphere as in § VI.3.d (this is shown in Figure 12, note that the upper corner of the polygon is the one used above and shown in Figure 11). □

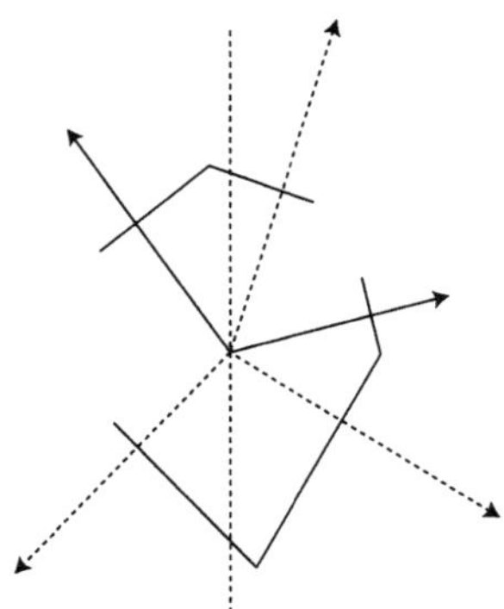

Figure 12

VIII.2.d. Karshon's uniqueness result. As we have noticed in Proposition VIII.2.8, some of the gradient spheres on our 4-manifold are only defined by the $\mathbf{S}^1$-action and do not depend on the invariant metric we have chosen: these are the spheres with stabilizer $\mathbf{Z}/k$ ($k \geqslant 2$). This is the basis of Karshon's construction. She associates, with any 4-manifold endowed with a periodic Hamiltonian, a graph

- the vertices of which correspond to the critical submanifolds (there are two kinds of vertices, those corresponding to isolated fixed points (ordinary vertices) and those corresponding to critical surfaces (that I will draw as horizontal segments))
- the edges of which correspond to the $\mathbf{Z}/k$-spheres.

The vertices are labelled by the values of the function (and the genus and symplectic area of the surface if they correspond to a non-isolated critical point). The edges are labelled by the order k of the stabilizer. Notice that the labels of the vertices keep track of the symplectic form, while the labels of the edges are defined by the action. Notice also (and this is something rather unpleasant) that there is no reason why such a graph should be connected (see Figures 13 and 14). For instance, the graphs associated, respectively, to the $\mathbf{S}^1$-actions on $\mathbf{P}^2(\mathbf{C})$

- by $u \cdot [x, y, z] = [ux, y, z]$; the bottom vertex corresponding to a sphere ($x = 0$, the minimum) and the top one to an isolated critical point ($[1, 0, 0]$, the maximum), there is no $\mathbf{Z}/k$-sphere,
- by $u \cdot [x, y, z] = [u^m x, u^n y, z]$, m and n coprime ($2 \leqslant m < n$); the three critical points are the minimum $[0, 0, 1]$, mapped to 0, the intermediate critical point $[1, 0, 0]$, mapped to m, and the maximum $[0, 1, 0]$. The three complex projective lines joining these points are the gradient spheres (and they are $\mathbf{Z}/k$-spheres as indicated),

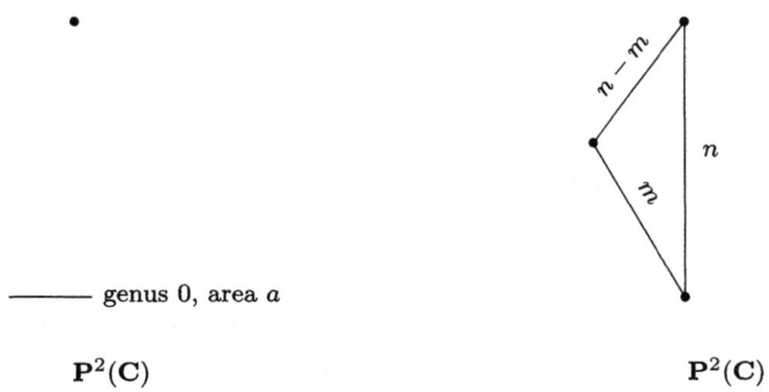

Figure 13

are shown in Figure 13, while Figure 14 shows the graphs associated with the $\mathbf{S}^1$-actions

$\mathbf{S}^2 \times \mathbf{S}^2$ $\mathbf{P}(L \oplus 1)$

Figure 14

- by $u \cdot (v, w) = (u \cdot v, u^k \cdot w)$ on $\mathbf{S}^2 \times \mathbf{S}^2$, the dot standing for the rotations about the z-axis, there are four isolated critical points and two $\mathbf{Z}/k$ gradient spheres,
- by rotations in the fibers of $\mathbf{P}(L \oplus \mathbf{1}) \to B$ (here the Euler class of L is e and λ denotes the total variation of H).

There is an obvious notion of isomorphism of such graphs.

The subtle observation which allows Karshon to prove her uniqueness result is that the gradient spheres which are not $\mathbf{Z}/k$-spheres, that is, the occurrence of which is not forced by the $\mathbf{S}^1$-action, are unimportant, since they can be avoided by a small perturbation of the (invariant and compatible) metric. She proves indeed that a generic invariant compatible metric has only $\mathbf{Z}/k$-spheres as gradient spheres relating critical points of index 2.

Remark VIII.2.13. Notice however that there *are* two gradient spheres meeting at any critical point of index 2. Let us consider the Karshon graph on the right of Figure 15 (this corresponds to a periodic Hamiltonian on $\mathbf{S}^2 \times \mathbf{S}^2$, see Figure 23). The next graph shows a possible arrangement of gradient spheres descending from the two critical points of index 2. A metric giving such a graph is not generic; for a generic metric, the gradient spheres would descend to the minimum, as shown in the last part of the figure.

After having defined the graphs, Karshon proves the uniqueness result her graphs were designed for: the graph determines the symplectic manifold and the action.

Theorem VIII.2.14 **(Uniqueness theorem of Karshon [83]).** *Let (W, ω) and (W', ω') be two compact connected symplectic manifolds of dimension 4 endowed with periodic Hamiltonians H and H'. Assume the corresponding graphs are isomorphic. Then there is a symplectic diffeomorphism $\varphi : W \to W'$ such that $H' \circ \varphi = H$.*

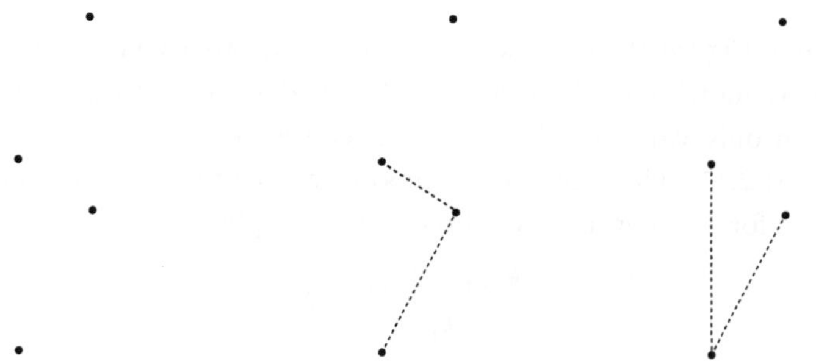

Figure 15

As for the symplectic aspects of Delzant's uniqueness theorem (Theorem IV.4.20), I will not give the proof and rather send the readers to the original paper [**83**].

Notice that, together with Theorem VIII.2.9, Theorem VIII.2.14 gives the nice following corollary.

Corollary VIII.2.15 **([83]).** *Any symplectic 4-manifold endowed with a periodic Hamiltonian is Kähler.* □

This is very nice but leaves us with the following problem: to prove an existence theorem, since Theorem VIII.2.14 does not tell us which graphs indeed appear. There are obviously obstructions coming from the Duistermaat–Heckman theorem, which gives restrictions on the areas of the surfaces under consideration. There is also the obvious fact that there are only two gradient spheres meeting at a point of index 2. Some obstructions are more subtle. Consider for instance the graph shown on the left part of Figure 16. I claim that this graph can only exist

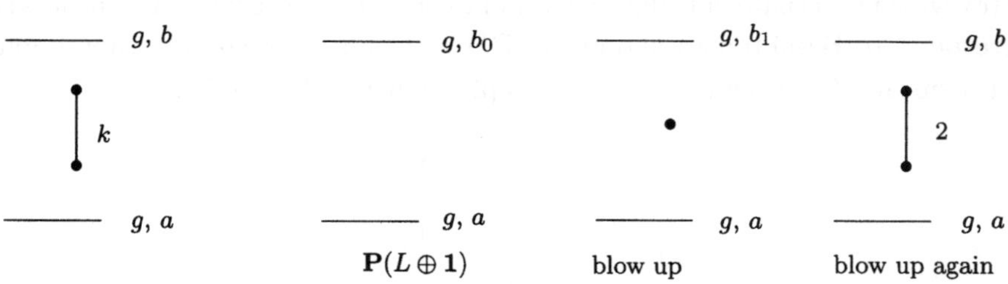

Figure 16

if $k = 2$. This is a consequence of Theorem VIII.2.9, which tells us that such a graph would come from a manifold obtained from a $\mathbf{P}(L \oplus \mathbf{1})$ by a sequence of blowing ups. So we should start from the graph on the right of Figure 14 and blow up fixed points. Figure 16 shows what happens and proves our assertion(5). Notice that this proves more generally that an isolated edge connecting two critical points of index 2 can only carry the label 2. This can also be viewed as a consequence of Lemma VIII.2.11, which of course imposes a lot of constraints on our graphs, since it forces, for any consecutive three gradient spheres, that

$$\frac{k_{i-1}+k_{i+1}}{k_i} \in \mathbf{N}.$$

VIII.2.e. Graphs associated with actual periodic Hamiltonians. According to Theorem VIII.2.9, the graphs that are actually obtained from Hamiltonian circle actions on 4-manifolds are obtained by blowing up the simple examples of $\mathbf{P}(L \oplus \mathbf{1})$ and $\mathbf{P}^2(\mathbf{C})$. We thus need

- to consider the graphs associated with the simple examples
- and to understand the effect of a blow up.

When both the maximum and the minimum are reached at isolated fixed points, the graph is divalent and the weights of the vertices can be characterized using the same ideas as in the proof of Proposition VIII.2.12 as those from which a toric surface can be constructed. I will thus concentrate on the case where at least one of the extrema is reached along a surface (and I will assume this is the case for the minimum). The graphs will thus have a "basis" that I will label by the genus g of the surface and the opposite b of the Euler class of its normal bundle (this number can be deduced from the data present on Karshon's graph, using Duistermaat–Heckman, on the other hand, we lose the information on the size of the surface with respect to the symplectic form).

Basic graphs. As announced above, we concentrate on graphs with a bottom vertex which is actually an edge, corresponding to a fixed surface. The basic graphs correspond to Hamiltonian actions on $\mathbf{P}(L \oplus \mathbf{1})$ for some complex line bundle L over a surface B, hence they are the graphs shown in Figure 17.

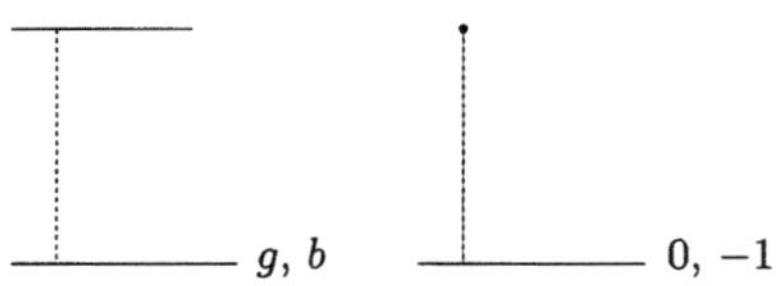

Figure 17. Basic graphs

(5) I shall be more explicit on the effect of blowing ups on graphs in § VIII.2.e

Blowing up a vertex. Let us turn now to the question of blowing up. From now on, it will be more convenient to use a slightly modified version of the graphs, taking Remark VIII.2.13 into account and adding to the graph a (dotted) edge labelled 1 relating the lowest vertex (corresponding to a critical point of index 2) of each connected component to the minimum, and similarly between the top vertices and the maximum. The effect of a blow up is shown in Figure 18.

Figure 18. Blowing up fixed points

Given a Karshon graph (completed as explained above), we will look at the complement Γ of whatever vertex corresponds to the maximum. The graph Γ has the shape of a star. Notice that it is equivalent to label its edges with new labels which are minus the self-intersection number of the sphere it represents: according to Lemma VIII.2.11, this replaces k_i by

$$b_i = \frac{k_{i-1} + k_{i+1}}{k_i} \cdots$$

which is an integer (the case of an isolated minimum is different, see Exercise VIII.5). Now we have a star-shaped graph the edges of which are labeled with the self-intersection numbers of the corresponding gradient spheres. Let us add to the picture minus the self-intersection number of the minimum surface (it will be determined from the rest of the graph later). The new labels are shown in Figure 19. The main advantage of this new labelling, which is equivalent to the

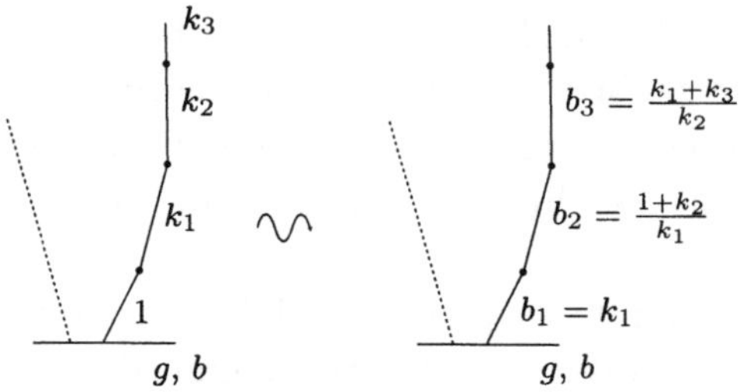

Figure 19

previous one, is that it is completely transparent with respect to blow ups: the labels are (minus) the self-intersection numbers, so that to blow up a fixed point produces a sphere of self-intersection number -1 and hence a new edge labelled 1.

Plumbing. These new graphs describe a construction called *plumbing*. This is a classical construction in low dimensional topology—also used in singularity theory, see for example [**115, 68**][(6)]. This is one of the many reasons to present this construction here. Some other ones are

- the relations with the Seifert manifolds investigated in Chapter I,
- the fact that plumbing describes quite accurately the 4-manifolds considered here,
- the fact that this allows one to determine which graphs are actual graphs of symplectic 4-manifolds endowed with periodic Hamiltonians,

... the main reason being the fact that I like plumbings and continued fractions.

Let me explain now how to plumb two complex line bundles on surfaces. Let $E_1 \to B_1$ and $E_2 \to B_2$ be the disc bundles of two complex line bundles on surfaces. Choose two discs D_1 and D_2 respectively in B_1 and B_2. We know that the bundles E_1 and E_2 are trivializable, over D_1 (*resp.* D_2) as well as over $\overline{B_1 - D_1}$ (*resp.* $\overline{B_2 - D_2}$). Choose trivializations

$$E_1|_{D_1} \cong D_1 \times B^2 \cong B^2 \times B^2$$
$$E_2|_{D_2} \cong D_2 \times B^2 \cong B^2 \times B^2$$

using which it is possible to glue, inverting factors. The result is an almost com-

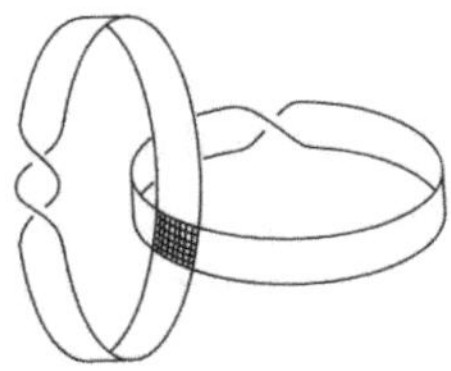

Figure 20. Plumbing

plex 4-manifold with boundary (and corners, that can easily be smoothed), which contains the two surfaces B_1 and B_2, with normal bundles E_1 and E_2. This is what Figure 20 schematically shows. The diffeomorphism type of the plumbed manifold is well-defined by that of B_1, B_2 and by the isomorphism type of the vector bundles.

Then we can iterate the process (and this is indeed something we have already done, see Figure 9). Given a graph with integral labels on its edges, we consider the line bundles having (minus) the label as Euler class on the corresponding surface (if no genus is indicated, the surface is a sphere) and we plumb the two bundles together if the corresponding edges have a common edge[(7)].

---

(6) Notice that the usual plumbing graphs are duals to the graphs used here.

(7) The usual convention in differential topology is to use the dual graph to the one we have here.

Consider for instance the graph associated with our periodic Hamiltonian. Remove the top vertex. What we have then is a star-shaped graph that we can consider as a plumbing graph. The 4-manifold with boundary that we obtain by plumbing along this graph is diffeomorphic with the complement of a neighborhood of the maximum in our original manifold.

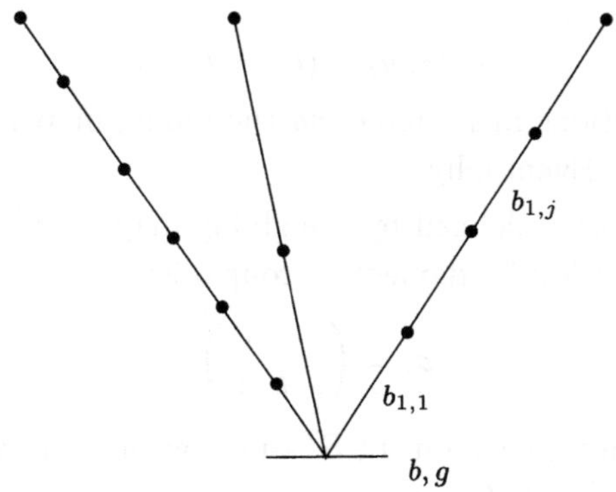

Figure 21

Notice that the abstract 4-manifold with boundary obtained by plumbing along a star-shaped graph can be endowed with an $\mathbf{S}^1$-action.

- If the basis B of the disc bundle E is not a sphere, we endow it with the trivial $\mathbf{S}^1$-action and make the circle act by rotation in the fibers of E (assumed to be Hermitian).
- Otherwise, namely if B is a sphere, we write it as the union of two hemispheres $\mathbf{S}^2 = \mathbf{S}^2_+ \cup \mathbf{S}^2_-$, viewed as discs

$$\mathbf{S}^2_+ = \{z \in \mathbf{C} \mid |z| \leqslant 1\}$$

and

$$\mathbf{S}^2_- = \{z \in \mathbf{C} \cup \infty \mid |z| \geqslant 1\} \cong \{z \in \mathbf{C} \mid |z| \leqslant 1\}.$$

Then the bundle $E \to B$ can be obtained by gluing two copies of the trivial bundle:

$$E = (\mathbf{S}^2_+ \times B^2) \cup_\varphi (\mathbf{S}^2_- \times B^2)$$

by

$$\begin{aligned} \varphi : \mathbf{S}^1 \times B^2 &\longrightarrow \mathbf{S}^1 \times B^2 \\ (z, u) &\longmapsto (\overline{z}, \varphi_z(u)) \end{aligned}$$

over $\mathbf{S}^1 = \mathbf{S}^2_+ \cap \mathbf{S}^2_-$ for some $\varphi_z \in \mathrm{SO}(2)$ for all z. The isomorphism class of the bundle is well-defined by the homotopy class of the mapping

$$\begin{aligned} \mathbf{S}^1 &\longrightarrow \mathrm{SO}(2) \\ z &\longmapsto \varphi_z \end{aligned}$$

which we may assume to have the form $\varphi_z(u) = \overline{z}^k u$ for some integer k. The bundle obtained this way is denoted $E(k)$.

Let $\mathbf{S}^1$ act on each piece, by

$$t \cdot (z, u) = (t^{m_1} z, t^{n_1} u)$$

on $\mathbf{S}^2_+ \times B^2$ and by

$$t \cdot (z, u) = (t^{m_2} z, t^{n_2} u)$$

on $\mathbf{S}^2_- \times B^2$. Both actions must agree *via* the gluing map φ, that is: $m_2 = -m_1$ and $n_2 = -km_1 + n_1$. Eventually,

(1) The bundle $E(k)$ is defined by the gluing map $\varphi : \mathbf{S}^1 \times \mathbf{S}^1 \to \mathbf{S}^1 \times \mathbf{S}^1$, the matrix of which (on fundamental groups) is:

$$\varphi_\star = \begin{pmatrix} -1 & 0 \\ -k & 1 \end{pmatrix}.$$

(2) If the $\mathbf{S}^1$-action is given, on the $\mathbf{S}^2_+$-side, by the integers (m_1, n_1), it will be given, on the $\mathbf{S}^2_-$-side by (m_2, n_2), with

$$\begin{pmatrix} m_2 \\ n_2 \end{pmatrix} = \begin{pmatrix} -1 & 0 \\ -k & 1 \end{pmatrix} \begin{pmatrix} m_1 \\ n_1 \end{pmatrix}.$$

The boundary $\partial W(\Gamma)$ is thus endowed with a fixed point free $\mathbf{S}^1$-action, in other words, this is a Seifert manifold. The next theorem allows us to compute the Seifert invariants of this manifold. Before stating it, we need to define a (signed) version of continued fractions. We write

$$[b_1, \dots, b_s] = b_1 - \cfrac{1}{b_2 - \cfrac{1}{b_3 - \cdots}}.$$

A good general reference on continued fractions is the book [**66**]. The fact that we have signs here does not make things very different. If we require that all b_i's are $\geqslant 2$, any rational number $\dfrac{n}{q} > 1$ can be written $[b_1, \dots, b_s]$ as one can easily show using a slight modification of the usual algorithm of Euclid[(8)]. The reduced fractions we get here are nonincreasing, and with standard notation

$$[b_1, \dots, b_n] = \frac{N_n}{D_n},$$

we have

$$N_n D_{n+1} - N_{n+1} D_n = 1$$

for any n, which can be considered more convenient than the analogous relation for continued fractions with "plus" signs.

---

(8) See also Exercise VIII.8.

Theorem VIII.2.16. *If* $\Gamma = (g \mid b, (b_{1,1}, \ldots, b_{1,s_1}), \ldots, (b_{r,1}, \ldots, b_{r,s_r}))$ *and if all the reduced fractions of the continued fractions*

$$[b_{j,1}, \ldots, b_{j,s_j}] = \frac{\alpha_j}{\alpha_j - \beta_j}$$

have a nonzero rational value, then

$$\partial W(\Gamma) = (g \mid \beta, (\alpha_1, \beta_1), \ldots, (\alpha_r, \beta_r))$$

for some integer β.

We have used here the notation for Seifert manifolds introduced in Theorem I.3.7.

Remarks VIII.2.17

(1) The invariants so obtained are not necessarily normalized in the sense that β_i is not assumed to be less than α_i.
(2) The determination of β in terms of b and $b_{i,j}$, that we do not make explicit here can be considered by the readers as an exercise.

Example VIII.2.18. If $\Gamma = (g \mid b)$, $\partial W(\Gamma)$ is a disc bundle of the line bundle with Euler class $-b$ on the genus g surface B. Glue a disc bundle of the line bundle with Euler class b to get the manifold $\mathbf{P}(L \oplus \mathbf{1})$ (with $e(L) = \pm b$).

The proof of the theorem will use the following lemma (an assertion on the manifold $\partial W(\Gamma)$ rather than on an $\mathbf{S}^1$-action).

Lemma VIII.2.19. *If* $\Gamma = (b_1, \ldots, b_s)$, *then the manifold* $\partial W(\Gamma)$ *is the lens space* $L(p, q)$ *where* $p/q = [b_1, \ldots, b_s]$.

Proof. We must glue the bundles $E(-b_1), \ldots, E(-b_s)$. One checks by an easy induction on s that this is the same as gluing the two solid tori $S_{1+} \times \mathbf{S}^1$ (where S_{1+} is the first hemisphere in the first sphere S_1) and $S_{s-} \times \mathbf{S}^1$ by the map

$$\partial S_{1+} \times \mathbf{S}^1 \longrightarrow \partial S_{s-} \times \mathbf{S}^1$$

with matrix

$$A_s = \begin{pmatrix} -1 & 0 \\ b_s & 1 \end{pmatrix} \begin{pmatrix} 0 & 1 \\ 1 & 0 \end{pmatrix} \cdots \begin{pmatrix} 0 & 1 \\ 1 & 0 \end{pmatrix} \begin{pmatrix} -1 & 0 \\ b_1 & 1 \end{pmatrix}.$$

Using reduced fractions, we can write

$$\begin{pmatrix} 0 & 1 \\ 1 & 0 \end{pmatrix} A_s = \begin{pmatrix} b_s & 1 \\ -1 & 0 \end{pmatrix} \cdots \begin{pmatrix} b_1 & 1 \\ -1 & 0 \end{pmatrix} = \begin{pmatrix} N_s & D_s \\ -N_{s-1} & -D_{s-1} \end{pmatrix}$$

where

$$\frac{N_i}{D_i} = [b_1, \ldots, b_i]$$

(by induction). Then

$$A_s = \begin{pmatrix} 0 & 1 \\ 1 & 0 \end{pmatrix} \begin{pmatrix} N_s & D_s \\ -N_{s-1} & -D_{s-1} \end{pmatrix} = \begin{pmatrix} -N_{s-1} & -D_{s-1} \\ N_s & D_s \end{pmatrix}.$$

To use less cumbersome notation, put $N_s = p$, $D_s = q$ (this is the notation in the statement), $N_{s-1} = u$, $D_{s-1} = v$ (hence $qu - pv = 1$), in such a way that we have

$$A_s = \begin{pmatrix} -u & -v \\ p & q \end{pmatrix}$$

and that the gluing map is:

$$\begin{array}{ccc} \partial S_{1+} \times \mathbf{S}^1 & \longrightarrow & \partial S_{s-} \times \mathbf{S}^1 \\ (a, z) & \longmapsto & (a^{-u} z^{-v}, a^p z^q). \end{array}$$

We recognize the lens space $L(p, q)$ (see §I.3.5). □

Proof of the theorem. We apply the lemma in each branch of the graph. We know, as the action is trivial at the "head", that we have to glue the lens space $L(p, q)$ with the trivial action in the fibers over S_{1+}: the action on $S_{1+} \times B^2$ is

$$t \cdot (a, z) = (ta, z).$$

At the end of this branch, we know that the $\mathbf{S}^1$-action is

$$t \cdot (a, z) = (t^{m'_s} a, t^{n'_s} z)$$

with

$$\begin{pmatrix} m'_s \\ n'_s \end{pmatrix} = A_s \begin{pmatrix} 1 \\ 0 \end{pmatrix} = \begin{pmatrix} -u & -v \\ p & q \end{pmatrix} \begin{pmatrix} 1 \\ 0 \end{pmatrix} = \begin{pmatrix} -u \\ p \end{pmatrix} = \begin{pmatrix} -N_{s-1} \\ N_s \end{pmatrix}.$$

The exceptional orbit is the central ($a = 0$) one, the order of its stabilizer is $n'_s = N_s = p$. To get the Seifert invariants (p, β) it is enough, up to normalization, to find some β such that $m'_s \beta = +1 \bmod n'_s$. But $m'_s = -N_{s-1} = -u$, we thus have to solve

$$-u\beta = 1 \mod p.$$

As we have $pu - qv = 1$, $\beta = p - q$ works. □

Eventually, the algorithm to decide whether a given graph is indeed the graph of some Hamiltonian action on a 4-manifold is the following:

- add to the graph the descending gradient spheres as explained above,
- relabel the edges,
- compute the continued fractions.

If the result is the inverse of an integer, we are done.

Proposition VIII.2.20. *Let Γ be a weighted graph. This is the graph describing a periodic Hamiltonian on a compact connected symplectic 4-manifold if and only if it is isomorphic to a graph, the complement of the top vertex of which*

– *is connected and star-shaped,*
– *the bottom vertex represents a surface of genus g and all the other vertices are di-valent,*
– *on each branch of the star, the weights satisfy*

$$[b_1, \ldots, b_s] = \frac{1}{\gamma_s}$$

(the inverse of an integer). □

Exercises

Exercise VIII.1. Let $f : \mathbf{S}^1 \to W$ be a smooth map into the symplectic manifold (W, ω). Assume W is endowed with a symplectic $\mathbf{S}^1$-action, the fundamental vector field of which is denoted X. Consider the mapping

$$\begin{aligned} F : \mathbf{S}^1 \times \mathbf{S}^1 &\longrightarrow W \\ (z, x) &\longmapsto z \cdot f(x). \end{aligned}$$

Check that

$$(F^\star\omega)_{(z,x)}\left(\left(\lambda\frac{d}{dt}, \xi\right), \left(\lambda'\frac{d}{dt}, \xi'\right)\right) = \lambda i_X \omega_{f(x)}(\mathsf{T}_x f(\xi')) - \lambda' i_X \omega_{f(x)}(\mathsf{T}_x f(\xi)),$$

(here $z = \exp(2i\pi t)$), that is,

$$F^\star\omega = dt \wedge f^\star i_X\omega.$$

Deduce that

$$\int_{\mathbf{S}^1\times\mathbf{S}^1} F^\star\omega = \int_{\mathbf{S}^1} f^\star i_X\omega.$$

Exercise VIII.2. Let H be a periodic Hamiltonian on a compact connected manifold W of dimension $2n$. Assume that H has only two critical values. Assume also that the minimum is reached at an isolated point. Prove that the maximum is reached on a symplectic manifold of dimension $2n-2$, which is diffeomorphic with $\mathbf{P}^{n-1}(\mathbf{C})$. Prove that W is equivariantly diffeomorphic with $\mathbf{P}^n(\mathbf{C})$, endowed with the $\mathbf{S}^1$-action by

$$u \cdot [x_0, x_1 \ldots, x_n] = [x_0, ux_1, \ldots, ux_n].$$

This is part of a result of Delzant in [**39**].

Exercise VIII.3. Prove that the graphs obtained from periodic Hamiltonians on the complex projective plane are as shown in Figure 22 (and their images by the symmetry about the horizontal axis).

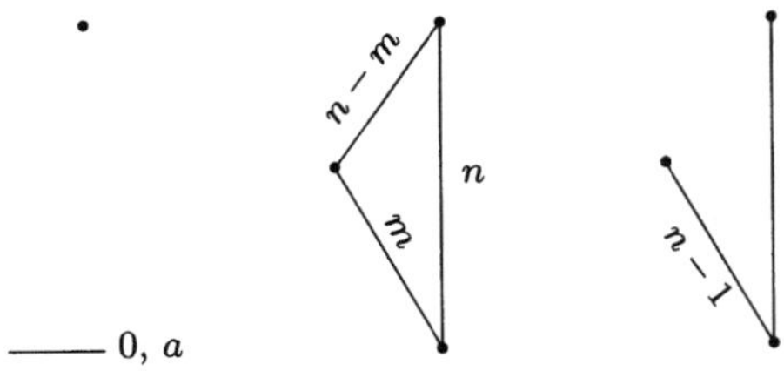

Figure 22. Graphs for $\mathbf{P}^2(\mathbf{C})$

Exercise VIII.4. Check that, if B is a surface of positive genus, the associated graph is that shown in the right part of Figure 14. Prove that, if B is a sphere, the Hamiltonians are the projections on the rational lines of the momentum mapping of the standard $\mathbf{T}^2$-action on the corresponding Hirzebruch surface (shown on the left of Figure 23) and that the graphs are as shown in Figure 23.

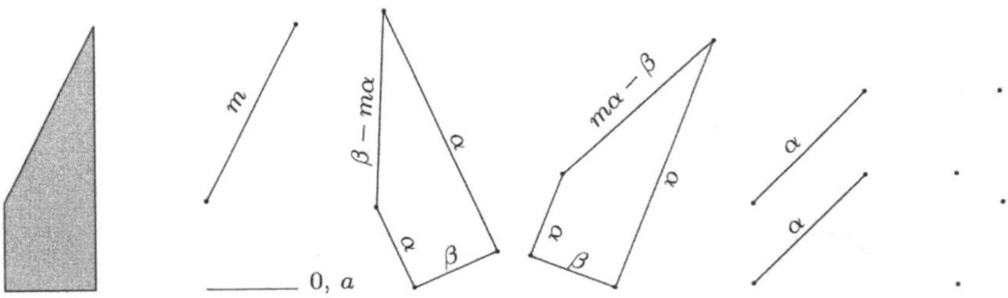

Figure 23. Graphs for $\mathbf{P}(\mathcal{O}(m) \oplus \mathbf{1})$

Is there any contradiction between the presence of the first graph and the remark we have made above (see Figure 16 and the discussion around it), forbidding an isolated edge with a label different from 2?

Prove that the penultimate graph can only occur for $m = 0$ (and α taking any value, we have already met this graph for an action on $\mathbf{S}^2 \times \mathbf{S}^2$, see Figure 14) and for $m = 1$ (the Hirzebruch surface is just $\mathbf{P}^2(\mathbf{C})$ blown up at a point) and $\alpha = 2$.

Prove that the very last graph corresponds to the case

$$\alpha = \beta = 1, \text{ and } \beta - m\alpha = \pm 1$$

and that it appears only for $m = 0$ or 2.

Exercise VIII.5. In the case of a minimum which is an isolated point, prove that the self-intersection numbers of the gradient spheres are the opposites of the numbers b_i, b'_i shown in Figure 24.

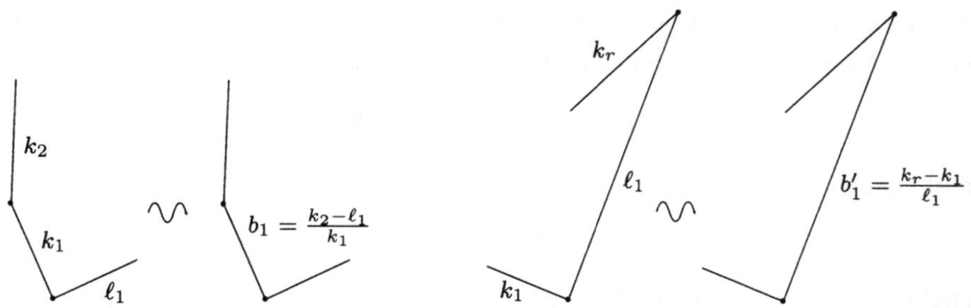

Figure 24

Exercise VIII.6. What are the self-intersection numbers of the gradient spheres corresponding to the middle graph in Figure 22? In a graph for which the top and bottom vertices correspond to isolated fixed points, do the intersection numbers determine the orders of the corresponding stabilizers?

Exercise VIII.7. Prove that to blow up an isolated minimum of a periodic Hamiltonian on a 4-dimensional manifold transforms its Karshon graphs as shown on Figure 25, in which dotted edges correspond to gradient spheres with trivial stabilizers and it is assumed that $\alpha > \beta$.

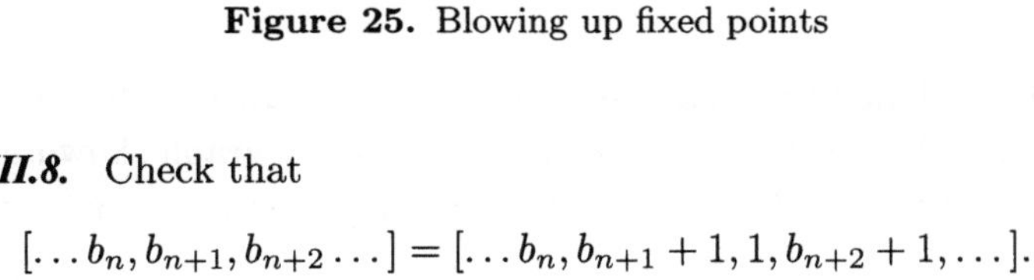

Figure 25. Blowing up fixed points

Exercise VIII.8. Check that

$$[\dots b_n, b_{n+1}, b_{n+2}\dots] = [\dots b_n, b_{n+1}+1, 1, b_{n+2}+1, \dots].$$

What is the relation with blow ups? Prove that $b_1, \dots, b_n \geqslant 2$ if and only if $[b_1, \dots, b_n] > 1$. Considering a star-shaped graph such that

$$[b_{i,1}, \dots, b_{i,s_i}] = \frac{1}{\gamma_i}$$

(the inverse of some integer) for all i, deduce that this graph is obtained by a sequence of blow ups starting from a graph $(g \mid b)$. Deduce a proof of Theorem VIII.2.9 in the case of periodic Hamiltonians on 4-manifolds having a surface as one of the extrema.

Exercise VIII.9. Consider the case of a 4-manifold endowed with a semi-free $\mathbf{S}^1$-action. Prove that, in each branch of the associated graph, one has

$$b_1 = 1, \quad b_2 = \cdots = b_j = 2,$$

and that

$$[b_1, \ldots, b_j] = \frac{1}{j}.$$

Deduce that a symplectic 4-manifold endowed with a semi-free Hamiltonian $\mathbf{S}^1$-action is obtained by a sequence of blow ups

$$\pi_i : W_i \longrightarrow W_{i-1}$$

in which each π_i is the blow up at some points of the surface on which the minimum is reached in W^{i-1}.

Exercise VIII.10. Prove that any (oriented) Seifert manifold (with oriented basis) is a level of some periodic Hamiltonian on a compact 4-dimensional symplectic manifold.

Exercise VIII.11. Let W be a compact symplectic manifold of dimension 4 endowed with a Hamiltonian $\mathbf{S}^1$-action. Assume that there exists a Hamiltonian $\mathbf{T}^2$-action extending the given action (for some way of embedding $\mathbf{S}^1$ as a subgroup in $\mathbf{T}^2$). Prove that the two following conditions hold:

- one of the regular levels has a sphere $\mathbf{S}^2$ as symplectic reduction
- no level of H meets more than two gradient manifolds.

Conversely, assume that W is endowed with an $\mathbf{S}^1$-action satisfying these two conditions. Prove that it extends to a Hamiltonian $\mathbf{T}^2$-action and that this action is unique.

***Exercise VIII.12* (Plumbing along polygonal graphs).** In this exercise, we plumb complex line bundles over spheres along the polygonal graph shown in Figure 26. The result is a 4-manifold with boundary. We try now to endow it with an $\mathbf{S}^1$-

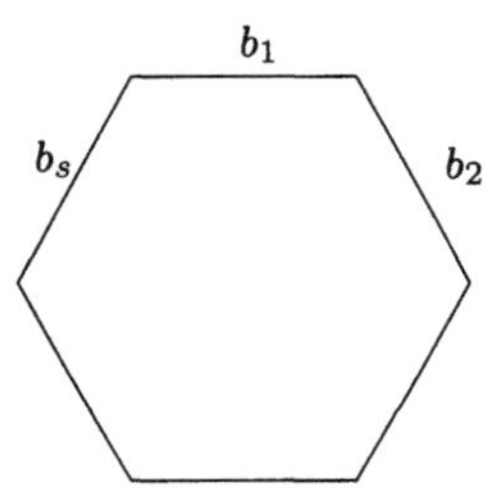

Figure 26

action. We use the notation of § VIII.2.e, so that, over the hemisphere S_{1+}, the action is given by

$$t \cdot (z, a) = (t^{m_1} z, t^{n_1} a)$$

and that over S_{s-} it is given by

$$t \cdot (z, a) = (t^{m'_s} z, t^{n'_s} a)$$

where

$$\begin{pmatrix} m'_s \\ n'_s \end{pmatrix} = A_s \begin{pmatrix} m_1 \\ n_1 \end{pmatrix}.$$

Prove that, in order that the actions agree on the last plumbing, it is necessary that the matrix

$$B_s = \begin{pmatrix} 0 & 1 \\ 1 & 0 \end{pmatrix} A_s = \begin{pmatrix} b_s & 1 \\ -1 & 0 \end{pmatrix} \cdots \begin{pmatrix} b_1 & 1 \\ -1 & 0 \end{pmatrix} \in \mathrm{SL}(2; \mathbf{Z})$$

has 1 as eigenvalue. Conversely, let B be a parabolic matrix, that is,

$$B = \begin{pmatrix} a & b \\ c & d \end{pmatrix} \quad \text{with } a, b, c, d \in \mathbf{Z},\ ad - bc = 1 \text{ and } a + d = 2.$$

Prove that there exists integers s, $b_1, \ldots, b_s$ such that

$$\begin{pmatrix} a & b \\ c & d \end{pmatrix} = \begin{pmatrix} N_s & D_s \\ -N_{s-1} & -D_{s-1} \end{pmatrix} = \begin{pmatrix} b_s & 1 \\ -1 & 0 \end{pmatrix} \cdots \begin{pmatrix} b_1 & 1 \\ -1 & 0 \end{pmatrix}.$$

Let $\begin{pmatrix} m \\ n \end{pmatrix}$ be a nonzero vector, fixed by B and such that m and n are relatively prime. Prove that the manifold got by plumbing along the graph in Figure 26 may be endowed with an (effective) $\mathbf{S}^1$-action, determined by m and n on the sphere number 1.

(1) If $B_s = \pm\mathrm{Id}$, check that there is even a $\mathbf{T}^2$-action on our plumbed manifold.
(2) Notice that

$$\underbrace{[2, \ldots, 2]}_{s \text{ times}} = \frac{s+1}{s}$$

and thus that a polygonal graph weighted by 2's gives

$$B_s = \begin{pmatrix} N_s & D_s \\ -N_{s-1} & -D_{s-1} \end{pmatrix} = \begin{pmatrix} s+1 & s \\ -s & -(s-1) \end{pmatrix}$$

which is a parabolic matrix. Deduce that, for any s, $(b_1, \ldots, b_s) = (2, \ldots, 2)$ give a solution.
(3) In the same way, check that

$$\underbrace{[-2, \ldots, -2]}_{s \text{ times}} = \frac{(-1)^s (s+1)}{(-1)^{s+1} j}$$

and that the unimodular matrix is

$$B_s = (-1)^s \begin{pmatrix} s+1 & -s \\ s & -(s-1) \end{pmatrix}$$

which is parabolic if and only if s is even.

(4) The case where $s = 3$. Prove that all the triangles which have two vertices weighted by -1 are solutions. When does the plumbed manifold have a $\mathbf{T}^2$-action? Prove that, in this case, the plumbed manifold may then be identified to a neighborhood of the union of the three coordinate lines $x = 0$, $y = 0$, and $z = 0$ in $\mathbf{P}^2(\mathbf{C})$.

(5) The case where $s = 4$. Prove that all the quadrilaterals weighted by $(0, a, 0, b)$ are solutions and that there is a $\mathbf{T}^2$-action if and only if $b = -a$. What is the plumbed manifold in this case?

***Exercise VIII.13* (Plumbing along polygonal graphs (continuation)).** We consider a 4-dimensional manifold W with an $\mathbf{S}^1$-action as constructed in Exercise VIII.12. Prove that the boundary of W is endowed with an action of $\mathbf{S}^1$ without exceptional orbit. Prove that the quotient of this action is a torus $\mathbf{T}^2$ (hint: the plumbing along the linear graph $(b_1, \ldots, b_s)$ gives a lens space, fibered on a 2-sphere (Lemma VIII.2.19), gluing the spheres number 1 and s amounts to gluing two small discs in this sphere).

Determine the Euler class of W as a principal $\mathbf{S}^1$-bundle over $\mathbf{T}^2$ in terms of the matrix B_s of Exercise VIII.12.

Compactify W into a closed 4-manifold endowed with an $\mathbf{S}^1$-action that admits no invariant symplectic form.

Example: Parabolic Inoue surfaces. This is the case of a polygon, all the weights of which are equal to 2. Recall that the matrix B_s is

$$\begin{pmatrix} s+1 & s \\ -s & -(s-1) \end{pmatrix} \sim \begin{pmatrix} 1 & s \\ 0 & 1 \end{pmatrix}.$$

Check that the $\mathbf{S}^1$-action on the closed 4-manifold constructed this way[9] has s isolated fixed points and a fixed surface which is a torus $\mathbf{T}^2$.

[9] There exists an analytic complex surface, endowed with a $\mathbf{C}^\star$-action with these properties. The spheres we used to plumb constitute a cycle of rational curves, and the torus $\mathbf{T}^2$ is an elliptic curve. This surface (*parabolic Inoue surface*) has remarkable properties, but has no nonconstant meromorphic functions (see for example [**98**] and [**114**]). For instance, it has no symplectic form defining the orientation we consider (*a fortiori* calibrating the complex structure). In fact the intersection pairing is negative definite (the proof of this fact is easy and thus left as an exercise).

BIBLIOGRAPHY

[1] K. AHARA & A. HATTORI – 4-dimensional symplectic S^1-manifolds admitting moment map, *J. Fac. Sci. Univ. Tokyo Sect. IA Math.* **38** (1991), p. 251–298.

[2] A. ALEKSEEV – On Poisson actions on compact Lie groups on symplectic manifolds, *J. Differential Geom.* **45** (1997), p. 241–256.

[3] A. ALEKSEEV & A. MALKIN – Symplectic structure of the moduli space of flat connection on a Riemann surface, *Comm. Math. Phys.* **169** (1995), p. 99–119.

[4] F. APÉRY – *Models of the real projective plane*, Vieweg, Braunschweig, 1987.

[5] V. I. ARNOLD – *Mathematical methods in classical mechanics*, Springer, 1978.

[6] V. I. ARNOLD & A. B. GIVENTAL – Symplectic geometry, in *Dynamical systems*, Encyclopædia of Math. Sci., Springer, 1985.

[7] M. F. ATIYAH – Convexity and commuting Hamiltonians, *Bull. London Math. Soc.* **14** (1982), p. 1–15.

[8] ———, Angular momentum, convex polyhedra and algebraic geometry, *Proceedings Edinburgh Math. Soc.* **26** (1983), p. 121–138.

[9] ———, *The geometry and physics of knots*, Lezioni Lincee, Cambridge University Press, 1990.

[10] M. F. ATIYAH & R. BOTT – The Yang-Mills equations over Riemann surfaces, *Phil. Trans. R. Soc. Lond.* **308** (1982), p. 523–615.

[11] ———, The moment map and equivariant cohomology, *Topology* **23** (1984), p. 1–28.

[12] M. Audin – Symplectic and almost complex manifolds, *in* [**21**], p. 41–74.

[13] ———, Hamiltoniens périodiques sur les variétés symplectiques compactes de dimension 4, in *Géométrie symplectique et mécanique* (C. Albert, ed.), Lecture Notes in Math., vol. 1416, Springer, 1990.

[14] ———, *The topology of torus actions on symplectic manifolds*, Progress in Math., Birkhäuser, 1991.

[15] ———, *Spinning tops, a course on integrable systems*, Cambridge University Press, 1996, Traduction en russe, Regular and chaotic dynamics, Moscou, 1999, traduction en japonais, Kyoritsu, 2000.

[16] ———, Écrire des mathématiques, `http://smf.emath.fr/Publications/InfoAuteurs`, 1997.

[17] ———, Lectures on gauge theory and integrable systems, in *Séminaire de Mathématiques Supérieures de Montréal*, Kluwer, 1997.

[18] ———, *Les systèmes hamiltoniens et leur intégrabilité*, Cours Spécialisés, vol. 8, Société Mathématique de France & EDP Sciences, 2001.

[19] ———, Hamiltonian monodromy *via* Picard-Lefschetz theory, *Commun. Math. Phys.* **229** (2002), p. 459–489.

[20] M. Audin, A. Cannas da Silva & E. Lerman – *Symplectic geometry of integrable Hamiltonian systems*, Birkhäuser, 2003.

[21] M. Audin & J. Lafontaine (eds.) – *Holomorphic curves in symplectic geometry*, Progress in Math., Birkhäuser, 1994.

[22] N. Berline & M. Vergne – Zéros d'un champ de vecteurs et classes caractéristiques équivariantes, *Duke Math. J.* **50** (1983), p. 539–549.

[23] P. Biran & K. Cieliebak – Symplectic topology on subcritical manifolds, *Comment. Math. Helv.* **76** (2001), p. 712–753.

[24] F. Bonahon & L. Siebenmann – The classification of Seifert fibred 3-orbifolds, in *Low dimensional topology* (R. Fenn, ed.), London Math. Soc. Lecture Notes Series, Cambridge University Press, 1997.

[25] N. Bourbaki – *Groupes et algèbres de Lie, chapitre 9*, Masson, Paris, 1982.

[26] A. Bouyakoub – Sur les fibrés principaux de dimension 4 en tores, munis de structures symplectiques invariantes et leurs structures complexes, *C. R. Acad. Sc. Paris* **306** (1988), p. 417–420.

[27] G. E. Bredon – *Introduction to compact transformation groups*, Pure and Applied Mathematics, vol. 46, Academic Press, New York, 1972.

[28] M. Brion – Points entiers dans les polyèdres convexes, *Ann. Scient. Éc. Norm. Sup.* **21** (1988), p. 653–663.

[29] M. Brion & C. Procesi – Action d'un tore dans une variété projective, in *Operator algebras, unitary representations, enveloping algebras, and invariant theory (Paris, 1989)*, Progress in Math., vol. 92, Birkhäuser Boston, Boston, MA, 1990, p. 509–539.

[30] J.-L. Brylinski – Éventails et variétés toriques, in *Séminaire sur les singularités des surfaces*, Lecture Notes in Math., vol. 777, Springer, 1980, p. 248–288.

[31] A. Cannas da Silva – Symplectic toric manifolds, *in* [**20**], p. 85–173.

[32] M. do Carmo – *Notas dum curso de grupos de Lie*, IMPA, Rio de Janeiro, 1974.

[33] H. Cartan – *Théorie élémentaire des fonctions analytiques d'une ou plusieurs variables complexes*, Hermann, Paris, 1961.

[34] M. Condevaux, P. Dazord & P. Molino – *Géométrie du moment (Séminaire Sud-Rhodanien)*, Publications du département de mathématiques, Université Claude Bernard-Lyon I, 1988.

[35] D. A. Cox – The homogeneous coordinate ring of a toric varieties, *J. Algebraic Geometry* **4** (1995), p. 17–50.

[36] ———, Recent developments in toric geometry, in *Algebraic geometry (Santa Cruz, 1995)*, Proc. Symp. Pure Math., vol. 62.2, Amer. Math. Soc., Providence, RI, 1997, p. 389–336.

[37] ———, Update on toric geometry, in *Geometry of toric varieties (Grenoble, 2000)*, Séminaires et Congrès, vol. 6, Société mathématique de France, Paris, 2002, p. 1–30.

[38] V. I. Danilov – The geometry of toric varieties, *Russian Math. Surveys* **33** (1978), p. 97–154.

[39] T. Delzant – Hamiltoniens périodiques et image convexe de l'application moment, *Bull. Soc. Math. France* **116** (1988), p. 315–339.

[40] A. Dold – Partitions of unity in the theory of fibrations, *Ann. of Math.* **78** (1963), p. 223–255.

[41] J. J. Duistermaat – Oscillatory integrals, Lagrange immersions and unfolding of singularities, *Comm. Pure Appl. Math.* **27** (1974), p. 207–281.

[42] ———, On global action-angle coordinates, *Comm. Pure Appl. Math.* **33** (1980), p. 687–706.

[43] ______, *Fourier integral operators*, Progress in Math., vol. 130, Birkhäuser Boston Inc., Boston, MA, 1996.

[44] J. J. DUISTERMAAT & G. J. HECKMAN – On the variation in the cohomology of the symplectic form of the reduced phase space, *Invent. Math.* **69** (1982), p. 153–158, and addendum, **72**, 1983, p. 259–269.

[45] F. EHLERS – Eine Klasse komplexer Mannigfaltigkeiten und die Auflösung einiger isolierter Singularitäten, *Math. Ann.* **218** (1975), p. 127–156.

[46] C. EHRESMANN – Sur les espaces fibrés différentiables, *C. R. Acad. Sc. Paris* **224** (1947), p. 1611–1612.

[47] ______, Les connexions infinitésimales dans un espace fibré différentiable, in *Colloque de topologie (espaces fibrés), Bruxelles, 1950*, Georges Thone, Liège, 1951, p. 29–55.

[48] R. FINTUSHEL – Classification of circle actions on 4-manifolds, *Trans. Amer. Math. Soc.* **242** (1978), p. 377–390.

[49] V. FOCK & A. ROSLYI – Poisson structure on moduli of flat connections on Riemann surfaces and r-matrices, preprint, 1992.

[50] ______, Flat connections and polyubles, *Teor. Mat. Fiz* **95** (1993), p. 228–238.

[51] T. FRANKEL – Fixed points on Kähler manifolds, *Ann. of Math.* **70** (1959), p. 1–8.

[52] I. M. GELFAND, M. M. KAPRANOV & A. V. ZELEVINSKY – *Discriminants, resultants and multidimensional determinants*, Birkhäuser, 1994.

[53] V. A. GINSBURG – Equivariant cohomologies and Kähler's geometry, *Funktsional. Anal. i Prilozhen.* **21** (1987), p. 271–283.

[54] V. L. GINZBURG – The topology of torus actions on symplectic manifolds, *Bull. Amer. Math. Soc.* **27** (1992), p. 315–320, review of [**14**].

[55] W. GOLDMAN – The symplectic nature of fundamental groups of surfaces, *Advances in Math.* **54** (1984), p. 200–225.

[56] ______, Invariant functions on Lie groups and Hamiltonian flows of surface group representations, *Invent. Math.* **85** (1986), p. 263–302.

[57] S. GRAFTON – *P is for peril*, Ballantine Books, New York, 2001.

[58] A. GRAMAIN – *Topologie des surfaces*, Presses Universitaires de France, Paris, 1971.

[59] H. B. GRIFFITHS – *Surfaces*, Cambridge University Press, Cambridge, 1976.

[60] P. A. GRIFFITHS & J. HARRIS – *Principles of algebraic geometry*, Wiley, 1978.

[61] V. GUILLEMIN – *Moment maps and combinatorial invariants of Hamiltonian T^n-spaces*, Progress in Math., vol. 122, Birkhäuser, 1994.

[62] V. GUILLEMIN, E. LERMAN & V. STERNBERG – *Symplectic fibrations and multiplicity diagrams*, Cambridge University Press, 1996.

[63] V. GUILLEMIN & S. STERNBERG – Convexity properties of the moment mapping, *Invent. Math.* **67** (1982), p. 491–513.

[64] ———, Birational equivalence in the symplectic category, *Invent. Math.* **97** (1989), p. 485–522.

[65] P. HALMOS – *A Hilbert space problem book*, Graduate Texts in Math., Springer, 1982.

[66] G. H. HARDY & E. M. WRIGHT – *An introduction to the theory of numbers*, fifth ed., The Clarendon Press Oxford University Press, New York, 1979.

[67] A. HATTORI – S^1-actions on unitary manifolds and quasi-ample line bundles, *J. Fac. Sci. Univ. Tokyo* **31** (1985), p. 433–486.

[68] F. HIRZEBRUCH, W. D. NEUMANN & S. S. KOH – *Differentiable manifolds and quadratic forms*, Marcel Dekker, Inc., New York, 1971.

[69] D. HUSEMOLLER – *Fibre bundles*, McGraw Hill, New York, 1966.

[70] P. IGLESIAS – Les SO(3)-variétés symplectiques et leur classification en dimension 4, *Bull. Soc. Math. France* **119** (1991), p. 371–396.

[71] K. JÄNICH – *Differenzierbare G-Mannigfaltigkeiten*, Lecture Notes in Math., vol. 59, Springer, 1968.

[72] L. JEFFREY – Extended moduli spaces of flat connections on Riemann surfaces, *Math. Annalen* (1994), p. 667–692.

[73] L. JEFFREY & F. KIRWAN – Localization for nonabelian group actions, *Topology* **34** (1995), p. 291–327.

[74] ———, On localization and Riemann-Roch numbers for symplectic quotients, *Quart. J. Math.* **47** (1996), p. 165–186.

[75] ———, Localization and the quantization conjecture, *Topology* **36** (1997), p. 647–693.

[76] L. JEFFREY & J. WEITSMAN – Bohr-Sommerfeld orbits in the moduli space of flat connections and the Verlinde dimension formula, *Commun. Math. Phys.* **150** (1992), p. 593–630.

[77] ———, Toric structures on the moduli space of flat connections on a Riemann surface: volumes and the moment map, *Advances in Math.* **106** (1994), p. 151–168.

[78] ———, Toric structures on the moduli space of flat connections on a Riemann surface. II. Inductive decomposition of the moduli space, *Math. Ann.* **307** (1997), p. 93–108.

[79] J. JURKIEWICZ – Chow ring of projective non-singular torus embedding, *Colloquium Mathematicum* **43** (1980), p. 261–270.

[80] ———, Torus embeddings, polyhedra, k^*-actions and homology, *Dissertationes Mathematicae* **236** (1985).

[81] M. V. KARASEV – The Maslov quantization conditions in higher cohomology and analogs of notions developed in Lie theory for canonical fibre bundles on symplectic manifolds I and II, *Selecta Math. Sov.* **8** (1989), p. 213–234 and 235–258, translation of a 1981 preprint.

[82] Y. KARSHON – An algebraic proof for the symplectic structure of the moduli space, *Proc. Amer. Math. Soc.* **116** (1992), p. 591–605.

[83] ———, *Periodic Hamiltonian flows on four dimensional manifolds*, Memoirs of the Amer. Math. Soc., vol. 141, 1999.

[84] A. A. KIRILLOV – *Éléments de la théorie des représentations*, MIR, Moscou, 1974.

[85] F. KIRWAN – *Cohomology of quotients in symplectic and algebraic geometry*, Princeton University Press, 1984.

[86] ———, Convexity properties of the moment mapping III, *Invent. Math.* **77** (1984), p. 547–552.

[87] B. KOSTANT – Quantization and representation theory I: prequantization, in *Lectures in modern analysis and applications III*, Lecture Notes in Math., vol. 170, Springer, 1970.

[88] ———, On convexity, the Weyl group and the Iwasawa decomposition, *Ann. Sci. Ec. Norm. Sup.* **6** (1973), p. 413–455.

[89] J.-L. KOSZUL – Sur certains groupes de transformations de Lie, *Invent. Math.* **52** (1953), p. 137–142.

[90] A. G. KUSHNIRENKO – Newton polygon and the number of solutions of a system of k equations and k unknowns (in Russian), *Uspekhi Math. Nauk* **30** (1975), p. 266–267.

[91] J. LAFONTAINE – *Introduction aux variétés différentielles*, Presses universitaires de Grenoble, 1996.

[92] F. LE LIONNAIS – Le second manifeste (1973), in *La bibliothèque oulipienne 2*, Ramsay, Paris, 1987, p. vii–xi.

[93] E. LERMAN – Symplectic cuts, *Math. Res. Lett.* **2** (1995), p. 247–258.

[94] E. LERMAN, E. MEINRENKEN, S. TOLMAN & C. WOODWARD – Nonabelian convexity by symplectic cuts, *Topology* **37** (1998), p. 245–259.

[95] E. LERMAN & S. TOLMAN – Hamiltonian torus actions on symplectic orbifolds and toric varieties, *Trans. Amer. Math. Soc.* **349** (1997), p. 4201–4230.

[96] P. LIBERMANN & C. M. MARLE – *Symplectic geometry and analytic mechanics*, Reidel, Boston, 1987.

[97] A. LICHNEROWICZ – Les variétés de Poisson et leurs algèbres de Lie associées, *J. Differential Geom.* **12** (1977), p. 253–300.

[98] E. J. N. LOOIJENGA – Rational surfaces with an anticanonical cycle, *Ann. of Math.* **114** (1981), p. 267–322.

[99] L. MANIVEL – *Fonctions symétriques, polynômes de Schubert et lieux de dégénérescence*, Cours Spécialisés, vol. 3, Société Mathématique de France, 1998.

[100] C.-M. MARLE – Modèle d'action hamiltonienne d'un groupe de Lie sur une variété symplectique, *Rendiconti del Seminarion Matematico* **43** (1985), p. 227–251.

[101] W. S. MASSEY – *Algebraic topology: an introduction*, Graduate Texts in Math., vol. 56, Springer-Verlag, 1967.

[102] D. MCDUFF – Examples of simply-connected symplectic non-Kählerian manifolds, *Journal Diff. Geom.* **20** (1984), p. 267–277.

[103] ———, The moment map for circle actions on symplectic manifolds, *Journal of geometry and physics* **5** (1988), p. 149–160.

[104] ———, The structure of rational and ruled symplectic 4-manifolds, *Journ. Amer. Math. Soc.* **3** (1990), p. 679–712.

[105] ———, Blow-ups and symplectic embeddings in dimension 4, *Topology* **30** (1991), p. 409–421.

[106] D. McDuff & D. Salamon – *Introduction to symplectic topology*, The Clarendon Press Oxford University Press, New York, 1995, Oxford Science Publications.

[107] J. Milnor – Construction of universal bundles I and II, *Ann. of Math.* **63** (1956), p. 272–284 and 430–436.

[108] ______, *Morse theory*, Princeton University Press, 1963.

[109] J. Milnor & J. Stasheff – *Characteristic classes*, Princeton University Press, 1974.

[110] H. Mineur – Réduction des systèmes mécaniques à n degrés de liberté admettant n intégrales premières uniformes en involution aux systèmes à variables séparées, *Journ. de Math.* **15** (1936), p. 385–389.

[111] J. Moser – On the volume elements on a manifold, *Trans. Amer. Math. Soc.* **120** (1965), p. 286–294.

[112] M. S. Narasimhan & S. Ramanan – Moduli of vector bundles on a compact Riemann surface, *Ann. of Math.* **89** (1969), p. 14–51.

[113] T. Oda – *Torus embeddings and applications*, Tata Institute of Fundamental Research Lectures on Mathematics and Physics, vol. 57, Tata Institute of Fundamental Research, Bombay, 1978, Based on joint work with Katsuya Miyake.

[114] ______, *Convex Bodies and algebraic geometry*, Springer, 2000.

[115] P. Orlik – *Seifert manifolds*, Lecture Notes in Math., vol. 291, Springer, 1972.

[116] G. Percé – *Tentative d'énumération de quelques objets qui pourraient se trouver dans ce livre*, Senais-Pavret, Paris, 2004.

[117] A. Pressley & G. Segal – *Loop groups*, The Clarendon Press Oxford University Press, New York, 1986, Oxford Science Publications.

[118] F. Raymond – Classification of the actions of the circle on 3-manifolds, *Trans. Amer. Math. Soc.* **19** (1968), p. 51–78.

[119] E. Reyssat – *Quelques aspects des surfaces de Riemann*, Progress in Math., vol. 77, Birkhäuser, 1990.

[120] I. Satake – On a generalization of the notion of manifold, *Proc. Nat. Acad. Sc.* **42** (1956), p. 359–363.

[121] I. Schur – Über eine Klasse von Mittelbildungen mit Anwendungen auf der Determinantentheorie, *Sitzungsberichte der Berliner Mathematischen Gesellschaft* **22** (1923), p. 9–20.

[122] H. SEIFERT – Topologie dreidimensionaler gefaserter Räume, *Acta Math.* **60** (1933), p. 147–238.

[123] J.-M. SOURIAU – *Structure des systèmes dynamiques*, Dunod, Paris, 1969.

[124] M. SPIVAK – *Differential geometry*, Publish or perish, 1975.

[125] I. VAISMAN – *Lectures on the geometry of Poisson manifolds*, Progress in Math., vol. 118, Birkhäuser, 1994.

[126] P. VANHAECKE – *Integrable systems in the realm of algebraic geometry*, Lecture Notes in Math., vol. 1638, Springer, 1996.

[127] S. VŨ-NGỌC – Mineur's proof of the Liouville-Arnold theorem, in preparation, 2004.

[128] A. WEINSTEIN – *Lectures on symplectic manifolds*, CBMS Regional Conference Series in Mathematics, vol. 29, Amer. Math. Soc., 1977.

[129] ———, Symplectic V-manifolds, periodic orbits of hamiltonian systems, and the volume of certain Riemannian manifolds, *Comm. Pure Appl. Math.* **30** (1977), p. 265–271.

[130] ———, The local structure of Poisson manifolds, *Journal Diff. Geom.* **18** (1983), p. 523–557.

INDEX

MIX
Papier aus verantwortungsvollen Quellen
Paper from responsible sources
FSC® C105338

If you have any concerns about our products,
you can contact us on
ProductSafety@springernature.com

In case Publisher is established outside the EU,
the EU authorized representative is:
Springer Nature Customer Service Center GmbH
Europaplatz 3, 69115 Heidelberg, Germany

Printed by Libri Plureos GmbH
in Hamburg, Germany